KV-350-143

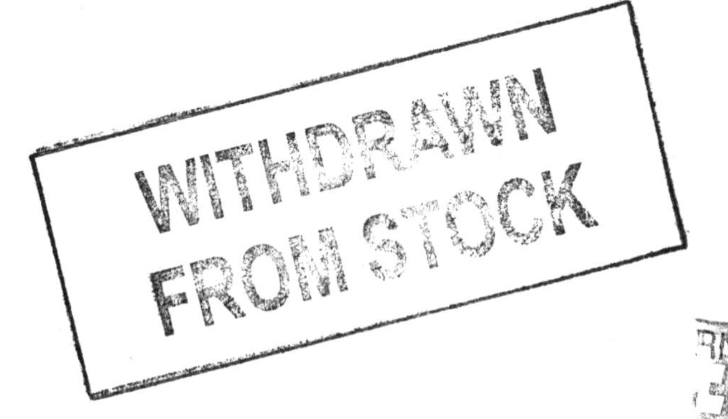
WITHDRAWN
FROM STOCK

Finite Element Modeling of Environmental Problems

Finite Element Modeling of Environmental Problems

Surface and Subsurface Flow and Transport

Edited by

Graham F. Carey
The University of Texas at Austin
U.S.A.

JOHN WILEY & SONS
Chichester • New York • Brisbane • Toronto • Singapore

Baffins Lane, Chichester
West Sussex PO19 1UD, England

National Chichester (01243) 779777
International (+44) 1243 779777

Other Wiley Editorial Offices

John Wiley & Sons, Inc., 605 Third Avenue,
New York, NY 10158-0012, USA

Jacaranda Wiley Ltd, 33 Park Road, Milton,
Queensland 4064, Australia

John Wiley & Sons (Canada) Ltd, 22 Worcester Road,
Rexdale, Ontario M9W 1L1, Canada

John Wiley & Sons (SEA) Pte Ltd, 37 Jalan Pemimpin #05-04,
Block B, Union Industrial Building, Singapore 2057

Library of Congress Cataloging-in-Publication Data

Finite element modeling of environmental problems: surface and
subsurface flow and transport / Graham F. Carey, editor.
p. cm.
Includes bibliographical references and index.
ISBN 0 471 95662 7
1. Water — Pollution — Mathematical models. 2. Soil pollution
— Mathematical models. 3. Finite element method. 1. Carey, Graham
F.
TD423.F556 1995
628.1'68'015118—dc20 95-7311
CIP

British Library Cataloguing in Publication Data

A catalogue record for this book is available from the British Library

Produced from camera-ready copy supplied by the author
Printed and bound in Great Britain by Biddles Ltd, Guildford and Kings Lynn.

This book is printed on acid-free paper responsibly manufactured from
sustainable forestation, for which at least two trees are planted for each one
used for paper production.

Contents

Preface

Rapid advances in microelectronics have dramatically expanded our capability for simulation in science and engineering. This has spurred interest in developing methodology and solution algorithms to solve practical problems. For example, the finite element method now enjoys widespread use in engineering analysis. It was initially developed for solution of problems in aircraft structural mechanics based on ideas from matrix analysis of structures. Subsequently, the method was placed on a rigorous mathematical foundation for approximate solution of boundary-value problems. Extensions of the methodology to transport applications and refinement of the solution algorithm have been topics of continuing research over the past two decades.

The main attributes of the method are its ability to model irregular geometries with irregular subregions or layers of differing material type, and its modular structure which lends itself to general-purpose programming. Consequently, the method is ideally-suited to simulation of complex real applications and it is now widely used in industry. In fact, there are several finite element software packages for stress and thermal analysis of structures and for industrial fluid flow problems. A number of texts detailing the method and its implementation are available.

The above features of the method also make it appropriate for modeling environmental problems. The rapid industrialization of society has placed a growing burden on the environment. This, together with an increasing awareness of the relationship of pollution to health and medical costs, has led to more stringent policies and requirements for environmental impact studies or remediation. Finite element simulation plays an important role here as it permits faithful, detailed characterization of the environmental problem in question. Of particular interest are shallow water and groundwater transport processes. For example, in coastal and estuarine problems very irregular coastlines can be accommodated easily as can extreme variations in bathymetry and the presence of numerous islands. Similarly, in groundwater pollution by contaminant transport through the soil, the variations in soil properties associated with differences in local geology can be modeled.

For these reasons the finite element method is increasingly being applied to surface water and soil transport problems and this is the focus of the present volume. We begin with a general discussion of the equations and main formulations for modeling surface water flow. This is followed by a development of a comprehensive model for a semi-enclosed coastal sea and a study of surface elevation and circulation in continental margin waters. The latter is particularly relevant in predicting the effect of hurricane storm surges. Two interesting new contributions to methodology for shallow water problems are then provided. The first is based on a fractional step approach which leads to a rational strategy for balancing dissipation terms. The second deals with a new entropy formulation and transformation to a symmetric flux Jacobian system. This leads to a new framework for developing approximate solution schemes. The development and performance of iterative solution strategies applied to the complex systems arising from the harmonic formulation of the shallow water equations is then considered. Various strategies are applied to representative applications and the effects of several preconditioners are compared. Some work initiated on a 3D model is also included. This is followed by three interesting application studies to the Po river delta flow, the Sydney offshore deepwater outfalls and Galveston Bay, Texas. In the Galveston Bay study the environmental impact of deepening the Ship Canal is investigated. The final two chapters in the shallow water treatment concern two different concepts: the method of sentinels and the use of fictitious domains. The former deals with the important issue of parameter identification in environmental systems that contain some unknown perturbations. For example, the data concerning a source of pollution and initial concentration may only be partially known. In the fictitious domain strategy, the effect of very shallow regions or islands can be accurately approximated by including artificial or fictitious elements. The above strategies are demonstrated in studies of the Finnish gulf and lake systems.

The analysis of transport of contaminants in soils and groundwater pollution is a problem of critical concern throughout the world. Our treatment of porous flow problems begins with the problem of modeling the long-term transport of radionuclides including the dominant geochemical processes. This is followed by two chapters dealing with mixed finite element methods. The first considers the inclusion of nonlinear, nonequilibrium adsorption kinetics using an upwind mixed scheme. The second describes a new numerical scheme for groundwater flow and transport problems with tensor coefficients using mixed methods on geometrically general domains. Another interesting new class of mixed methods based on least-squares is then proposed and some new theoretical results and estimates are summarized. The effects of dramatic changes in material properties are also considered.

Hybrid mixed methods are the basis of a 3D substructuring formulation and multilevel substructuring preconditioner described in the next chapter. A brief exposition of non-Darcy constitutive models and theoretical aspects including error estimates for the resulting nonlinear class of problems is included. The volume concludes with two interesting treatment technologies: *in situ* vitrification and soil venting. *In situ* vitrification is an exploratory technology for encapsulation of hazardous buried waste. The basic idea is to use electrical heating to melt the soil and waste which then solidifies to a glass-like impermeable mass. The problem involves finite element modeling of coupled fluid flow for the melt, phase change, electric field and electrical heating equations. In soil venting technology, air is pumped through the contaminated zone and the volatilized contaminant is removed by the gaseous phase flow. This necessitates modeling multiphase flow systems. In the present work a fractional flow formulation is introduced (motivated by the success of similar strategies for multiphase flow in petroleum reservoir simulation). The solution scheme involves a standard finite element formulation for the saturation equation coupled to a mixed formulation for the pressure equation.

The methodology, algorithms and application studies indicated here are representative of those encountered in finite element simulation of environmental problems. There are numerous other features of the methodology such as the use of adaptive mesh refinement and parallel algorithms that are not taken up here but are available in the literature and references cited. The volume has been arranged and edited to a cohesive treatment so that it is more accessible and is appropriate for a graduate course or seminar sequence on modeling environmental problems. It also provides a valuable reference source with current material on methodology and applications that is of interest to the scientific community.

I would like to express my appreciation first of all to the contributing authors who bring an international perspective to the book. Varis Carey and Yun Shen implemented the LaTeX organization of the chapters and Varis also assisted extensively in the editorial work. Pat Bozeman helped coordinate the interaction and correspondence with the contributing authors. I am also indebted to Bob McLay and our associates in the CFD Lab for their support.

G.F. Carey, Jan. 1995

Chapter 1

Modeling Surface Water Flow

R. A. Walters[1]

1.1 Introduction

There are many different factors involved in the mathematical statement of a problem in environmental fluid dynamics. In fact, there is such a wealth of possibilities that it is often difficult to decide how to formulate a particular model. The general problem is to calculate the velocity for a fluid held in a three-dimensional container with a free surface. The fluid is subject to external and internal forces, and may involve the transport of a solute or particulate species in the fluid. The applicable governing equations are a statement of the conservation of mass and momentum, subject to appropriate boundary and initial conditions.

A fundamental consideration is the determination of the time and space scales that the model must resolve. As a result, some form of time and/or space averaging is usually employed. Thus there are subscale processes that must be considered such as turbulence and dispersion. These concepts depend on the nature of the problem, and are pursued in more detail in the subsections that follow.

[1]U.S. Geological Survey,1201 Pacific Ave., Suite 600,Tacoma, WA, 98402, U.S.A.

1.2 Primitive Equation Formulation

A convenient starting point for surface water models is the continuity equation and the Navier-Stokes equations. These equations are averaged over turbulent time-scales to arrive at the Reynolds equations which contain terms with correlations between the velocity fluctuations that are known as Reynolds stresses. For the purpose of this treatment, there are three assumptions that we will use: the hydrostatic and Boussinesq approximations, and eddy viscosity closure of the stress terms.

Under the hydrostatic approximation, all vertical accelerations are small compared to gravitational acceleration. Thus the vertical momentum equation reduces to the hydrostatic equation and can be used to eliminate the pressure variable from the horizontal momentum equation. In practical terms, this limits the equations to shallow water theory, so deep and intermediate water waves cannot be considered. However, shallow water theory is applicable to most of the field-scale environmental flow problems encountered.

In the Boussinesq approximation, variations in fluid density are neglected where they affect the inertia terms, but are retained in the buoyancy terms where they provide the forcing for density-driven flows. It is a good approximation for flows encountered in fresh and salt water where the variation in density is small. It is not a good approximation for flows with heavy sediment loads where the variation in density is of the same order as the density.

The use of eddy viscosity closure has two implications regarding the applicability of the model: first it defines the form of the turbulence submodel, and secondly it defines the fluid as a Newtonian fluid; ie, there is a linear stress-strain relation. Examples of flows with non-Newtonian rheology are mud flows and glacier flows.

With these assumptions, the governing equations are easily formulated. First, the continuity equation for an incompressible fluid with horizontal velocity $\boldsymbol{u}$ and vertical velocity w, can be written

$$\boldsymbol{\nabla} \cdot \boldsymbol{u} + \frac{\partial w}{\partial z} = 0 \tag{1.1}$$

where $\boldsymbol{\nabla}$ is the horizontal gradient operator. Next, its two-dimensional vertical average is used as the governing equation for the free surface η relative to mean water level

$$\frac{\partial \eta}{\partial t} + \boldsymbol{\nabla} \cdot [(h+\eta)\overline{\boldsymbol{u}}] = 0 \tag{1.2}$$

where h is the depth from the mean water level and $H = h + \eta$ is the total depth. The horizontal components of the three-dimensional momentum equation are given by

$$\frac{\partial \boldsymbol{u}}{\partial t} + \nabla \cdot (\boldsymbol{u}\boldsymbol{u}) + \frac{\partial}{\partial z}(\boldsymbol{u}w) + \boldsymbol{f} \times \boldsymbol{u} + g\nabla\eta - \nabla \cdot (A_h \nabla \boldsymbol{u}) - \frac{\partial}{\partial z}(A_v \frac{\partial \boldsymbol{u}}{\partial z}) = \boldsymbol{F} \quad (1.3)$$

with coriolis vector $\boldsymbol{f}$, gravitational acceleration g and viscosity coefficients A_h, A_v. $\boldsymbol{F}$ denotes the applied body forces such as those due to density gradients. Finally, the vertical momentum equation reduces to the hydrostatic equation.

$$\frac{\partial p}{\partial z} = -\rho g \quad (1.4)$$

Here we have explicitly separated horizontal and vertical directions following the physics of the flows. If the hydrostatic assumption is not acceptable, then the full vertical momentum equation is retained. In this case, a pressure variable appears in the equations. (Usually a reduced pressure is used such that hydrostatic pressure is subtracted from the total pressure.)

The surface and bottom boundary conditions are

$$A_v \frac{\partial \boldsymbol{u}}{\partial z} = \frac{\tau_s}{\rho} \qquad (z = \eta) \quad (1.5)$$

$$A_v \frac{\partial \boldsymbol{u}}{\partial z} = \frac{\tau_b}{\rho} = C_D |\boldsymbol{u}| \boldsymbol{u} \qquad (z = -h), \quad (1.6)$$

where τ_s is the surface stress, τ_b is the bottom stress, ρ is density and C_D is the bottom stress coefficient. There are essential boundary conditions on η at open boundaries, and no normal flow on solid boundaries.

The viscous stress terms normally represent the effects of the Reynolds stresses. In the limit of low Reynolds number flow, the coefficients approach molecular viscosity coefficients. For shallow water, bottom stress generally greatly exceeds horizontal viscous stress and the latter can be neglected. However, fjords are one example where horizontal friction and the generation of turbulence on the nearly vertical sidewalls can be important.

The vertical eddy viscosity can be defined in several ways depending on whether the bottom boundary layer is to be resolved, and whether a turbulence closure scheme is used. For water that is shallower than the bottom boundary layer thickness, mixing-length theory can be used successfully. However, for cases where advection of turbulence is important, a separate turbulence closure model is required.

The value for $\overline{\boldsymbol{u}}$ can be calculated from a vertical average of the model results for $\boldsymbol{u}$ or from the vertical average of equation (1.3). The latter is performed in most standard texts on fluid flow (e.g. see [13]) and may be written

$$\frac{\partial(H\overline{\boldsymbol{u}})}{\partial t} + \nabla\cdot(H\overline{\boldsymbol{u}\boldsymbol{u}}) + \boldsymbol{f}\times(H\overline{\boldsymbol{u}}) + gH\nabla\eta - \nabla\cdot(HA_h\nabla\overline{\boldsymbol{u}}) - \frac{\tau_s}{\rho} + \frac{\tau_b}{\rho} = H\overline{\boldsymbol{F}} \tag{1.7}$$

Equation (1.7) is typically used in a two-dimensional (horizontal) model where the variations in the vertical are known or are insignificant. A coefficient α can be defined that takes into account the autocorrelation of the vertical variation in $\boldsymbol{u}$ and can be used to express the advection term as a function of $\overline{\boldsymbol{u}}$ with

$$\alpha_{ij} = \frac{\overline{u_i u_j}}{\overline{u}_i\,\overline{u}_j} \tag{1.8}$$

where $u_1 = u$ and $u_2 = v$.

In a primitive equation model, equations (1.1) to (1.3) are used as the governing equations for the variables η, u, v, and w, and discretized directly. In a two-dimensional model, equations (1.2) and (1.7) are used.

Primitive equation models typically use different order interpolation for η and $\boldsymbol{u}$ in order to suppress spurious computational modes [6,16,20]. Introducing test functions, the Galerkin weighted residual equations for the two-dimensional shallow water equations become

$$\int_\Omega \frac{\partial\eta}{\partial t}\psi d\Omega - \int_\Omega (h+\eta)\overline{\boldsymbol{u}}\cdot\nabla\psi d\Omega = -\oint \boldsymbol{Q}\cdot\boldsymbol{n}\psi d\Gamma, \tag{1.9}$$

where the divergence term in (1.2) has been integrated by parts, $\boldsymbol{Q} = H\boldsymbol{u}$, and

$$\int_\Omega \frac{\partial\overline{\boldsymbol{u}}}{\partial t}\phi d\Omega + \int_\Omega \overline{\boldsymbol{u}}\cdot\nabla\overline{\boldsymbol{u}}\;\phi d\Omega + \int_\Omega \boldsymbol{f}\times\overline{\boldsymbol{u}}\;\phi d\Omega + g\int_\Omega \nabla\eta\;\phi d\Omega$$
$$- \int_\Omega \frac{1}{H}\nabla\cdot(HA_n\nabla\overline{u})\,d\Omega - \int_\Omega \frac{\tau_s}{\rho H}\phi d\Omega + \int_\Omega \frac{\tau_b}{\rho H}\phi d\Omega = \int_\Omega \overline{\boldsymbol{F}}\phi d\Omega \tag{1.10}$$

where the equation is written in non-conservative form with $\alpha = 1$. Here Ω is the computational domain with boundary Γ.

For the continuity equation, essential conditions on η or the boundary integral for discharge are specified on all external boundaries. The natural boundary condition is that $\boldsymbol{Q}\cdot\boldsymbol{n} = 0$.

There are several options for the formulation of the momentum equation. The equation can be stated in conservative or non-conservative form by applying the continuity equation to the first two terms in (1.7). In a numerical sense, the equation is easier to solve in its non-conservative form although this is not an issue of great importance. In addition, the gravity term can be integrated by parts so that the boundary forcing for η appears as a line integral. Also, the conservative form of the advection term and lateral stress terms can be integrated by parts. Thus there appears a boundary term for the momentum transfer at the boundaries. A primary criterion for choosing these options is to provide a more convenient method to apply the boundary conditions.

The three-dimensional momentum equations are approximated in a similar manner as the two-dimensional equations, with the replacement of the test function ϕ by the three-dimensional set Φ. This procedure is discussed in more detail in the next section.

1.3 Wave Equation Formulation

A different form of the continuity equation is frequently used. The primary reason is that most finite element approximations to the primitive equations give rise to spurious oscillation modes also known as $2\Delta x$ oscillations [1,9,16,19]. Even the use of mixed-interpolation does not remove the modes in velocity [16]. However, the continuity and momentum equations can be combined to give a wave equation that replaces the continuity equation. This form of the equations does not support spurious modes and has computational advantages in the fact that the surface elevation and velocity solutions are decoupled [9,11,16]. Differentiating (1.2) with respect to time and replacing the second term with terms from the momentum equation,

$$\frac{\partial^2 \eta}{\partial t^2} - \boldsymbol{\nabla} \cdot (gH\boldsymbol{\nabla}\eta) = \boldsymbol{\nabla} \cdot \boldsymbol{R}, \tag{1.11}$$

$$\boldsymbol{R} = \boldsymbol{\nabla} \cdot (H\overline{\boldsymbol{u}\boldsymbol{u}}) + \boldsymbol{f} \times H\overline{\boldsymbol{u}} - \boldsymbol{\nabla} \cdot (HA_h \boldsymbol{\nabla}\overline{\boldsymbol{u}}) - \frac{\tau_s}{\rho} + \frac{\tau_b}{\rho} - H\overline{\boldsymbol{F}} \tag{1.12}$$

There are several variations of the equations above. Perhaps the most common is the addition of a penalty term that is a coefficient times the continuity equation [7,11]. Another is to formulate the advection terms so they are consistent in both the wave and momentum equations [7].

Suitable combinations of the equations above may then be discretized in space and time to form a numerical model for surface water flow. For a three-dimensional primitive equation model, (1.1) to (1.3) are used, whereas for a wave equation model equation (1.2) is replaced with (1.11). For the two-dimensional shallow water problem, (1.2) and (1.7) are used and for a wave equation model (1.11) and (1.7) are used.

The three-dimensional elements are constructed by defining a set of two-dimensional elements in the horizontal plane and then placing nodes along a vertical line beneath these nodes. The vertical coordinates can be level coordinates or terrain-following (σ coordinates) such that there is a fixed number of vertical nodes beneath each surface node. There are well-known problems with the non-orthogonal σ coordinate system (see [3,21]).

For the wave equation formulations, it is convenient to express the test functions as a tensor product of the horizontal and the vertical test functions of the form $\Phi = \phi(x,y)\zeta(z)$ [11]. Using the fact that (1.11) is not a function of z, ϕ is used as the weighting function in the wave equation for η and the weighted residual form of the equations may be expressed as

$$\int_\Omega \frac{\partial^2 \eta}{\partial t^2}\phi d\Omega + \int_\Omega gH\nabla\eta\cdot\nabla\phi d\Omega = -\int_\Omega \boldsymbol{R}\cdot\nabla\phi d\Omega - \oint \frac{\partial H\overline{\boldsymbol{u}}}{\partial t}\cdot\boldsymbol{n}\phi\, d\Gamma \quad (1.13)$$

where the divergence terms in (1.11) have been integrated by parts [11]. The wave equation is solved by replacing ϕ with Lagrange bases of order p and numerically integrating over the spatial domain with sufficiently high order that the integrals are exact.

With Φ as the weighting function, the momentum equation becomes

$$\begin{aligned}\int_\Omega \frac{\partial \boldsymbol{u}}{\partial t}\Phi d\Omega &+ \int_\Omega \left[\nabla\cdot(\boldsymbol{u}\boldsymbol{u}) + \frac{\partial(\boldsymbol{u}w)}{\partial z}\right]\Phi d\Omega + \int_\Omega \boldsymbol{f}\times\boldsymbol{u}\,\Phi dx \\ &+ \int_\Omega A_h\nabla\boldsymbol{u}\cdot\nabla\Phi d\Omega + \int_\Omega A_v\frac{\partial\boldsymbol{u}}{\partial z}\frac{\partial\Phi}{\partial z}d\Omega \\ &= -g\int_\Omega(\nabla\eta - \boldsymbol{F})\Phi d\Omega + \oint \boldsymbol{B}\cdot\boldsymbol{n}\phi\, d\Gamma \qquad (1.14)\end{aligned}$$

where the stress terms have been integrated by parts and the last term is the associated boudary integral for stress. As before, there are several options in manipulating these equations. Any of the divergence terms can be integrated by parts if this eases the specification of the boundary conditions. A detailed discussion of the time stepping form of these equations can be found in [11].

1.4 Harmonic Method in Time

For a wide range of problems that use steady or periodic forcing, such as astronomical and radiational tides, and steady flows, it is more convenient to solve the governing equations in the frequency domain than in the time domain. This approach can lead to much smaller computational effort and is especially useful for exploratory studies with complicated geometries. This is a technique commonly used in oceanography because many of the variables of interest are characterized by line spectra. The idea was developed in the context of numerical models independently by several groups [5,8,9,12,14]. In addition, a comparison between basic harmonic and time-stepping models for the two-dimensional equations is presented in [22].

In following this approach, the dependent variables are expressed as periodic functions

$$\eta(x,y,t) = \frac{1}{2} \sum_{n=-N}^{N} \eta_n(x,y)e^{-i\omega_n t} \tag{1.15}$$

$$\boldsymbol{u}(x,y,z,t) = \frac{1}{2} \sum_{n=-N}^{N} \boldsymbol{u}_n(x,y,z)e^{-i\omega_n t} \tag{1.16}$$

where ω is the angular frequency, and n is the index for the N tidal and residual constituents. Applying (1.15) and (1.16) to (1.2) and (1.3) and using harmonic decomposition [14], the governing equations become

$$-i\omega_n \eta_n + \boldsymbol{\nabla} \cdot (h\overline{\boldsymbol{u}}_n) = \boldsymbol{\nabla} \cdot \boldsymbol{W}_n \tag{1.17}$$

and

$$-i\omega_n \boldsymbol{u}_n + \boldsymbol{f} \times \boldsymbol{u}_n - \frac{\partial}{\partial z}(A_v \frac{\partial \boldsymbol{u}_n}{\partial z}) = -g(\boldsymbol{\nabla}\eta_n - \boldsymbol{T}_n) \tag{1.18}$$

where $\boldsymbol{W}$ is the continuity nonlinearity, and $\boldsymbol{T}$ contains the density forcing term and all the nonlinear terms in the momentum equation, including advection and terms arising from the time-dependent viscosity [17,18].

The treatment of the nonlinear continuity and advection terms is straitforward as they contain only simple products of the various frequencies. Thus they lead to sums and differences between the frequencies of the N constituents and contribute as source terms in the generation of overtides, compound tides, and low frequency tides. The treatment of the vertical friction term and bottom friction is more difficult due to the factor $|\boldsymbol{u}|$. However, these terms can be treated by a method developed in [14] for the two-dimensional shallow water equations and extended to three dimensions with the finite element method [17,18].

The depth-averaged form of equation (1.18) is

$$(-i\omega_n + \gamma)\overline{\boldsymbol{u}}_n + \boldsymbol{f} \times \overline{\boldsymbol{u}}_n = -g(\boldsymbol{\nabla}\eta_n - \overline{\boldsymbol{T}}_n) \tag{1.19}$$

where $\gamma(x,y)$ is the time-independent part of the bottom friction term $C_D|\boldsymbol{u}|$, and $\overline{\boldsymbol{T}}_n$ is the depth-average of $\boldsymbol{T}_n$ with the inclusion of surface and time-dependent part of bottom stress.

In a three-dimensional primitive equation model, (1.1), (1.17), and (1.18) are used as the governing equations and discretized directly. For a two-dimensional model, (1.17) and (1.19) are used.

The wave equation form of these equations becomes

$$-i\omega_n\eta_n - \nabla\cdot\left\{\left(\frac{gh}{q_n^2+f^2}\right)\left[q_n\left(\nabla\eta_n - \boldsymbol{T}_n\right) - \boldsymbol{f}\times\left(\nabla\eta_n - \boldsymbol{T}_n\right)\right]\right\} = \nabla\cdot\boldsymbol{W}_n \tag{1.20}$$

where $q_n = -i\omega_n + \gamma$.

This equation is derived by solving for $\overline{\boldsymbol{u}}$ in the depth-averaged momentum equation and substituting that expression into the divergence term in the continuity equation [10, 17]. This form of the equation has the same advantages as were mentioned earlier.

The three-dimensional elements for this formulation may be constructed by defining a set of two-dimensional elements in the horizontal plane and then placing nodes in planes along vertical lines beneath these nodes. As before, it is convenient to express the test functions as a tensor product of the horizontal and the vertical test functions as $\Phi = \phi(x,y)\zeta(z)$. Using the fact that (1.20) is not a function of z, ϕ is used as the weighting function in the Helmholtz equation for η and the weighted residual form may be expressed as

$$\begin{aligned}-i\omega_n\int_\Omega \eta_n\phi d\Omega + \int_\Omega\left\{\left(\frac{gh}{q_n^2+f^2}\right)\left[q_n\left(\nabla\eta_n - \boldsymbol{T}_n\right) - \boldsymbol{f}\times\left(\nabla\eta_n - \boldsymbol{T}_n\right)\right]\right\}\\ \cdot\nabla\phi d\Omega - \int_\Omega \boldsymbol{W}_n\cdot\nabla\phi d\Omega = -\oint \boldsymbol{Q}_n\cdot\boldsymbol{n}\phi d\Gamma\end{aligned} \tag{1.21}$$

where the divergence term in (1.20) has been integrated by parts, and $\boldsymbol{W}_n$ and $\boldsymbol{T}_n$ are treated as known functions. The wave equation is approximated by replacing ϕ with Lagrange bases of order p and numerically integrating over the spatial domain with sufficiently high order that the integrals are exact.

Similarly, with Φ as the weighting function, the momentum equation becomes

$$\begin{aligned}-i\omega_n\int_\Omega \boldsymbol{u}_n\Phi d\Omega + \boldsymbol{f}\times\int_\Omega \boldsymbol{u}_n\Phi d\Omega \quad &- \quad \int_\Omega A_v\frac{\partial\boldsymbol{u}_n}{\partial z}\frac{\partial\Phi}{\partial z}d\Omega\\ &= \quad -g\int_\Omega(\nabla\eta_n - \boldsymbol{T}_n)\Phi d\Omega\end{aligned} \tag{1.22}$$

1.5 Transport Equations

The transport equation is derived from the conservation of mass for a solute. As in the hydrodynamic equations, there are certain time and space scales that the model must resolve. In a similar manner, this leads to time and/or space averaging of the equation. The result is that dispersive terms appear that are similar to the Reynolds stress terms in the momentum equations. These terms are usually much larger than molecular diffusion and are written in the same form. The transport equation can then be written as

$$\frac{\partial C}{\partial t} + \nabla \cdot (\boldsymbol{u}C) + \frac{\partial}{\partial z}(wC) - \nabla \cdot (E_h \nabla C) - \frac{\partial}{\partial z}(E_v \frac{\partial C}{\partial z}) = S \tag{1.23}$$

where C is concentration, E_h is the horizontal dispersion coefficient, E_v is the vertical diffusion coefficient, and S represents source/sink terms. This equation is subject to the boundary conditions that flux or concentration is specified on all boundaries.

Following the development of the shallow water equations, the transport equation can be averaged in the vertical direction to give a two-dimensional form that can be used with the shallow water equations. The equation becomes

$$\frac{\partial H\overline{C}}{\partial t} + \nabla \cdot (H\overline{\boldsymbol{u}C}) - \nabla \cdot (E_c H \nabla \overline{C}) = H\overline{S} \tag{1.24}$$

where E_c is the two-dimensional horizontal dispersion coefficient. As before, a coefficient $\boldsymbol{\alpha}$ can be defined as

$$\alpha_i = \frac{\overline{u_i C}}{\overline{u}_i \overline{C}} \tag{1.25}$$

in order to express the advection term as a function of the depth-averaged variables.

Using the Galerkin method with ϕ as the weighting function, the depth-averaged equation becomes

$$\int_\Omega \frac{\partial \overline{C}}{\partial t}\phi d\Omega + \int_\Omega \overline{\boldsymbol{u}} \cdot \nabla \overline{C} \phi d\Omega + \int_\Omega E_c \nabla \overline{C} \cdot \nabla \phi d\Omega = \int_\Omega \overline{S} \phi d\Omega + \oint \overline{\boldsymbol{Q}}_s \cdot \boldsymbol{n} \phi d\Gamma \tag{1.26}$$

where the dispersion term is integrated by parts and $\overline{\boldsymbol{Q}}_s$ is the dispersive flux appropriate to the depth-averaged model. Suitable boundary conditions are essential conditions on $\overline{C}$ or a specified dispersive flux on all external boundaries. The natural boundary condition is no dispersive flux. A study of the propagation characteristics of this equation is presented in [2].

Using the three-dimensional elements as defined earlier, and the basis Φ, the transport equation becomes

$$\int_\Omega \frac{\partial C}{\partial t}\Phi d\Omega + \int_\Omega \left[\nabla\cdot(\boldsymbol{u}C) + \frac{\partial(wC)}{\partial z}\right]\Phi d\Omega + \int_\Omega E_h \nabla C\cdot\nabla\Phi d\Omega + \int_\Omega E_v \frac{\partial C}{\partial z}\frac{\partial \Phi}{\partial z} d\Omega = \int_\Omega S\Phi d\Omega + \oint \boldsymbol{Q}_s \cdot \boldsymbol{n}\Phi d\Gamma \quad (1.27)$$

where $\boldsymbol{Q}_s$ is the dispersive flux over the three-dimensional boundary. Following a procedure analogous to the development of (1.19), the transport equation can also be developed in terms of periodic motion.

1.6 Solution Strategies

All of the weighted residual statements discussed above lead to a system of equations for the nodal values of the dependent variables. The method of solution impacts the computational resources necessary to solve a particular problem such that there can be 2 or 3 orders of magnitude difference in the run times for different model equations and different matrix solution methods. Thus the solution method is of primary importance for model development.

The primitive equation formulation of the shallow water equations is much more computationally intensive than the wave equation formulation. One reason is that the former involves a simultaneous solution for all dependent variables η and the $\boldsymbol{u}$ components whereas with the latter η and $\boldsymbol{u}$ are decoupled which leads to smaller matrices. A common method of solving the non-linear system of equations arising from the primitive equations is to use an implicit finite difference time-stepping method coupled with a Newton-Raphson iteration on the resulting matrix. The resulting efficiency then depends on the matrix solution algorithm used. Three useful algorithms are frontal solvers, banded matrix solvers, and iterative solvers. All of these methods take advantage of the sparse nature of the matrix. Because the continuity equation is not necessarily diagonally dominant, some form of pivoting must be used to avoid an ill-conditioned matrix. This method of solving the shallow water equations is by far the most computationally intensive option. For periodic problems such as tides, there can be an order of magnitude or better improvement in run times when using the harmonic formulation of the problem because of the reduced number of iterations.

Using the wave equation formulation, η is solved for first followed by a back-calculation of $\boldsymbol{u}$. For the explicit time-stepping method [10], a matrix system is solved and no iteration is required. Moreover, for the case where node-point integration is used and horizontal friction is neglected, the matrix becomes diagonal and the velocity solution involves a simple 2×2 block diagonal matrix.

For periodic problems, an extremely useful approach is to use the harmonic form of the wave equation. First η is calculated from (1.21) and $\boldsymbol{u}$ is back-calculated from (1.22). The procedure is iterated until convergence. The convergence is oscillatory and can be shortened considerably with an under-relaxation factor for $\boldsymbol{u}$. Typical values range from 0.5 to 0.7. The wave equation is diagonally dominant so that pivoting is not necessary. A modified form of the frontal solver that reduces storage and run time has been used extensively for this problem [15]. In addition, iterative methods modified for complex matrices have been used successfully, although the break even point is rather high when compared with the frontal solver (see Chapter 6). Because of its high efficiency, it is sometimes useful to use this procedure to make exploratory calculations for a particular problem in order to understand the dynamics, then proceed to time-stepping methods for a more detailed study.

1.7 An Application

This volume contains other examples of several formulations and solution strategies indicated above. In this section we will focus on a study of salt-flux in a coastal plain estuary. The results presented here provide an illustration of the numerical methods and are not intended to be definitive or final results.

The estuary considered here is Delaware Bay, a large but shallow bay located between the states of Delaware, Pennsylvania, and New Jersey. This estuary is generally shallow (depths up to 5 m) but contains a deeper, relict river channel (depth over 10 m). The study area spans the distance from the Atlantic Ocean to the head of tide at Trenton, New Jersey, a distance of about 210 km. Water motions are dominated by tidal currents over most of the bay and the density is normally well-mixed vertically during late summer.

This study was prompted by a desire to predict changes in the salt distribution caused by a major alteration – a rise in sea level. As dispersion and mixing coefficients in a transport model are empirically derived, they can only be estimated from present conditions. Furthermore, they are scale-dependent and derived by sub-grid scale averaging of the advection terms to arrive at dispersion terms. The intention here is to simulate the hydrodynamics and salt dynamics with sufficient resolution that the major salt flux processes are treated explicitly. In this way, the use of empirical relations is minimized.

The requirements, then, are to formulate a coupled, three-dimensional flow and transport model with sufficient efficiency that a high resolution grid can be used. Toward this goal, the procedures for the harmonic wave equation model were used. The grid was formed using the automatic grid generation program of [4] and contains about 13000 nodes in a horizontal triangular grid and 25 nodes in the vertical. The highest resolution is placed in the channels near the end of the salt penetration. The model used terrain-following coordinates and the nodes in the vertical direction were arranged logarithmically in order to resolve the bottom boundary layer.

Details of the model formulation and the results for the hydrodynamics are reported in [18]. The model retained all non-linear interactions and included 5 primary tidal constituents (M_2, S_2, N_2, K_1, O_1), one overtide (M_4), and the residual (Z_0). The constituents were solved for sequentially with 5 to 10 iterations per constituent in an inner loop, and 3 iterations in an outer loop over all constituents. A frontal solver was used and typical run times were 136 sec per constituent per iteration on a 10 MFlop workstation.

The transport model is based on the harmonic version of (1.23) and (1.27). There are three kinds of salt flux processes in this type of model: tidal dispersion caused by correlations between tidal-period velocity and concentration, advection of the salt distribution by the residual velocity, and sub-grid scale dispersion. For this estuary and this level of resolution, the first two processes dominate. As an example, the interaction of the salt field and the residual circulation is illustrated here.

There are several mechanisms that drive the residual velocity, including river inflow, pressure gradient force from the density distribution, surface wind stress, and non-linear tidal processes. For the late summer period, river inflow does not contribute significantly to the residual velocity but is still important in the salt dynamics. Wind stress has been neglected. A plot of current speed normal to a transect through the upper part of Delaware Bay shows the characteristic distribution of the density driven circulation (Figure 1.1). The current tends to flow up-estuary in the deeper parts of the cross-section, and there is a compensating flow seaward in the more shoal parts. This pattern can be explained by the fact that the density forcing is stronger in the deeper parts. A major issue that is unresolved is whether the up-estuary flow in the channels extends all the way to the surface. The surface pressure gradient drives the surface current seaward,

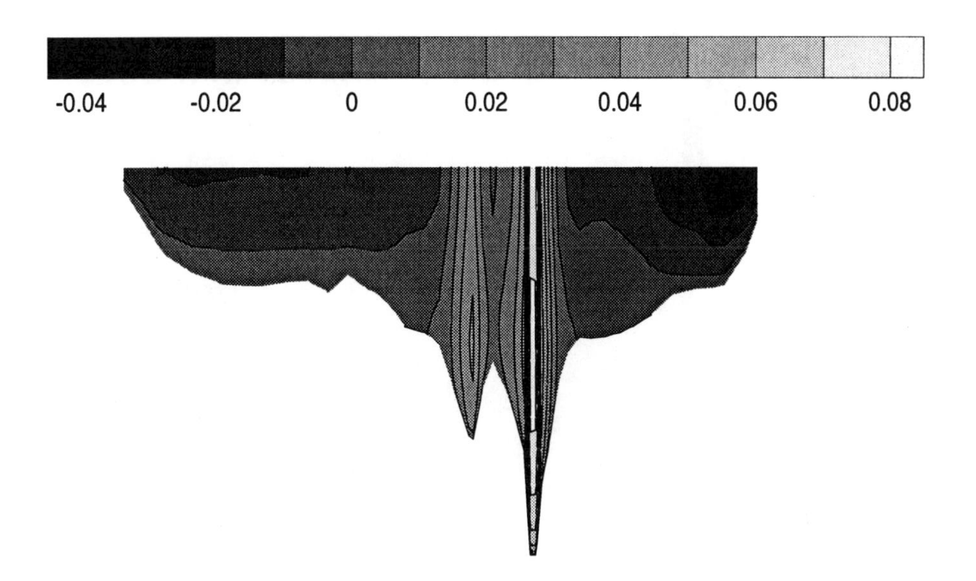

Figure 1.1: Residual current speed in a transect across upper Delaware Bay. The current is positive landward. Note the density current directed landward in the channels.

but the viscous coupling with the density flow drives it landward. This balance is sensitive to the details of the length scale used with the vertical eddy viscosity coefficient.

The salt distribution then responds to the residual flow pattern as shown in Figure 1.2. In particular, there is a landward displacement of the isohalines in the channels. This pattern is in agreement with observations that were made both longitudinally and laterally in the bay.

1.8 Conclusions

In summary, there are several ways to formulate the surface water flow problem. The details of the formulation depend on the spatial and temporal variations in the problem to be investigated. The methods outlined here are not complete, but reflect the range of applications considered in this volume. In addition, the choice of formulation has a great impact on computational efficiency – both array storage and run times. Traditional

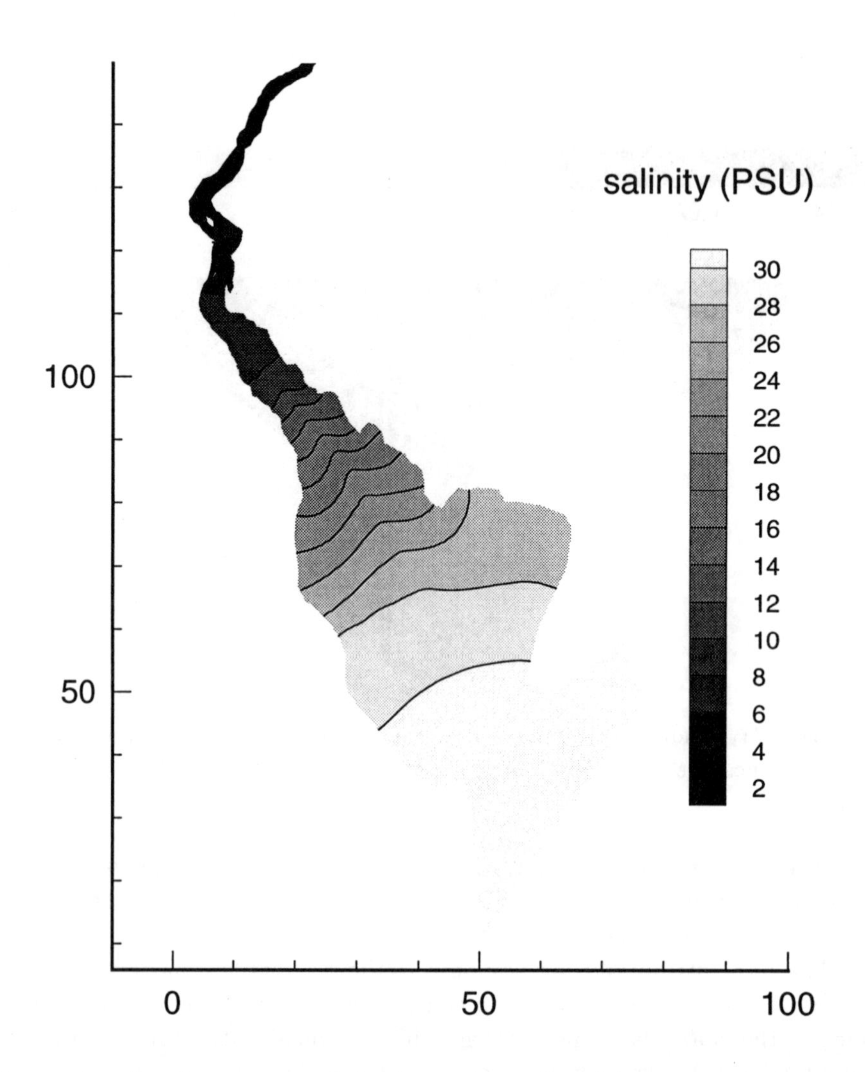

Figure 1.2: Depth-averaged salinity for Delaware Bay. The axes are in km. The displacement in the center of the isohalines indicates the location of the channel.

approaches with the primitive equations and Newton-Raphson iteration require by far the most computational resources. More recent developments with the wave equation approach have led to methods that do not support spurious oscillation modes and are much more computationally efficient.

Acknowledgments: This research was funded through the core research program of U.S. Geological Survey. The author is indebted to G.F.

Carey and E.J. Barragy for numerous discussions.

References

[1] Gray, W.G. "Some inadequacies of finite element models as simulators for two-dimensional circulation", *Advances in Water Resources*, 5, 171-177, 1982.

[2] Gray, W.G. and Pinder, G.F. "An analysis of the numerical solution of the transport equation", *Advances in Water Resources*, 12, 547-555, 1976.

[3] Haney, R.L. "On the pressure gradient force over steep topography in sigma coordinate ocean models", *J. Phys. Oceanogr.*, 21, 610-619, 1991.

[4] Henry, R.F. and Walters, R.A. "A geometrically-based, automatic generator for irregular triangular networks", *Communications in Numerical Methods in Engineering*, 9, 555-566, 1993.

[5] Kawahara, M. and K. Hasegawa, "Periodic Galerkin finite element method of tidal flow.",*IJNME*, 12, 115-127, 1978.

[6] King, I.P. and W.R. Norton, "Recent applications of RMA's finite element models to two-dimensional hydrodynamics and water quality, *Finite Elements in Water Resources*, Pentech Press, London, 1978.

[7] Kolar, R.L., Westerink, J.J., Catekin, M.E., and Blain, C.A. "Aspects of nonlinear simulations using shallow-water models based on the wave continuity equation", *Computers and Fluids*, 23, 523-538, 1994.

[8] Le Provost, C., G. Rougier and A. Poncet, "Numerical modeling of the harmonic constituents of the tides, with application to the English Channel", *J. Phys. Oceanography*, 11, 1123-1138, 1981.

[9] Lynch, D.R, and Gray, W.G. "A wave equation model for finite element tidal computations", *Computers and Fluids*, 7, 207-228, 1979.

[10] Lynch, D.R. and Werner F.E. "Three-dimensional hydrodynamics on finite elements. Part 1, Linearized harmonic model", *Int. J. Numerical Methods in Fluids*, 7, 871-909, 1987.

[11] Lynch, D.R. and Werner F.E. "Three-dimensional hydrodynamics on finite elements. Part 2, Non-linear time-stepping model", *Int. J. Numerical Methods in Fluids*, 12, 507-533, 1991.

[12] Pearson, C.E. and D.F. Winter, "On the calculation of tidal currents in homogeneous estuaries", *J. Phys Oceanography*, 7, 520-531, 1977.

[13] Pinder, G.F. and Gray, W.G. *Finite elements in surface and sub-surface hydrology*, 1977.

[14] Snyder, R.L., Sidjabat, M., and Filloux, J.H. "A study of tides, setup, and bottom friction in a shallow semi-enclosed basin. Part II, Tidal model and comparison with data", *J. Phys. Oceanog.*, 9, 170-188, 1979.

[15] Walters, R.A. "The frontal method in hydrodynamics simulations", *Computers and Fluids*, 8, 265-272, 1980.

[16] Walters, R.A. "Numerically induced oscillations in finite element approximations to the shallow water equations", *Int. J. Numerical Methods in Fluids* , 3, 591-604, 1983.

[17] Walters, R.A. "A three-dimensional, finite element model for coastal and estuarine circulation", *Continental Shelf Research*, 12, 83-102, 1992.

[18] Walters, R.A. "A model study of tidal and residual circulation in Delaware Bay and River", Submitted to *Journal of Geophysical Research*, 1994.

[19] Walters, R.A. and Carey, G.F. "Analysis of spurious oscillation modes in finite element approximations to the shallow water equations", *Computers and Fluids*, 11, 51-68, 1984.

[20] Walters, R.A. and R.T. Cheng, "Accuracy of an estuarine hydrodynamic model using smooth elements", *Water Resources Research*, 16, 187-195, 1980.

[21] Walters, R.A. and M.G.G. Foreman, "A three-dimensional, finite element model for baroclinic circulation on the Vancouver Island continental shelf", *Journal of Marine Systems*, 3, 507-518, 1992.

[22] Walters, R.A. and Werner, F.E., "A comparison of two finite element models using the North Sea data set", *Advances in Water Resources*, 12, 184-193, 1989.

Chapter 2

Environmental Hydrodynamics: Comprehensive Model for the Gulf of Maine

D.R. Lynch[1], *J.T.C. Ip*[1], *F.E. Werner*[2] *and E.M. Wolff*[1]

2.1 Introduction

The Gulf of Maine is a semi-enclosed coastal sea on the eastern North American shelf, stretching from the Cape Cod Islands eastward to the Bay of Fundy and the Nova Scotian shelf, and from the coast of Maine/New Brunswick seaward to the shelf break on the southern edge of Georges Bank. The Gulf represents a major international resource, with heightened interest in recent decades in tidal power, oil and gas, and fisheries. The availability of a valid, comprehensive circulation model for the Gulf is prerequisite to understanding and managing the Gulf ecosystem. The need for such a model has been articulated in several forums [22,25,28]. Herein we describe such a comprehensive model in general, and illustrate its performance relative to the Gulf circulation. In so doing, we seek to display state-of-the-art simulation methods for the generic class of environmental problems driven by coastal and shelf-scale circulation.

The basic Gulf-wide circulation is depicted in Figure 2.1 – a composite schematic summarizing numerous observational, theoretical, and modeling

[1]Dartmouth College, Hanover, NH.
[2]University of North Carolina, Chapel Hill, NC.

studies, beginning with the seminal work of Bigelow [1]. What is needed from computational models today is a quantitative, three-dimensional description of the transport, transit times, and branch points illustrated; their variability at decadal, seasonal, and shorter time scales; and a clearer picture of the controlling physical processes.

Generally, all important shelf processes are operative in the Gulf and contribute in various blends to the features shown in Figure 2.1. These include two primary and distinct inflows across the Scotian Shelf and through the Northeast Channel; deepwater formation in the three major Gulf basins; tides, tidal mixing, and tidal rectification; wind; stratification and frontal circulation; freshwater inflow along the coast; and local estuarine processes. These processes are all inextricably interwoven at various length- and time-scales and cannot be easily separated. Therefore, any comprehensive modeling strategy must ultimately be capable of simultaneous simulation of all of them.

The resolution demands of conventional numerical methods can be overwhelming in this context. For example, tidal rectification and frontal circulation can demand resolution of order 2 km, while properly equilibrated inflows from the Scotian Shelf demand upstream spatial coverage of order 4 shelf widths, i.e. horizontal coverage of order 1000 km. Similarly in the vertical, proper resolution of surface and bottom heat and momentum transfers demands vertical resolution of order 1 m, with total depth reaching 300 m in the Gulf basins and 1000 m or more at the shelf break. Uniform gridding over these ranges produces an estimated 250,000 horizontal cells, with perhaps on average 200 vertical cells. Coupling these spatial needs with the requirements of simulating nonlinear dynamics in tidal time produces a formidable computational burden for conventional modeling strategies.

The Finite Element Method (FEM) is uniquely suited to this task. Its basic strength lies in its freedom to use unstructured, nonuniform computational grids (i.e. with no *a priori* orientations or connections) of variable size triangular elements. This allows high local resolution of topographic (banks, coast) and physical (fronts, jets) features without necessarily sacrificing the horizontal extent of the model. In this way, fine local detail is embedded into the far-field or broad scale context in a natural and mathematically rigorous manner. It is this context which provides our working definition of a *Comprehensive Model*:

> A *Comprehensive Model* is one which provides shelf-scale geographic coverage; adequate local resolution of topography, coast, and flow features; and internal physics sufficient to capture all important shelf processes in tidal time.

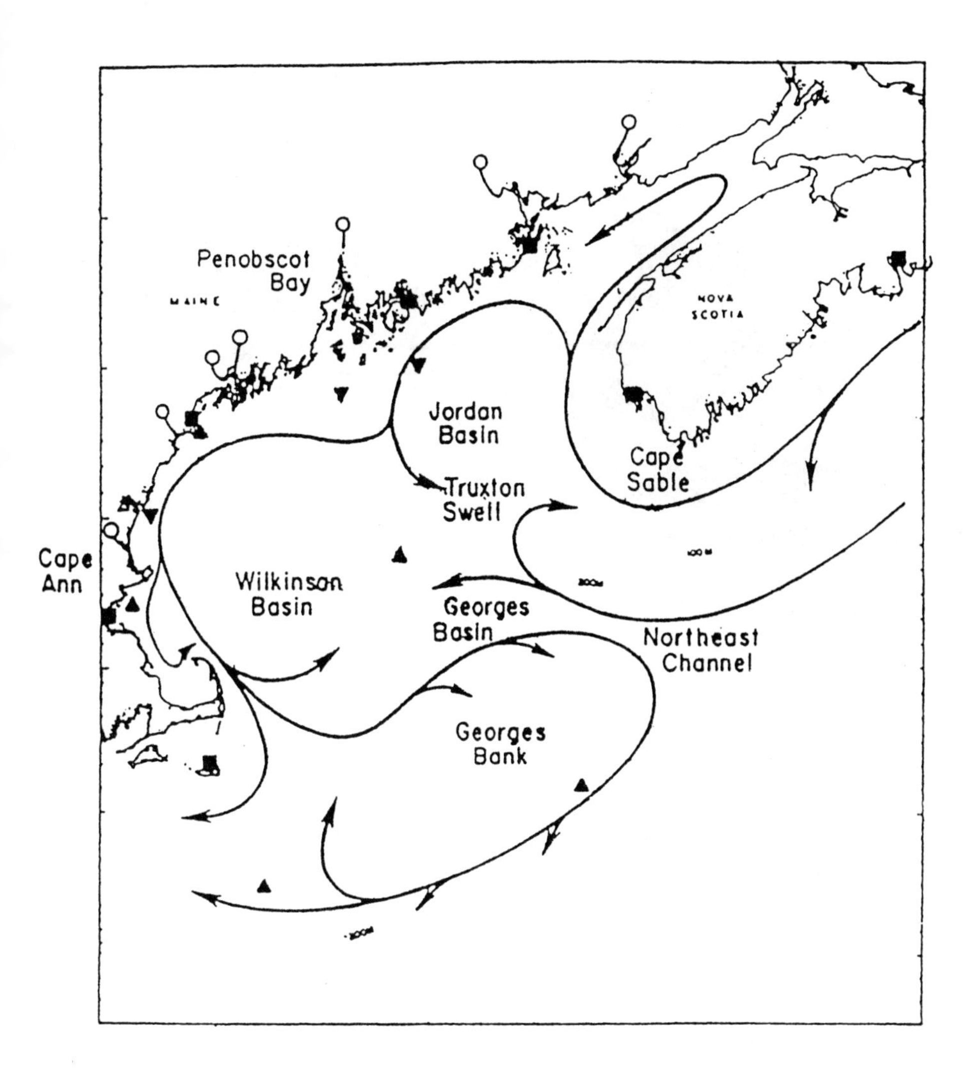

Figure 2.1: The general pattern of non-tidal surface circulation in the Gulf of Maine emphasizing inflow from the Scotian Shelf, general Gulf-scale counterclockwise flow, interior basin scale flow, and branch points in the coastal flow south of Penobscot, and off Cape Ann and Cape Cod [22]. Deeper exchanges through the Northeast Channel are not shown.

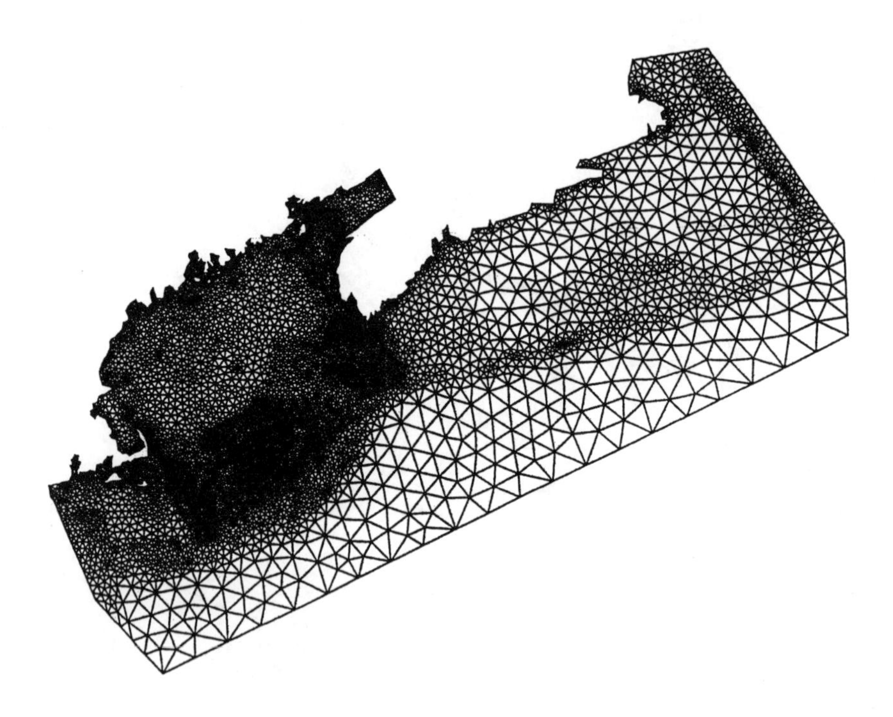

Figure 2.2: Horizontal finite element mesh discretizing the simulation domain of interest in the Gulf of Maine using linear triangles.

Such a FEM mesh appears in Figure 2.2. It includes 6756 horizontal nodes and 12877 triangles. Coverage extends from the Laurentian Channel to the western tip of Long Island, and seaward to roughly the 1000 m isobath, beyond which it terminates in a gently-sloping idealized ocean. Resolution on Georges Bank, for example, is of order 3 km, and approaches 2 km across the steep northern flank.

Several studies of the Gulf of Maine have now been completed with this mesh. Lynch and Naimie [14] studied three-dimensional tidal rectification; Ridderinkhof and Loder [21] examined Lagrangian trajectories in the tidally-rectified flow; Greenberg [7] examined the dynamic response to wind at various frequencies; and Naimie et al. [19] studied composite seasonal mean circulation fields for six two-month periods using climatological wind and density data. The flow fields from these studies have in turn been used in basic studies of advective influences on early life stages for cod [12,26,27]. All of these studies use the basic diagnostic FEM approach of Lynch et al. [16] and employ iteration in the frequency domain to achieve a nonlinear solution

[20]. The speed and simplicity of this approach has allowed rapid evaluations of basic physical and biological relationships in the Gulf which have proven extremely insightful. The diagnostic simplification of the physics carries, of course, a concomitant penalty in physical realism.

Herein we describe the results of a fully nonlinear, prognostic, time-domain model on the above mesh, with advanced turbulence closure incorporated. This model is comprehensive both geographically and physically, embodying most of the physical processes thought to be operative on the shelf.

2.2 Prognostic Time-Domain Model

The nonlinear time-domain model follows the algorithmic approach of Lynch and Werner [15], with improvements [8]. It is a free-surface, tide-resolving model based on the conventional 3-D shallow water equations. Temperature and/or salinity are transported in tidal time and the density field is closed prognostically via an equation of state. Vertical mixing of momentum, heat and mass is represented by a level 2.5 turbulence closure scheme [17]. This approach provides stratification- and shear-dependent mixing coefficients which evolve with the simulation and requires the prognostic evaluation of two additional macroscopic state variables, turbulent kinetic energy and mixing length. The improvements in Galperin et al. [5] have been found to be very important in avoiding unphysical situations on Georges Bank and have been adopted as standard. In the horizontal, Smagorinsky-type closure provides shear- and mesh scale-dependent eddy viscosity [23].

The governing equations are the 3-D, Reynolds-averaged Navier-Stokes and transport equations with Boussinesq and hydrostatic assumptions. The Reynolds stress (transport) is represented in eddy viscosity (diffusivity) form, parameterized in terms of stratification and two subgrid-scale quantities – turbulent kinetic energy and mixing length – which evolve at the macroscale. Using standard notation, we have the continuity, in 3-D and vertically integrated forms:

$$\frac{\partial w}{\partial z} = -\nabla_{xy} \cdot \mathbf{v} \tag{2.1}$$

$$\frac{\partial \zeta}{\partial t} = -\nabla_{xy} \cdot \int_{-h}^{\zeta} \mathbf{v} dz \tag{2.2}$$

momentum, heat and salt conservation:

$$\frac{d\mathbf{v}}{dt} + \mathbf{f} \times \mathbf{v} + g\nabla\zeta - \frac{\partial}{\partial z}\left(qls_m \frac{\partial \mathbf{v}}{\partial z}\right) = -\frac{g}{\rho_0}\int_z^{\zeta} \nabla\rho dz + \mathbf{F}_m \tag{2.3}$$

$$\frac{dT}{dt} - \frac{\partial}{\partial z}\left(qls_h \frac{\partial T}{\partial z}\right) = F_T \tag{2.4}$$

$$\frac{dS}{dt} - \frac{\partial}{\partial z}\left(qls_h \frac{\partial T}{\partial z}\right) = F_S \tag{2.5}$$

and transport equations for the turbulent kinetic energy $q^2/2$ and mixing length l:

$$\frac{dq^2}{dt} - \frac{\partial}{\partial z}\left(qls_q \frac{\partial q^2}{\partial z}\right) =$$
$$2\left[qls_m\left(\left(\frac{\partial u}{\partial z}\right)^2 + \left(\frac{\partial v}{\partial z}\right)^2\right) + \frac{g}{\rho_0} qls_h \frac{\partial \rho}{\partial z}\right] - 2\left[\frac{q^3}{B_1 l}\right]$$
$$\frac{dq^2 l}{dt} - \frac{\partial}{\partial z}\left(qls_q \frac{\partial q^2 l}{\partial z}\right) =$$
$$lE_1\left[qls_m\left(\left(\frac{\partial u}{\partial z}\right)^2 + \left(\frac{\partial v}{\partial z}\right)^2\right) + \frac{g}{\rho_0} qls_h \frac{\partial \rho}{\partial z}\right] - lW\left[\frac{q^3}{B_1 l}\right] \tag{2.6}$$

wherein E_1 and B_1 are experimental constants [17]; W is a wall proximity function [3]; s_m, s_h are algebraic stability functions dependent on the local stratification [5]; s_q is fixed at the value 0.2 [17]; and the density is related to temperature and salinity by the equation of state [6] $\rho = \rho(T, S)$. Finally, the non-advective horizontal exchanges $\mathbf{F}_m$, F_T, etc. are all expressed in Laplacian forms, e.g.

$$\mathbf{F}_m = \nabla_{xy} \cdot (A\nabla_{xy}\mathbf{v}) \tag{2.7}$$

with A given in terms of the local shear and grid scale as in Smagorinsky [23].

The depth-averaged continuity equation is expressed in "wave equation" form. The Galerkin weighted residual method is used to obtain the weak forms, which are discretized on simple linear elements (triangular in the horizontal, linear in the vertical) with all variables expressed in the same linear basis. A general terrain-following vertical coordinate is used, with a flexible FEM approach to vertical resolution. This provides continuous tracking of the free surface and proper resolution of surface and bottom boundary layers. The baroclinic pressure gradient is computed on level surfaces as in Naimie et al. [19].

The method of solution is essentially unchanged from Lynch and Werner [15]. A semi-implicit time-stepping procedure is used to solve the implicit wave equation for elevation (a 2-D calculation), followed by tridiagonal solution for velocity with vertical shear and bottom stress terms implicit in time. The discretization of the T, S, q^2, and q^2l equations is analogous to that for velocity and requires only tridiagonal matrix solution in each time step. Diffusivities and viscosity are represented vertically as element-wise constants which are recomputed in each time step. The decay terms in the q^2 and q^2l equations are treated in each time step as quasi-linear first-order decays, and handled implicitly, along with vertical diffusion in general, to avoid stability constraints. Based on scaling arguments, we neglect advection of q^2 and q^2l.

Illustrative results for some key nonlinear processes follow.

- *Tidal Rectification*
  In Figure 2.3 we show the Eulerian circulation on Georges Bank under M_2 tidal forcing alone, averaged over one tidal cycle. As discussed in Lynch and Naimie [14] and several previous studies, this is a basic point of departure, being a year-round barotropic contributor to the Bank circulation. The agreement with previous model results [14] is remarkably good, confirming that the earlier nonlinear frequency-domain and present time-domain results are equivalent. This is especially interesting since the present results use more advanced closure – which is evidently not an issue here in the absence of wind and stratification. Recent studies with this model reveal that enhanced resolution is required (mesh length 1-2 km) in the jet on the northern flank of Georges Bank, where topographic gradients are steepest [13].

- *Stratification*
  The above results are quite sensitive to stratification in the frontal region of Georges Bank [19]. In Figure 2.4 we show the equivalent of Figure 2.3, except with surface heating sufficient to force a $3\sigma_t$ stratification in the bank area. Apparent is a significant southwestward extension of the recirculation zone in the vicinity of the Great South Channel. Tremblay et al. [26] and Werner et al. [27] have highlighted the importance of seasonally-intensified recirculation in this area to the retention of cod and scallop larvae on the Bank.

  In addition to enhanced retention, there is a significant increase in speed, especially on the northern flank of the Bank. Peak velocities in this location increase from 0.125 cm/s in the barotropic case to 0.35

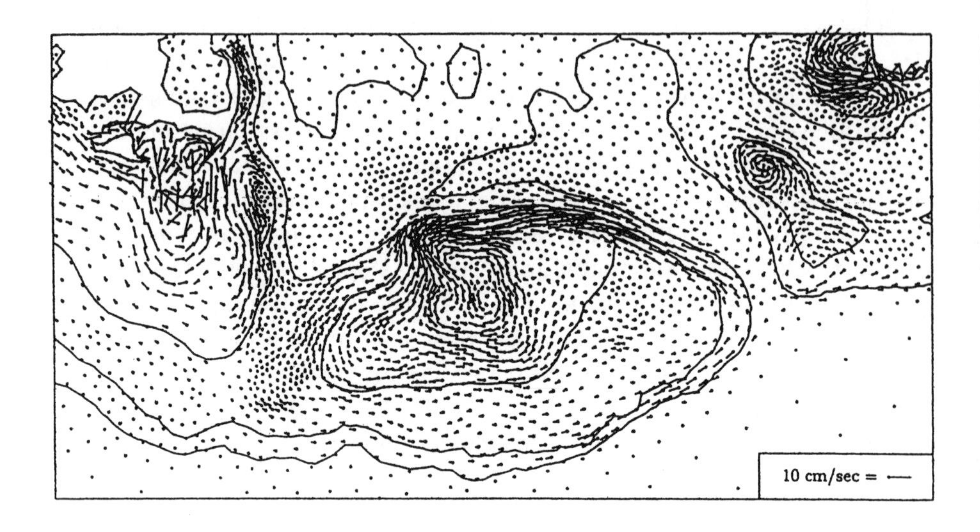

Figure 2.3: Vertically averaged tidal residual current at Georges Bank obtained from the simulation results using the prognostic time-domain model and employing a Mellor-Yamada level 2.5 turbulence closure scheme with Galperin modification. Simulation is forced with tide only.

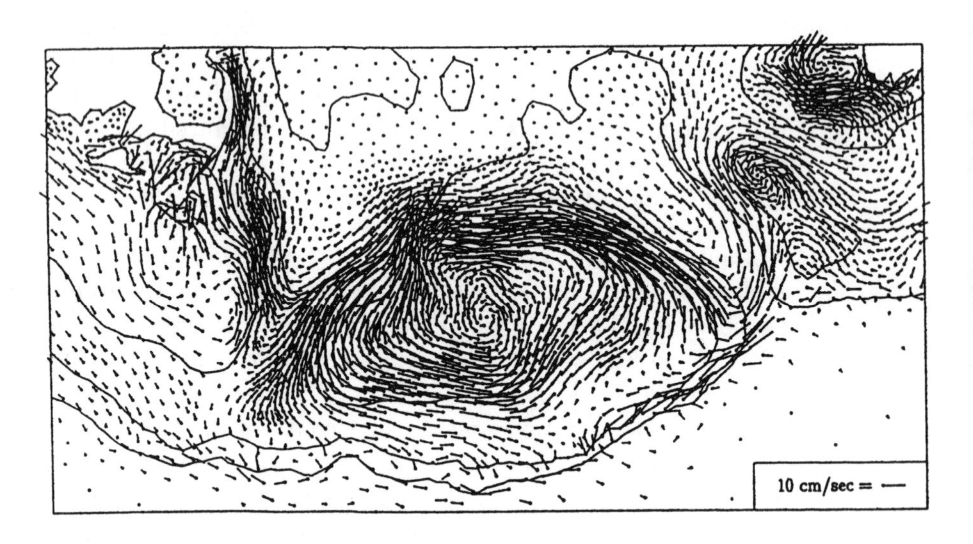

Figure 2.4: Same as Figure 2.3, but forced with tide and surface heating.

cm/s in the stratified case and observational evidence of seasonal intensification of this feature is well-established [9,11]. Figure 2.5 shows a transect across the northern flank and reveals a complex frontal structure, changing from stratified conditions off-bank to well-mixed conditions on the bank top where tidal velocities are highest and the resulting turbulent mixing is strongest. The location of the front is in good agreement with observation and theory [10,14]. A complex cross-bank velocity pattern accompanies the intensified along-bank jet. Figure 2.6a and 2.6b show the turbulent energy and resulting eddy viscosity profiles on this transect at one point in tidal time. The increasing vertical penetration of the bottom boundary layer with position on-bank is evident, with turbulent energy peaking at the bottom. As expected, eddy viscosity peaks above the bottom, below which it decays to zero in an asymptotically linear manner. Finally, Figure 2.7 shows, again on the same transect, the density structure at two points in tidal time, separated by 6.2 hours. The propagation of internal waves from the northern flank is evident. Corresponding to the tidal excursion, there is an approximately 10 km cross-bank migration of isopycnals on the bank top. On the sides of the bank this generates vertical isopycnal motion of order 20-30 m.

- *A Seasonal Composite*
  In Figure 2.8 we show the Gulf-wide tidal residual circulation pattern in early spring (March-April), forced by the M_2 tide and climatological wind stress. Initial conditions for this simulation were from Naimie et al. [19], including the climatological density field for this period and the velocity field derived therein with the earlier diagnostic model. These initial conditions were allowed to evolve prognostically; surface heating was handled by a first-order "nudging" of the surface density toward its climatological value, with a time constant of five days. There are several important features of this simulation. There is the familiar partly-open gyre surrounding Georges Bank, reflecting the dominance of tidal rectification in this lightly-stratified season. There is upwelling off southwestern Nova Scotia, which is present in essentially all model runs and thought to be the year-round result of topographic tidal rectification [24]. Over Jordan and Georges Basins there is a large-scale, counterclockwise gyre reminiscent of baroclinic circulation [1]. Finally, there is a well-organized coastal current along the northern and western Gulf with three important branch points. The first is south of Penobscot Bay – there the current is drawn off-

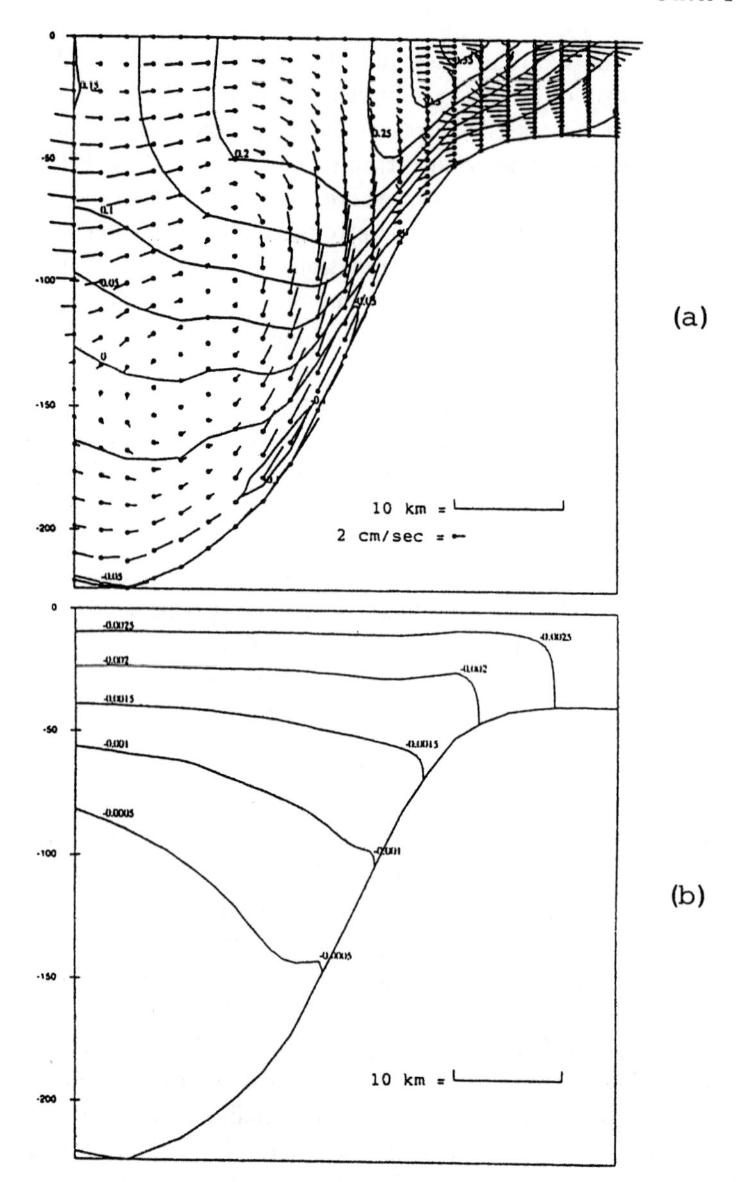

Figure 2.5: Tidally averaged vertical transect across the northern flank from the stratified case of Figure 2.4, showing the density profile changing from stratified off-bank to well-mixed on the bank top where tidal turbulence is strongest. (a) The contour lines illustrate the current speed, and the origin-marked vectors show the cross-bank velocity on the transect (top figure). (b) The contour lines show density (bottom figure).

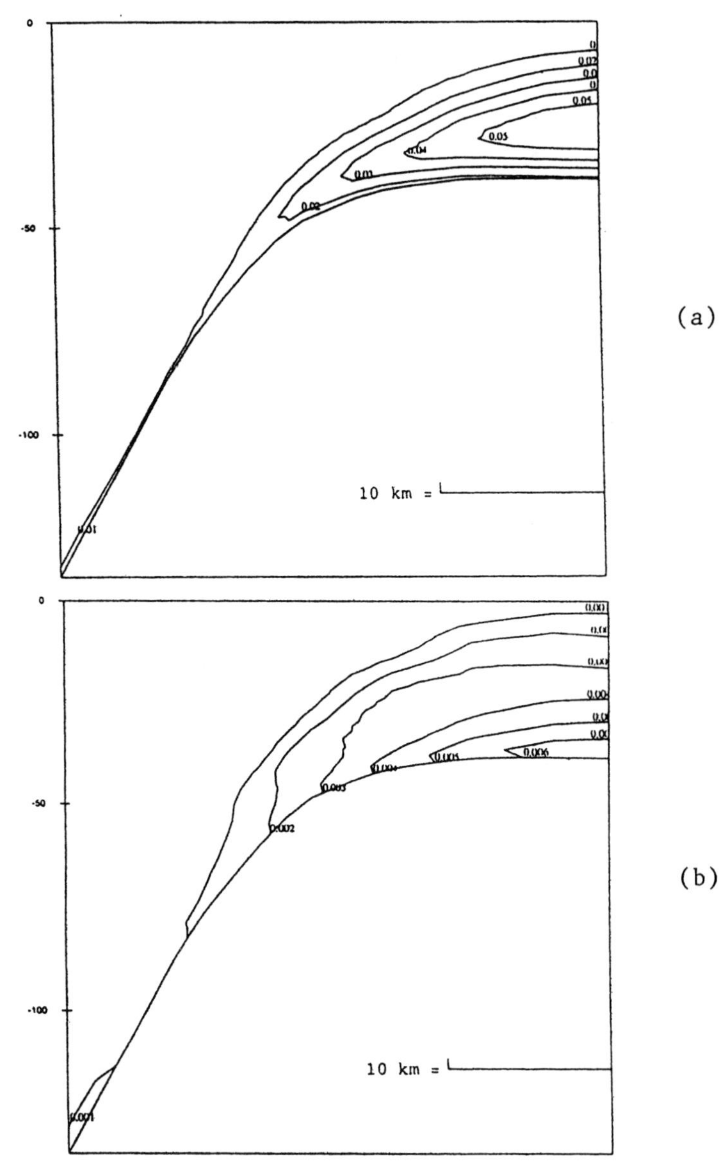

Figure 2.6: Tidal time vertical transect across the northern flank from the baroclinic simulation showing the increasing vertical penetration of the bottom boundary layer with position on-bank. (a) The contour lines show vertical eddy viscosity increasing from the surface to bottom, reaching a maximum and dropping off in the bottom boundary layer where mixing length is limited (top figure). (b) The contour lines show turbulent kinetic energy increasing from the surface to bottom (bottom figure).

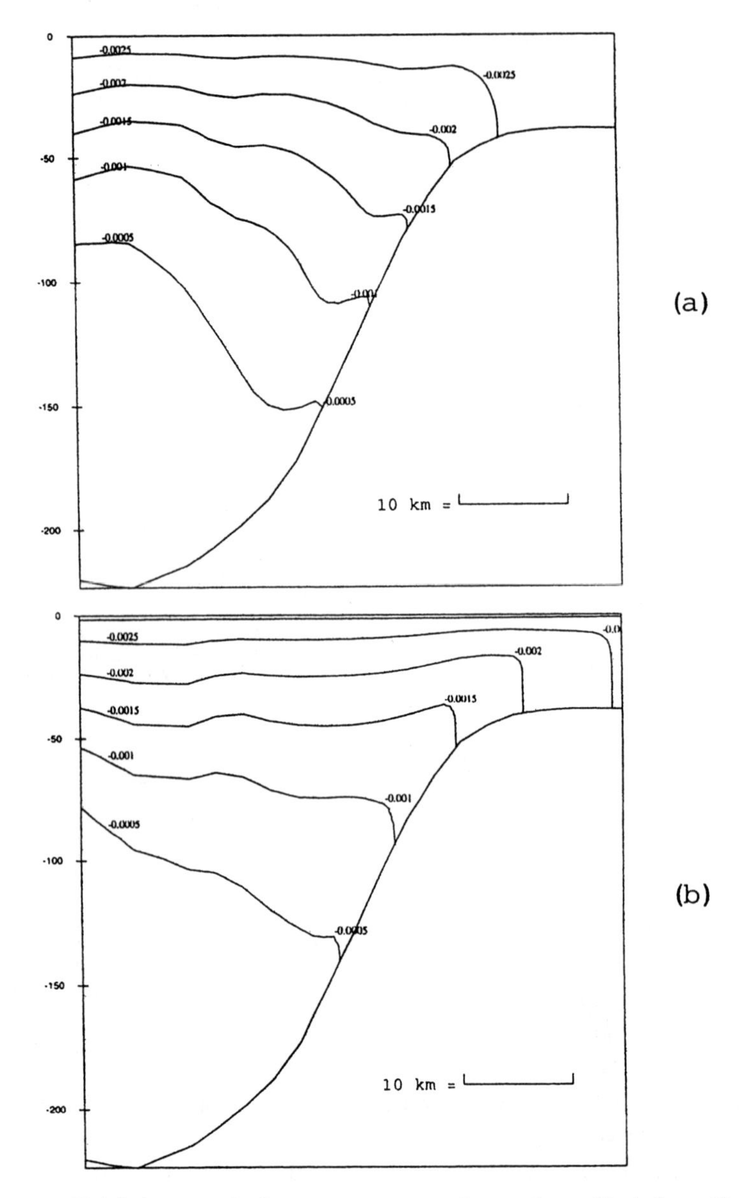

Figure 2.7: Tidal time vertical transects across the northern flank from the stratified simulation showing advection of density structures. The two figures are separated by 6.2 hours.

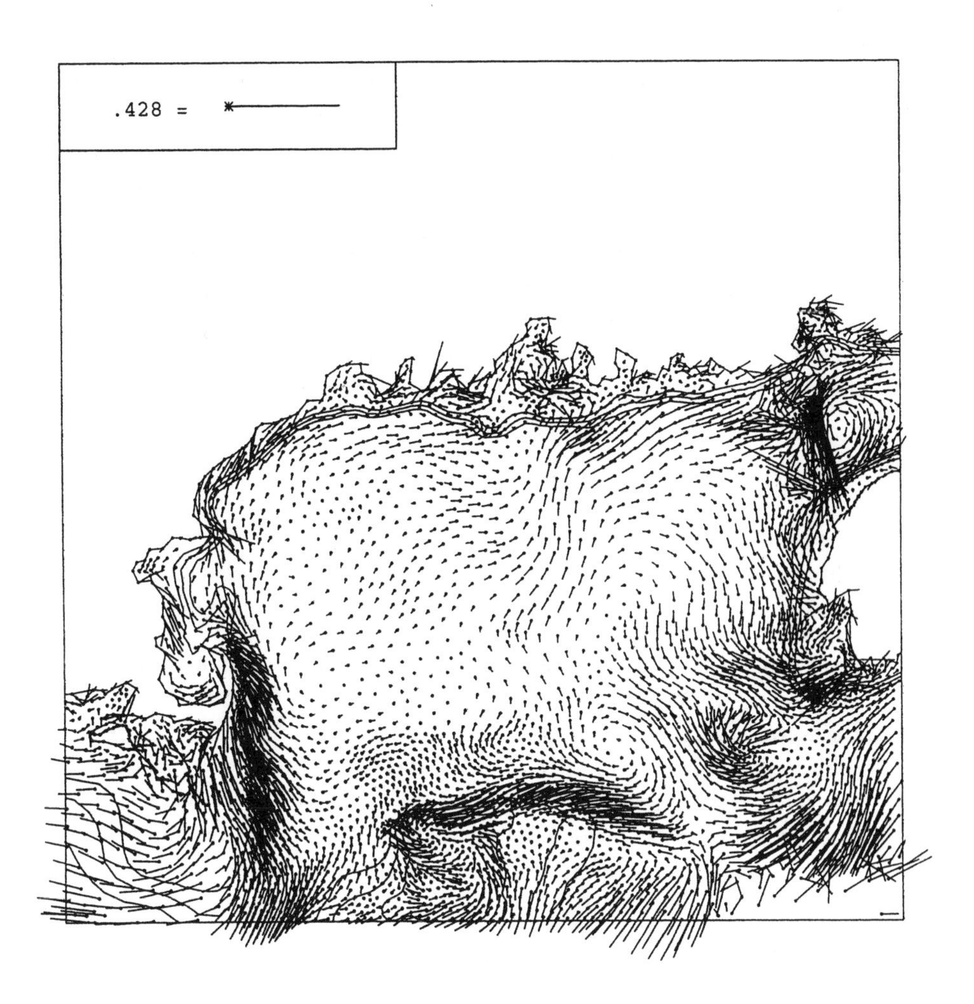

Figure 2.8: The baroclinic Gulf-wide tidal residual circulation pattern in early spring (March-April), forced by the M_2 tide and climatological wind stress. Initial conditions for this simulation were from Naimie et al. [19], including the climatological density field for this period and the velocity field derived therein.

shore, a portion merging with the Jordan Basin gyre and a portion of flow rejoining the coastal current further to the west. Following the coastal current west, one encounters a second branch point at Cape Ann, where a portion enters and circuits the Massachusetts and Cape Cod Bays. Further downstream, there is a third branch point east of Cape Cod, where a portion of the flow is diverted east to Georges Bank. These features are qualitatively very realistic by comparison with Figure 2.1. Interestingly, there is no source of freshwater from coastal rivers in this simulation; the coastal current is driven solely by the large-scale baroclinic circulation which is set up by the climatological initial conditions and the Gulf-wide influence of the wind and through-flows at the Scotian Shelf and Northeast Channel.

2.3 Transport Models

The computed flow fields described above support two different types of models for following the motion and interpreting its significance.

- *Lagrangian Particle Tracking*
  Some insight into the 3-D nature of the coastal current can be obtained by tracking particles in the computed flow field. Some representative trajectories appear in Figure 2.9, based on the 3-D March-April flow field computed in Naimie et al. [19]. Passive drifters were released at several locations in the western Gulf at 20 m below the surface, and tracked numerically using a 4th order Runge-Kutta algorithm [2]. For the conditions simulated, this depth is below the Ekman layer and the drifters generally are recruited north, below the influence of the prevailing southeastward wind, to the coast, where they upwell and then join the coastal current. The composite of these different trajectories illustrates both the complexity of 3-D behavior in the presence of topography and the branch point structure described above. These particle-tracking calculations have been used to advantage in the study of cod and scallop early-life history and are the basis of our present efforts at building and interpreting more advanced individual-based models of these species and their prey.

- *Advection-Diffusion-Reaction Model*
  Finally, the conventional approach to transport of nutrients, phytoplankton, etc. is to construct advection-diffusion-reaction models driven by the flow fields of interest. In Figure 2.10 we show such a cal-

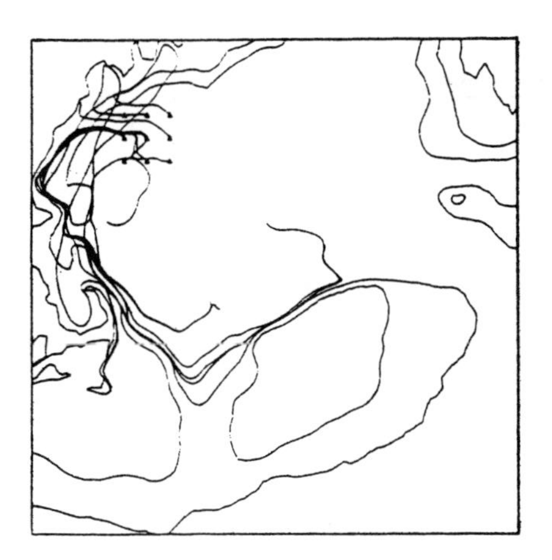

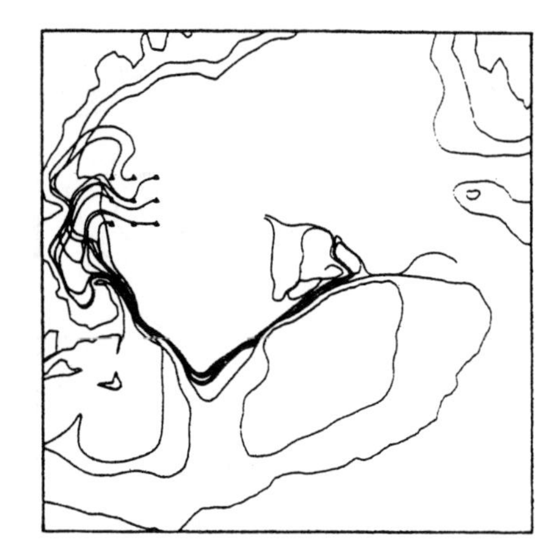

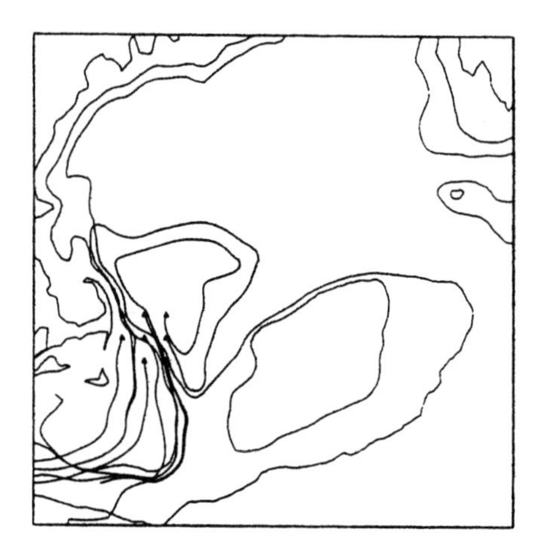

Figure 2.9: Simulated movement of a passive tracer in the western Gulf for March and April. The triangles mark the starting points of particles launched at various sites near the New England coastline and west of Georges Bank at 20 m depth. The particles were allowed to drift passively for 120 days. The particle paths were calculated using the program `DROG3D.f` [2]

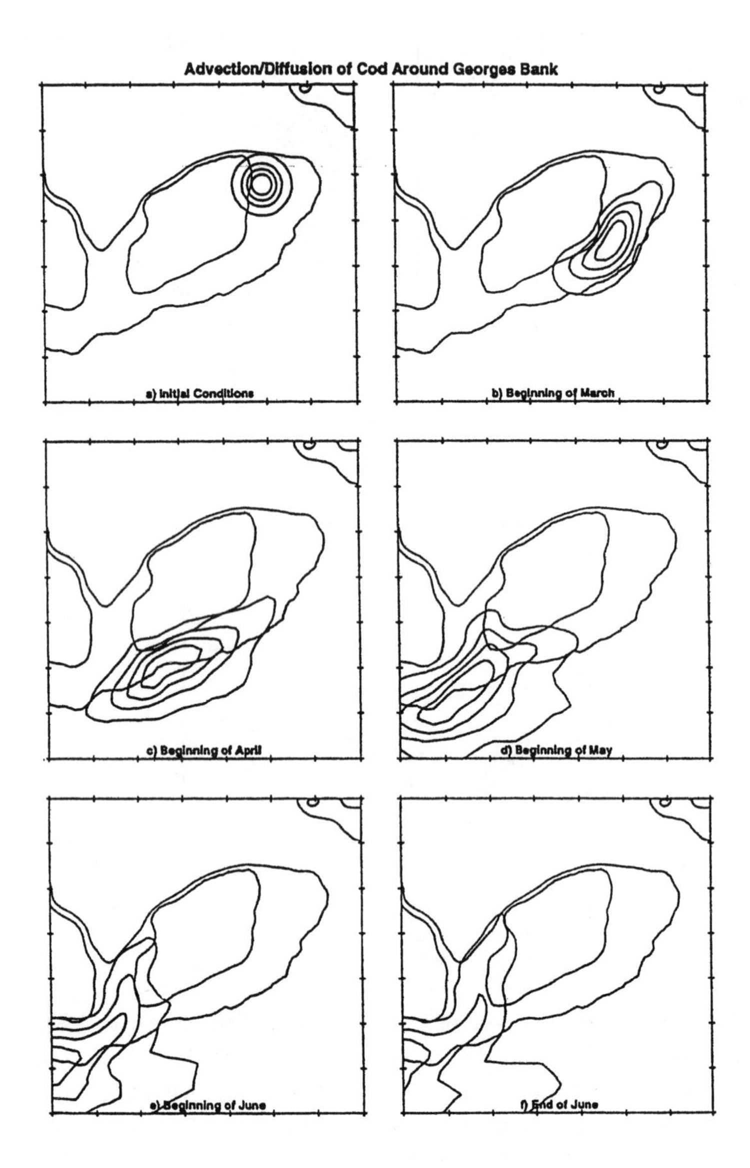

Figure 2.10: Transport of an initial Gaussian distribution of a conservative tracer from the cod spawning grid [27], from the beginning of February to the end of June. A monthly climatologically-derived flow field was averaged over depth to provide the advective velocities in tidal time.

culation based on depth-averaged monthly flow fields from Naimie [18] computed as in Werner et al. [27] A Gaussian distribution of a passive tracer is introduced near the primary spring spawning site for cod; its tidal time evolution over subsequent months illustrates the dominant transport effect of advection in the residual flow field, and the dispersive effect of the tidal-time excursion. The partial recirculation of the distribution in this case is highly sensitive to the specific location of spawning; see Figure 2.11 for a comparison. In additional, more detailed studies of both the depth-averaged 2-D and the complete 3-D advection-diffusion-reaction models coupled with nutrients-phytoplankton-zooplankton population dynamics driven by the diagnostic flow fields of interest have been undertaken. These provide important insights into the mechanism of shear dispersion in the complex 3-D flow on the bank [29].

2.4 Discussion and Directions

We have demonstrated the validity of a comprehensive coastal/shelf circulation model. The FEM provides the essential spatial ingredients: broad geographic coverage with flexible local resolution both horizontally and vertically. The incorporation of turbulence closure and fully prognostic mass field evolution in tidal time brings the internal physical processes to state-of-the-art for hydrostatic shelf modeling. The examples demonstrate a broad range of circulation features representing important nonlinear interactions among shelf physics and detailed topography.

The results to date for the Gulf of Maine illustrated several important circulation features of the region which embody a diverse set of critical nonlinear processes. There are many detailed aspects of these circulation fields which will require careful scrutiny in the coming years. Specific ongoing activities center on three general themes:

- *Coastal Current*
  This involves a study of increased resolution; the interaction of local (i.e. local wind and river discharge) versus large-scale (i.e. Jordan Basin gyre) physical features; and the posing of limited-area-model boundary conditions from more comprehensive model results.

- *Georges Bank*
  Investigation of feeding, mortality, and advective losses of larval fish and zooplankton through the use of advanced individual-based models

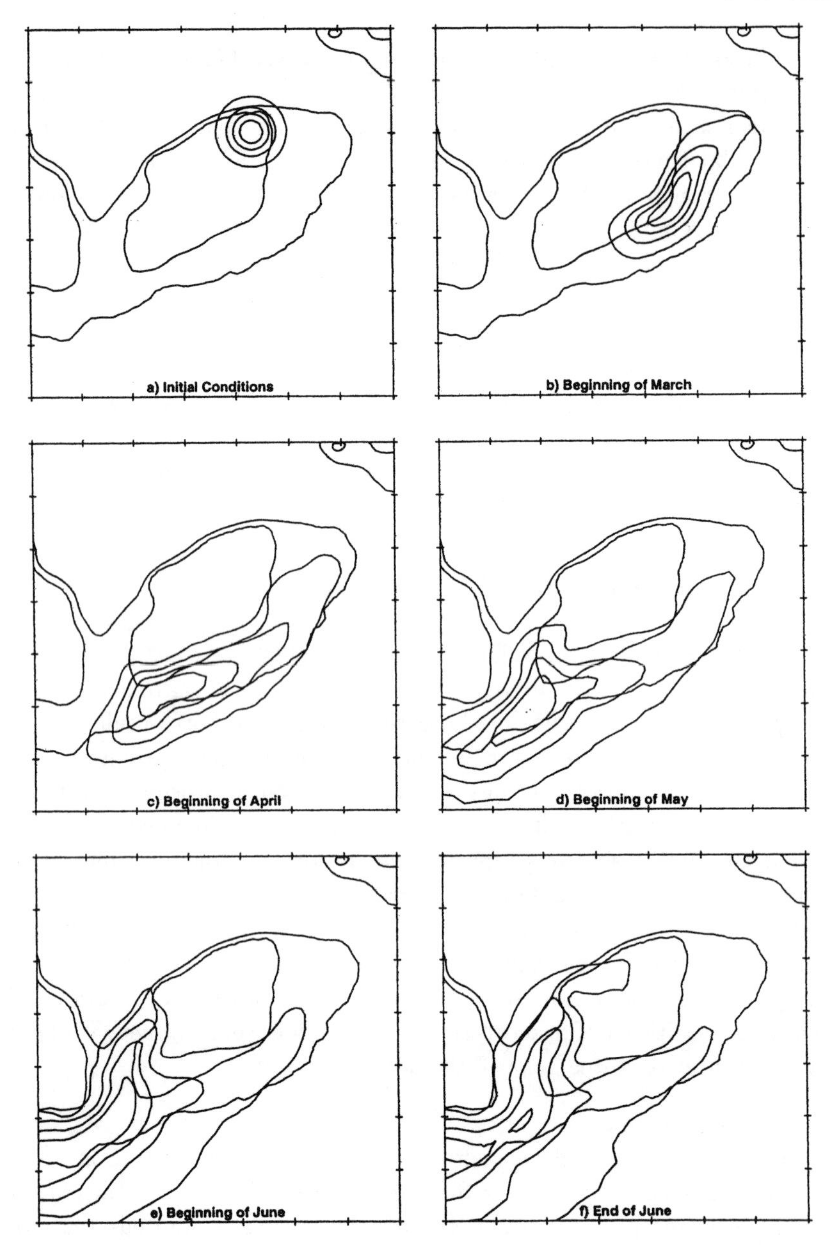

Figure 2.11: When the initial Gaussian distribution is moved on-bank, the increase in on-bank circulation and retention is clearly visible. The figure depicts the advection and diffusion of this concentration from early February through the end of June. The flow field is the same as in Figure 2.10.

in conjunction with Lagrangian particle-tracking; investigation of population dynamics of the zooplankton community through advective-diffusive models; and study of the interactions among nutrients, phytoplankton, and zooplankton with those models. All of this implies a detailed ongoing effort to understand the Bank's frontal structure.

- *Deep Basins*
  Investigation of the processes of deep water intrusion and water mass evolution, and the role of these processes in controlling the large-scale circulation in the Gulf. Implied is the investigation of the processes governing the two primary inflows at the Scotian Shelf and Northeast Channel.

We look forward to progress in these matters, and to the continued evolution of generic circulation models based on detailed site-specific studies in the Gulf and elsewhere.

Acknowledgments: We thank our colleagues David Brooks, Wendell Brown, David Greenberg, Charles Hannah, Matthias Johnsen, John Loder, Christopher Naimie, Peter Smith, and David Townsend for their generous help in innumerable ways. Financial support from the National Science Foundation, the Office of Naval Research, the Joint NSF-NOAA GLOBEC program, the Gulf of Maine Regional Marine Research Program, and the New Hampshire Sea Grant College Program is gratefully acknowledged. This is Contribution No. 29 of the U.S. GLOBEC Program. An earlier version of this paper appeared in the RARGOM workshop proceedings [4].

References

[1] Bigelow, H.B., "Physical Oceanography of the Gulf of Maine", *Bull. U.S. Bur. Fish.*, 40, 511-1027, 1927.

[2] Blanton, B. *Drogues.f and Drogdt.f User's Manual for 2-Dimensional Drogue Tracking on a Finite Element Grid with Linear Finite Elements*, Skidaway Institute of Oceanography, Savannah, Georgia, 1992.

[3] Blumberg, A. F., B. Galperin, and D. J. O'Connor, "Modeling vertical structure of open-channel flows", *ASCE, J. Hydraulic Engg.*, 118, 1119–1134, 1992.

[4] Braasch, E. (ed.), *Gulf of Maine Circulation Modeling: Workshop Proceedings*, RARGOM Report no. 94-1, Dartmouth College, Hanover, NH, 101, 1994.

[5] Galperin, B., L.H. Kantha, S. Hassid, and A. Rosati, "A quasi-equilibrium turbulent energy model for geophysical flows", *J. Atmos. Sci.*, 45, 55–62, 1988.

[6] Gill, A.E., *Atmosphere-Ocean Dynamics*, Academic Press, 599–603, 1982.

[7] Greenberg, D.A., "Model Progression in the Gulf of Maine", in Proc. RARGOM Circulation Modeling Workshop, Dartmouth College, Hanover, New Hampshire, Oct. 1993, RARGOM Report #94-1, 1994.

[8] Ip, J.T.C., and D.R. Lynch, *Three-Dimensional Shallow-Water Hydrodynamics on Finite Elements: Nonlinear Time-Stepping Prognostic Model, QUODDY User's Manual, Numerical Methods Laboratory Report NML-94-1*, Thayer School of Engineering, Dartmouth College, Hanover, New Hampshire, 1994.

[9] Loder, J.W., D. Brickman, and E.P.W. Horne "Detailed structure of currents and hydrography on the northern side of Georges Bank", *J. Geophys. Res.*, 97, C9, 14331-14351, 1992.

[10] Loder, J.W., and D.A. Greenberg, "Predicted positions of tidal fronts in the Gulf of Maine region", *Continental Shelf Res.*, 6, 397-414, 1986.

[11] Loder, J.W., and D.G. Wright, "Tidal rectification and frontal circulation on the sides of Georges Bank", *J. Marine Res.*, 43, 581-604, 1985.

[12] Lough, R.G., W.G. Smith, F.E. Werner, J.W. Loder, C.G. Hannah, C.E. Naimie, F.H. Page, R.I. Perry, M.M. Sinclair, and D.R. Lynch, "The influence of advection processes on the interannual variability in cod egg and larval distributions on Georges Bank: 1982 vs 1985", *ICES Mar. Sci. Symp.*, 198, 1994 (in press).

[13] Lynch, D.R., J.T.C. Ip, F.E. Werner, and C.E. Naimie, "Convergence studies of tidally-rectified circulation on Georges Bank", in *Quantitative Skill Assessment for Coastal Ocean Models*, Editors: D.R. Lynch and A.M. Davies, Coastal and Estuarine Series, American Geophysical Union, 1994 (in press).

[14] Lynch, D.R., and C.E. Naimie, "The M2 tide and its residual on the outer banks of the Gulf of Maine", *J. Phys. Oceanogr.*, 23, 2222-2253, 1993.

[15] Lynch, D.R., and F.E. Werner, "Three-dimensional hydrodynamics on finite elements. Part II: Non-linear time-stepping model", *Intl. J. Numer. Meth. Fluids*, 12, 507–533, 1991.

[16] Lynch, D.R., F.E. Werner, D.A. Greenberg, and J.W. Loder, "Diagnostic Model for Baroclinic and Wind-Driven Circulation in Shallow Seas", *Continental Shelf Res.*, 12, 37-64, 1992.

[17] Mellor, G. L., and T. Yamada, "Development of a turbulence closure model for geophysical fluid problems", *Reviews of Geophys. Space Phys.*, 20, 851–875, 1982.

[18] Naimie, C.E., "Seasonal variation of the three-dimensional low-frequency circulation in the vicinity of Georges Bank", *PhD Thesis*, Dartmouth College, Hanover, New Hampshire, 1995.

[19] Naimie, C.E., J.W. Loder, and D.R. Lynch, "Seasonal variation of the 3-D residual circulation on Georges Bank", *J. Geophys. Res.*, 99, C8, 1184-1200, 1994.

[20] Naimie, C.E., and D.R. Lynch, "Three-Dimensional Diagnostic Model for Baroclinic, Wind-Driven and Tidal Circulation in Shallow Seas – FUNDY5 Users' Manual", *Numerical Methods Laboratory Report NML-93-1*, Dartmouth College, Hanover, New Hampshire, March 1993.

[21] Ridderinkhof, H., and J.W. Loder, "Lagrangian characterization of circulation over submarine banks with application to the outer Gulf of Maine", *J. Phys. Oceanogr.*, 24,1184-1200, 1994.

[22] RMRP, "Gulf of Maine Research Plan" Gulf of Maine Regional Marine Research Program, University of Maine, June 1992.

[23] Smagorinsky, J., "General circulation experiments with the primitive equations I. The basic experiment", *Monthly Weather Review*, 91, 99–164, 1963.

[24] Tee, K.-T., P.C. Smith, and D. LeFaivre, "Topographic upwelling off southwest Nova Scotia", *J. Phys. Oceanogr.*, 23, 1703-1726, 1993.

[25] Townsend, D.W., "The Implications of Slope Water Intrusions in the Gulf of Maine", Proceedings of a Workshop, March 6-7 1989, Editor: D.W. Townsend, sponsored by ARGO-Maine, Bigelow Laboratory for Ocean Sciences, 1989.

[26] Tremblay, J.M., J.W. Loder, F.E. Werner, C.E. Naimie, F.H. Page, and M.M. Sinclair, "Drift of sea scallop larvae on Georges Bank: a model study of the roles of mean advection, larval behavior and larval origin", *Deep Sea Research II*, 41, 7-49, 1994.

[27] Werner, F.E., F.H. Page, D.R. Lynch, J.W. Loder, R.G. Lough, R.I. Perry, D.A. Greenberg, and M.M. Sinclair, "Influences of mean advection and simple behavior on the distribution of cod and haddock early life stages on Georges Bank", *Fish. Oceanogr.*, 2, 43-64, 1993.

[28] Wiggin, J., and C.N.K. Mooers, "Proceedings of the Gulf of Maine Scientific Workshop", Urban Harbors Institute, Boston, December 1992.

[29] Wolff, E.M., "An advection-diffusion-reaction model with NPZ (nutrients, phytoplankton, Zooplankton) relationship applied to Georges Bank", *M.S. Thesis*, Dartmouth College, Hanover, New Hampshire, 1994.

Chapter 3
Surface Elevation and Circulation in Continental Margin Waters

J.J. Westerink[1], *R. A. Luettich Jr.*[2], *C.A. Blain*[1] *and S.C.Hagen*[1]

3.1 Introduction

In the development of models to predict tidal and hurricane driven surface water elevation and currents in continental margin waters along the U.S. east and Gulf coasts, we have found it highly advantageous to define a domain which encompasses a large expanse of the deep ocean in addition to the continental margin regions of interest. In fact, our Western North Atlantic Tidal (*WNAT*) model domain, shown in Figure 3.1, encompasses a significant portion of the Atlantic Ocean, as well as the entire Gulf of Mexico and Caribbean Sea. The advantages of this domain are related to the geometry and location of the open ocean boundary which significantly simplify the task of boundary condition specification.

For tidal computations, Westerink et al. [23] demonstrate the advantages of locating the open ocean boundary in deep waters along the 60°W meridian where tides vary more gradually and nonlinear constituents are much smaller than on the continental shelf. Further benefits of the specified open ocean boundary include its geometric simplicity, avoiding complicated cross shelf

[1]Department of Civil Engineering and Geological Sciences, University of Notre Dame
[2]University of North Carolina at Chapel Hill, Institute of Marine Sciences

boundaries and the ability to couple the *WNAT* model to global tidal models which tend to be the most accurate in deep waters (in particular global models under current development).

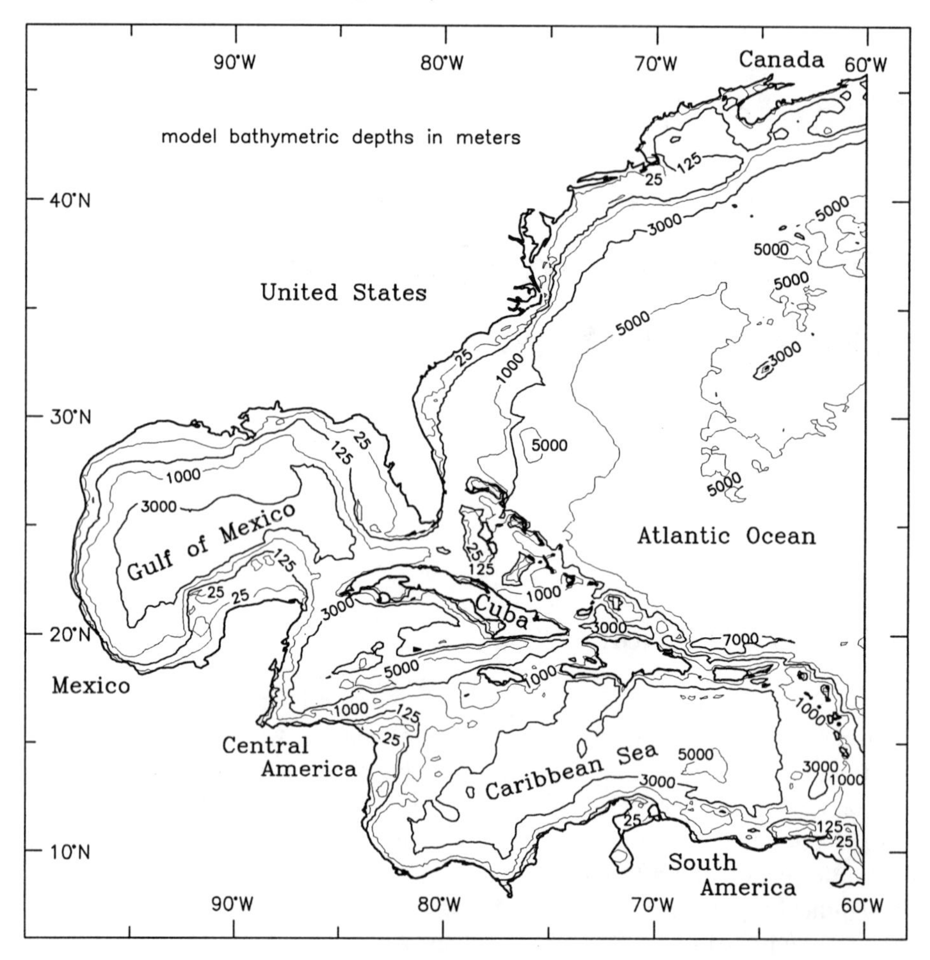

Figure 3.1: The model domain including bathymetry (in meters).

For hurricane storm surge calculations, it has been shown that not placing open ocean boundaries on shelf regions located in the vicinity of the storm leads to significantly more accurate storm surge predictions [1]. This is related to the fact that surface elevation response varies slowly in deep water and can be approximated well by an inverted barometer related to the pressure deficit in the center of the hurricane while in shallow shelf waters the inverted barometer disappears and is replaced by a rapid variation in elevation response which is highly dependent on the shape of the shoreline as well as local bathymetry. Furthermore, it is important that the open ocean boundary be placed outside of and away from the Gulf of Mexico and Caribbean within which storms can excite basin wide resonant modes/hurricane forerunner which are extremely difficult to correctly represent in a boundary forcing function [1].

The *WNAT* model domain covers an area of more than $8.347 \times 10^2 km^2$. The same reasons that make the large *WNAT* domain attractive in terms of simplifying the specification of boundary conditions for both tidal and hurricane storm surge calculations, make it attractive to apply unstructured grids with large variability in node-to-node spacing. The significant variability in hydrodynamic response as well as response gradients and structure (which are functions of the large variability in bathymetry and coastline features as well), will require localized grid refinement in specific regions. Unstructured graded grids will allow the domain to be optimally discretized and will result in the most accurate solutions for the fewest nodes. The focus of this paper is to examine the grid resolution requirements for both tidal and hurricane storm surge calculations. In particular, we examine the importance of providing a high level of localized resolution in regions with steep bathymetric gradients (i.e. the continental shelf break and slope) when developing unstructured graded grids for tidal computations. We then examine the influence of variable grid structure for hurricane storm surge predictions.

3.2 Hydrodynamic Model ADCIRC-2DDI

The computations described in this chapter were performed using *ADCIRC-2DDI*, the depth integrated option of a system of two- and three-dimensional hydrodynamic codes named *ADCIRC* [10]. *ADCIRC-2DDI* uses the depth integrated equations of mass and momentum conservation, subject to the incompressibility, Boussinesq and hydrostatic pressure approximations. Using the standard quadratic parameterization for bottom stress and neglecting baroclinic terms and lateral diffusion/dispersion effects leads to the following set of conservation statements in primitive non-conservative form expressed in a spherical coordinate system [7]:

$$\frac{\partial \zeta}{\partial t} + \frac{1}{R\cos\phi}\left(\frac{\partial UH}{\partial \lambda} + \frac{\partial (VH\cos\phi)}{\partial \phi}\right) = 0 \tag{3.1}$$

$$
\begin{aligned}
\frac{\partial U}{\partial t} + \frac{1}{R\cos\phi}U\frac{\partial U}{\partial \lambda} + \frac{1}{R}V\frac{\partial U}{\partial \phi} - (\frac{\tan\phi}{R}U + f)V \\
= -\frac{1}{R\cos\phi}\frac{\partial}{\partial\lambda}[\frac{p_s}{\rho_0} + g(\zeta - \eta)] + \frac{\tau_{s\lambda}}{\rho_0 H} - \tau_* U \qquad (3.2)
\end{aligned}
$$

$$
\begin{aligned}
\frac{\partial V}{\partial t} + \frac{1}{R\cos\phi}U\frac{\partial V}{\partial \lambda} + \frac{1}{R}V\frac{\partial V}{\partial \phi} + (\frac{\tan\phi}{R}U + f)U \\
= -\frac{1}{R}\frac{\partial}{\partial\phi}[\frac{p_s}{\rho_0} + g(\zeta - \eta)] + \frac{\tau_{s\phi}}{\rho_0 H} - \tau_* V \qquad (3.3)
\end{aligned}
$$

where t = time, ζ is the free surface elevation relative to the geoid, U, V are depth averaged horizontal velocities, $H = \zeta + h$ is the total water column and h is the bathymetric depth relative to the geoid. The Coriolis parameter $f = 2\Omega\sin\phi$ and Ω is the angular speed of the Earth(radius R). The atmospheric pressure at the free surface is denoted p_s and λ,ϕ are degrees longitude and latitude; g is the acceleration due to gravity, η is the effective Newtonian equilibrium tide potential, ρ_0 is the reference density of water, τ_{sx}, τ_{sy} are applied free surface stress components, and

$$
\tau_* = C_f \frac{(U^2 + V^2)^{1/2}}{H} \qquad (3.4)
$$

with bottom friction coefficient C_f.

A practical expression for the effective Newtonian equilibrium tide potential is [13]

$$
\eta(\lambda, \phi, t) = \sum_{n,j} \alpha_{jn} C_{jn} f_{jn}(t_0) L_j(\phi) \cos[\frac{2\pi(t - t_0)}{T_{jn}} + j\lambda + \nu_{jn}(t_0)] \qquad (3.5)
$$

where C_{jn} is a constant characterizing the amplitude of tidal constituent n of species j, α_{jn} is the effective earth elasticity factor for tidal constituent n of species j, f_{jn} is the time-dependent nodal factor and ν_{jn} is the time dependent astronomical argument for j = 0, 1, 2 = tidal species (j = 0, declinational; j = 1, diurnal; j = 2, semi-diurnal) and $L_0 = 3\sin^2\phi - 1$, $L_1 = \sin(2\phi)$, $L_2 = \cos^2\phi$; T_{jn} is the period of constituent n of species j and t_0 is the reference time.

Values for C_{jn} are presented in [13]. We note that the value for the effective elasticity factor is typically taken as 0.69 for all tidal constituents [5,14] although its value has been shown to be slightly constituent dependent [17,24].

To facilitate a finite element solution to equations (3.1) - (3.3), these equations are mapped from spherical form into a rectilinear coordinate system using a Carte Parallelogrammatique (CP) projection [7]

$$x' = R(\lambda - \lambda_0)\cos\phi_0 \tag{3.6}$$

$$y' = R\phi \tag{3.7}$$

where λ_0, ϕ_0 is the center point of the projection.

Applying the CP projection to Equations (3.1) - (3.3) gives the shallow water equations in primitive non-conservative form expressed in the CP coordinate system

$$\frac{\partial \zeta}{\partial t} + \frac{\cos\phi_0}{\cos\phi}\frac{\partial(UH)}{\partial x'} + \frac{1}{\cos\phi}\frac{\partial(VH\cos\phi)}{\partial y'} = 0 \tag{3.8}$$

$$\begin{aligned}\frac{\partial U}{\partial t} &+ \frac{\cos\phi_0}{\cos\phi}U\frac{\partial U}{\partial x'} + V\frac{\partial U}{\partial y'} - (\frac{\tan\phi}{R}U + f)V \\ &= -\frac{\cos\phi_0}{\cos\phi}\frac{\partial}{\partial x'}[\frac{p_s}{\rho_0} + g(\zeta - \eta)] + \frac{\tau_{s\lambda}}{\rho_0 H} - \tau_* U\end{aligned} \tag{3.9}$$

$$\begin{aligned}\frac{\partial V}{\partial t} &+ \frac{\cos\phi_0}{\cos\phi}U\frac{\partial V}{\partial x'} + V\frac{\partial V}{\partial y'} + (\frac{\tan\phi}{R}U + f)U \\ &= -\frac{\partial}{\partial y'}[\frac{p_s}{\rho_0} + g(\zeta - \eta)] + \frac{\tau_{s\phi}}{\rho_0 H} - \tau_* V\end{aligned} \tag{3.10}$$

Utilizing the finite element method to resolve the spatial dependence in the shallow water equations in their primitive form gives inaccurate solutions with severe artificial (near $2\Delta x$) modes [3]. However, reformulating the primitive equations into a Generalized Wave Continuity Equation (GWCE) form gives highly accurate, noise-free, finite element solutions to the shallow water equations [4,6,12,18,19]. The GWCE is derived by combining a time differentiated form of the primitive continuity equation and a spatially differentiated form of the primitive momentum equations recast into conservative form, reformulating the convective terms into non-conservative form and adding the primitive form of the continuity equation multiplied by a constant in time and space, τ_0 [6,8,11,12]. The GWCE in the CP coordinate system is:

$$\frac{\partial^2 \zeta}{\partial t^2} + \tau_0\frac{\partial \zeta}{\partial t} + \frac{\cos\phi_0}{\cos\phi}\frac{\partial}{\partial x'}(U\frac{\partial \zeta}{\partial t} - \frac{\cos\phi_0}{\cos\phi}UH\frac{\partial U}{\partial x'} - VH\frac{\partial U}{\partial y'} +$$

$$
\begin{aligned}
&(\frac{\tan\phi}{R}U + f)VH - H\frac{\cos\phi_0}{\cos\phi}\frac{\partial}{\partial x'}(\frac{p_s}{\rho_0} + g(\zeta - \eta)) - (\tau_* - \tau_0)UH + \frac{\tau_{s\lambda}}{\rho_0}) \\
&\quad + \frac{\partial}{\partial y'}(V\frac{\partial\zeta}{\partial t} - \frac{\cos\phi_0}{\cos\phi}UH\frac{\partial V}{\partial x'} \\
&\quad -VH\frac{\partial V}{\partial y'} - (\frac{\tan\phi}{R}U + f)UH - H\frac{\partial}{\partial y'}(\frac{p_s}{\rho_0} + g(\zeta - \eta)) \\
&\quad -(\tau_* - \tau_0)VH + \frac{\tau_{s\phi}}{\rho_0}) - \frac{\partial}{\partial t}[\frac{\tan\phi}{R}VH] - \tau_0[\frac{\tan\phi}{R}VH] = 0 \qquad (3.11)
\end{aligned}
$$

Equation (3.11) is solved in conjunction with the primitive momentum equations in non-conservative form, (3.9) and (3.10).

The high accuracy of this formulation is a result of its excellent numerical amplitude and phase propagation characteristics. In fact, Fourier analysis indicates that in constant depth water and using linear interpolation, a linear tidal wave with 25 nodes per wavelength is more than adequately resolved over the range of Courant numbers, $C = \sqrt{gh}\Delta t/\Delta x \leq 1.0$ [10]. Furthermore, the monotonic dispersion behavior of this generalized wave equation approach avoids generating artificial (near $2\Delta x$) modes which plague primitive based finite element solutions. We note that the monotonic dispersion behavior of GWCE-based finite element solutions is very similar to that associated with staggered finite difference solutions to the primitive shallow water equations [20]. GWCE-based finite element solutions to the shallow water equations allow for extremely flexible spatial discretizations which result in a highly effective minimization of the discrete size of any problem [2,9,16,21].

The details of ADCIRC, our implementation of this generalized wave equation are described in [7,8,10]. As in most GWCE based finite element codes, ADCIRC employs 3-node C^0 linear triangles for surface elevation, velocity and depth. Furthermore, the decoupling of the time and space discrete form of the GWCE and momentum equations, together with time independent and/or tri-diagonal system matrices, full vectorization of all major loops and the use of a preconditioned conjugate gradient solver results in a highly efficient code (running at approximately 115 megaflops on a CRAY YMP 6128).

3.3 Gridding Studies: Tidal Computations

Often the criterion to establish grids for tidal computations is based on the one-dimensional, linear, frictionless, constant topography wavelength to grid size ratio

$$\frac{\lambda}{\Delta x} = \frac{\sqrt{gh}}{\Delta x} T \tag{3.12}$$

where h is the waterdepth and T is the tidal period.

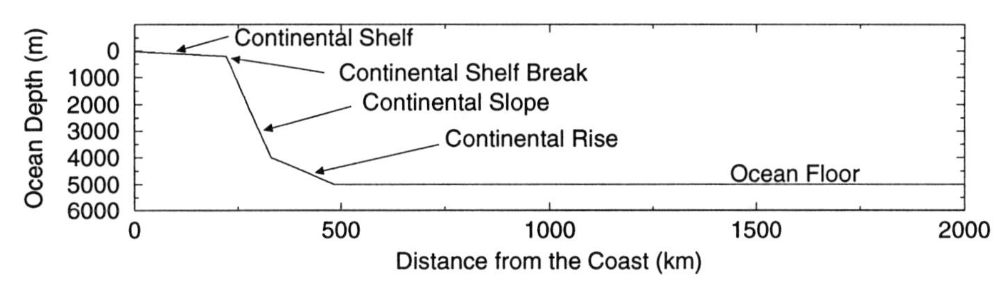

Figure 3.2: One dimensional idealized bathymetry for truncation error analysis.

We have found that this criterion does not, of itself, lead to unstructured graded grids which perform satisfactorily. Among the shortcomings of this criterion are its inability to identify the two-dimensional structure of the waves associated with flow features such as amphidromes and with circulation forced by two-dimensional topography and/or coasts. In these cases the actual wavenumber content of the response is much greater than that predicted by the one-dimensional criterion and we require additional grid resolution. Another significant shortcoming of the one-dimensional criterion is that it is based on uniform waves propagating in constant depth water. Therefore it does not take into consideration the rate of change of wavelength or the associated significant gradients in response which occur as the waves propagate over steep topographic changes. Nonetheless, as is shown in Figure 3.1, very steep topographic changes are a wide-spread feature of the *WNAT* model domain and occur in the vicinity of the continental shelf break and the continental slope.

Westerink et al. [22] examine the question of grid resolution in the vicinity of the continental shelf break and slope using numerical experiments in the context of a one-dimensional idealized slice of ocean. From this it is noted that the $\lambda/\Delta x$ criterion has to be significantly exceeded in the vicinity of the continental shelf break and slope. In fact, in order to obtain highly accurate unstructured graded grid solutions which matched uniform fine grid solution, at least 20 nodes had to be placed across the shelf break/slope region.

The results of the one-dimensional numerical experiments in [22] are readily confirmed by performing a local truncation error analysis of the one-dimensional linearized form of the harmonic shallow water equations in

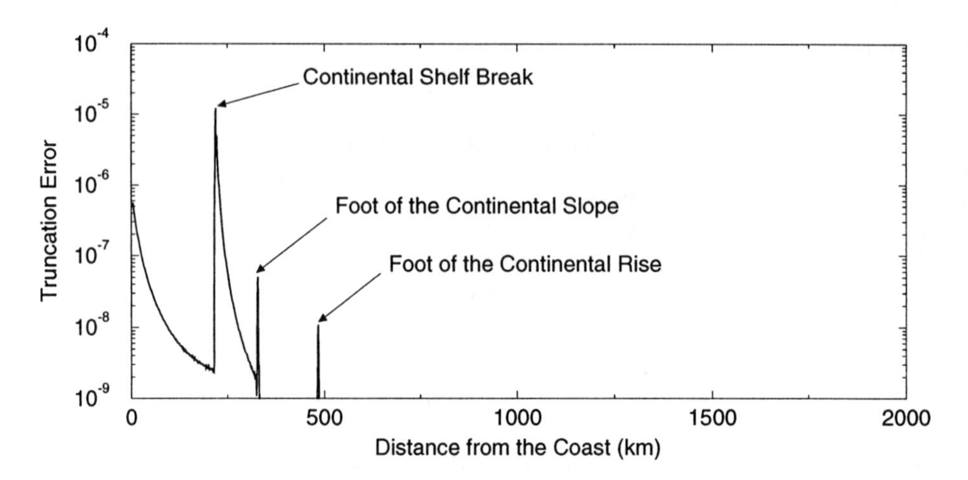

Figure 3.3: Local truncation error for the one-dimensional momentum equation.

GWCE form. Utilizing Taylor series expansions in the generic discrete nodal momentum and GWCE equations leads to expressions for nodal truncation error. Developing a very fine grid ($\Delta x = 1\ km$) "truth" solution allows us to estimate the spatial derivatives of elevation and velocity in the expressions for the local truncation errors. Applying this analysis for a typical slice of our *WNAT* model topography, shown in Figure 3.2, leads to the localized truncation error in the momentum equation (with up to second order terms and an assumed uniform grid) shown in Figure 3.3. We note that the predominant local truncation errors occur near the coast on the continental shelf and near the continental shelf break and slope. The local truncation errors for the GWCE are very similar to those of the momentum equation. Since a high local truncation error will adversely affect the accuracy of the entire solution, we want to reduce the local truncation error in these areas by providing a much higher level of grid refinement very near the coast and in the vicinity of the shelf break than on most of the shelf or in the deep ocean. We note that the $\lambda/\Delta x$ criterion does not indicate the need for a locally high level of grid refinement in the vicinity of the shelf break and slope.

We now present a sequence of two-dimensional numerical experiments which examine the actual *WNAT* domain and the influence of grid resolution in the vicinity of the continental shelf break by comparing responses obtained using a uniform grid and two unstructured graded grids (grids SS4, T1 and T2) to a "truth" response computed using a very fine uniform grid (SSS5). The characteristics of these four grids are summarized in Table 3.1.

Table 3.1: *WNAT* Model Grid Characteristics

Grid	Nodes	Grid Structure	Grid Size		Resolution	
			$(\frac{degrees}{min})$	(km)	$(\frac{\lambda}{\Delta x})_{max}$	$(\frac{\lambda}{\Delta x})_{min}$
SS4	24255	uniform	$12'$	19	704	11.3
SSS5	95999	uniform	$6'$	9	1322	22.6
T1	11712	graded	$5' \rightarrow 0.8°$	$7 \rightarrow 70$	1184	16.0
T2	28889	graded	$5' \rightarrow 0.8°$	$7 \rightarrow 70$	1013	16.0

For simplicity and to ensure identical boundaries, all grids approximate the *WNAT* domain using a coarse boundary.

Grid SS4 uniformly discretizes the domain with a resolution of approximately $12' \times 12'$. Bathymetry is interpolated onto the grid using the ETOPO5 data base with a minimum defined depth of between 3 m (for the Gulf of Mexico and Caribbean Sea) and 7 m (for the northeast U.S. and Canadian coasts) to avoid drying of elements during maximum ebb. Grid SSS5 is derived from grid SS4 by splitting each element into four smaller triangular elements effectively doubling the resolution to $6' \times 6'$. Bathymetry from the ETOPO5 data base was interpolated onto grid SSS5 and minimum depth values were again set. Responses from grid SSS5 are used as our "truth" solution against which other solutions are compared.

Grid T1, shown in Figure 3.4a, is an unstructured graded grid with element sizes varying between $5'$ and $0.8°$ and is based on a target minimum $\frac{\lambda}{\Delta x}$ criterion of between 25 and 150. The higher $\frac{\lambda}{\Delta x}$ criterion is predominantly applied in deep water regions where two-dimensional flow features require additional resolution. The actual values achieved for the T1 grid are somewhat higher in some regions due to adjacent element size ratio and element skewness restrictions. (That is, elements should not change in size too rapidly). We note that grid T1 does not provide a very high level of grid refinement over the shelf break. Bathymetry for grid T1 is obtained by interpolating directly from grid SSS5 depths.

Grid T2, shown in Figure 3.4b, is developed from grid T1 by providing additional refinement over the shelf break region (typically $5'$ resolution between 100 m and 500 m depth) and the continental slope (typically $10'$ resolution between 500 m and 4000 m). This results in a grid with a very high level of resolution in the vicinity of the shelf break and slope with $\frac{\lambda}{\Delta x}$ ratios ranging between 250 to 500. Again bathymetry is interpolated onto grid T2 from grid SSS5.

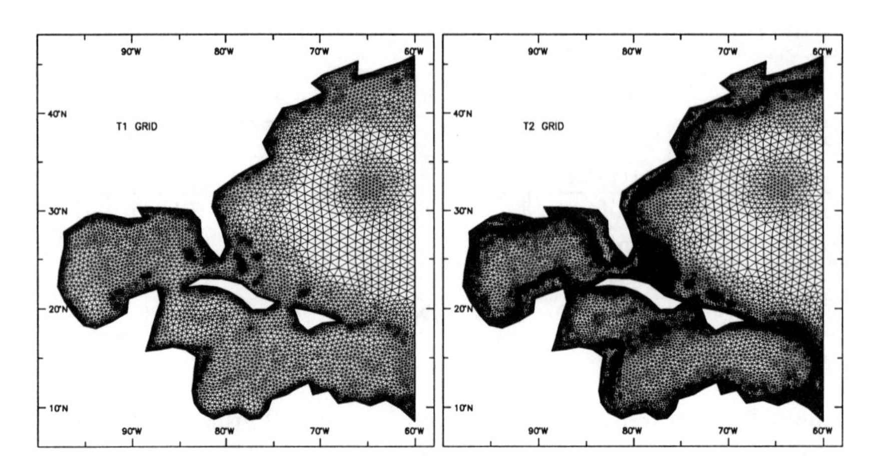

Figure 3.4: Grid T1 (a) and grid T2 (b).

All four grids are identically forced using an M_2 tide on the open ocean boundary as well as an M_2 tidal potential function within the interior domain [23]. The fully nonlinear form of the shallow water equation is used and the nonlinear bottom friction coefficient is set to $C_f = 0.003$ throughout the entire domain. The simulations are run for 55 days in order for all start up transients to dissipate and to reach a dynamic steady state. Harmonic analysis of the stationary elevation responses provides elevation amplitudes and phases at all nodes and establishes a basis for comparing the results obtained with these four grids.

We now compare elevation amplitude and phase responses for grids SS4, T1, and T2 to our "truth" solution obtained with grid SSS5. This is accomplished by interpolating the amplitude and phase responses from grids SS4, T1, and T2 onto grid SSS5 and then subtracting off the nodal response values obtained with grid SSS5. We then summarize these error distributions by defining the fraction of the domain area which has levels of under- or overprediction which exceed a set limit. This results in the cumulative error distribution curves for absolute elevation amplitude error, relative amplitude error (the absolute elevation errors are normalized with the "truth" amplitude values) and elevation phase error shown in Figure 3.5. We note that a domain median under- or overprediction error corresponds to a cumulative fraction approximately equal to 2.5×10^{-1} while the maximum domain errors correspond to the end points of the curves. Furthermore, we note that a cumulative area fraction of 5×10^{-6} (0.0005% of the *WNAT* domain) equals the area of one finite element in the SSS5 grid ($\sim 28\ km^2$). Therefore the lower portions of the under- and overprediction curves correspond to very small fractions of the *WNAT* model domain. Some of the information in the cumulative error distribution curves is summarized for convenience in Table 3.2 which indicates maximum errors for all three criteria as well as the percentage of the domain exceeding pre-defined threshold error values (± 1 *cm* for elevation amplitude, ± 0.05 for relative elevation amplitude and $\pm 5°$ for elevation phase).

We first examine the cumulative area error distribution of the absolute elevation amplitude, shown in Figure 3.5a. For grid SS4 we note that over 97.4% of the domain underprediction errors do not exceed -1 *cm* and over 96.3% of the domain overprediction errors are less than 1 *cm*. Regions with greater than 1 *cm* error are typically located adjacent to the coastline. The extreme underprediction error equals -16.4 *cm* and the extreme overprediction error equals 7.7 *cm*. Both these extreme errors occur in the Gulf of Maine, a region with high M_2 amplitude values and gradients. We note that the underprediction errors are in general more severe than the overprediction errors. For grid T1, the absolute elevation amplitude errors are significantly greater than for grid SS4 over almost the entire domain. In fact for this grid only 91.3% of the domain underprediction errors do not exceed 1 *cm* and only 80.6% of the domain overprediction errors are less than 1 *cm*. Regions with amplitude errors exceeding 1 *cm* now extend over the entire Atlantic and Florida shelves. Furthermore, while the extreme underprediction error is less than for grid SS4, the extreme overprediction error is about the same as for grid SS4. The extreme underprediction error now occurs near an amphidrome located off of Haiti while the extreme overprediction error is still located in the Gulf of Maine. For grid T2, the error levels are less than for grid SS4 over most of the domain. Like grid SS4, error levels in excess of 1 *cm* are again restricted to regions very near the Atlantic coast. However the extreme underprediction errors have been dramatically reduced as compared to grid SS4. Furthermore the amplitude error levels of this unstructured graded grid with a high level of resolution over the shelf break are dramatically better than for grid T1 over the entire domain. Finally we note that there is no bias in the under and overprediction errors with these curves being much more symmetrical about the zero amplitude error axis than results for grids SS4 or T1.

We now examine the cumulative area distribution of the relative elevation amplitude error, shown in Figure 3.5b. We note that the patterns and especially relative positioning of the three plotted curves are somewhat different for the relative amplitude error as compared to the absolute amplitude error (since the relative errors tend to weigh errors in regions where the actual amplitude is small). For grid SS4 over 98.5% of the domain

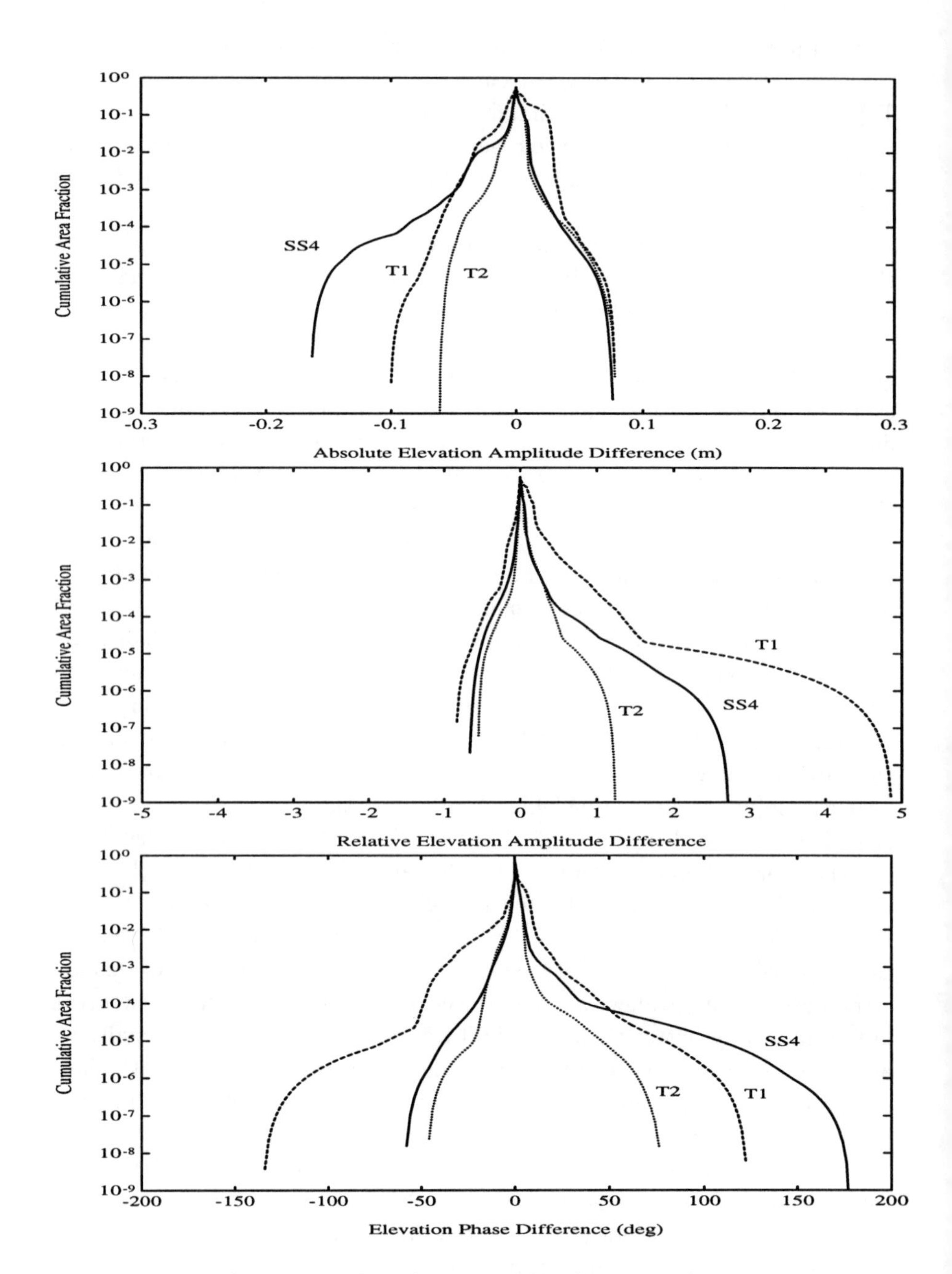

Figure 3.5: Cumulative area fraction error curves

Table 3.2: Error Measures for Grids SS4,T1,T2 compared to Grid SSS5

ERROR MEASURE	SS4	T1	T2
ELEVATION AMPLITUDE (absolute)			
Largest negative difference	−16.4 *cm*	−10.1 *cm*	−6.1 *cm*
Largest positive difference	+7.7 *cm*	+7.8 *cm*	+7.8 *cm*
Percent Area exceeding −1 *cm* difference	2.6%	8.7%	1.8%
Percent Area exceeding +1 *cm* difference	3.7%	19.4%	0.3%
ELEVATION AMPLITUDE (relative)			
Largest negative difference	−0.67	−0.85	−0.55
Largest positive difference	+2.7	+4.9	+1.2
Percent Area exceeding −0.05 difference	1.5%	5.2%	0.4%
Percent Area exceeding +0.05 difference	11.3%	35.1%	3.3%
ELEVATION PHASE (degrees)			
Largest negative difference	−59.2°	−135.4°	−47.2°
Largest positive difference	+178.3°	+123.7°	+77.8°
Percent Area exceeding −5° difference	0.7%	2.9%	1.0%
Percent Area exceeding +5° difference	1.6%	12.5%	0.4%

underprediction errors don't exceed a factor of −0.05 and 88.7% of the overprediction errors don't exceed a factor of 0.05. These 0.05 factor error levels are located predominantly in the Gulf of Mexico and Caribbean Sea in the vicinity of amphidromes. The extreme underprediction error equals a factor of −0.67 and the extreme overprediction error equals a factor of 2.7. Both these extreme errors are located within amphidromes off Haiti. We note that the very low amplitude values within amphidromes tend to bias the absolute error there, resulting in large relative errors. For grid T1, the relative amplitude errors are always significantly greater than for grid SS4 over the entire domain. The 0.05 factor error levels now extend over most of the Gulf and Caribbean. For grid T2, the relative amplitude errors are always less than for grid SS4 and dramatically less than for grid T1. The 0.05 factor error levels extend over smaller parts of the Gulf and Caribbean than grid SS4 and are again located in the vicinity of amphidromes. The extreme overprediction error as compared to grid SS4 has in particular dropped to a factor of 1.2 and is still located near an amphidrome off Haiti.

Finally, we examine the cumulative area distribution of the phase error, shown in Figure 3.5c. For grid SS4, over 99.3% of the domain underpredictions in phase do not exceed −5° and over 98.4% of the domain overpredictions in phase do not exceed 5°. The 5° phase error regions are located primarily within the Gulf and Caribbean and are typically in the vicinity of amphidromes. Amphidromes have associated high gradients in phase. Extreme errors are significant with an extreme underprediction in phase of $-59°$ occurring in the Caribbean and an extreme overprediction in phase of $178°$ occurring in the Gulf on the Florida shelf. Both these extreme errors occur very near amphidromic points. For grid T1, phase errors are generally significantly greater than for the SS4 grid except for the extreme overprediction errors. Phase errors exceed $5°$ in large portions of the Gulf and Caribbean. Finally the T2 grid results in phase errors which are always less than grid SS4 and significantly less than grid T1. Again the $5°$ phase error regions are limited to regions near amphidromes in the Gulf and Caribbean as was the case for grid SS4. The extreme overprediction error has significantly dropped as compared to grid SS4 and now equals $78°$ and is located in the Caribbean adjacent to an amphidrome.

Thus the regions where errors remained the highest in grid T2 are almost entirely located either very near the Atlantic coast (absolute elevation amplitude errors) or near amphidromes (relative elevation amplitude and phase errors). Both regions near the coast and near amphidromic points often have high gradients in amplitude and phase associated with them. Based on examining contour maps of error distributions for the simulations presented in this paper as well as others [11], we have found that the errors which remain in grid T2 are extremely sensitive to the level of resolution provided over regions of steep bathymetric change. This is consistent with the findings of the one dimensional numerical experiments in [22] which indicated that at least 20 nodes were required over the shelf break and slope. Since we lack this level of grid refinement over all but the gentlest continental slopes, we must provide additional grid resolution in these regions to reduce the remaining localized errors in grid T2.

3.4 Studies of Hurricane Storm Surge Computations

In the second study, we examine gridding requirements for hurricane storm surge calculations by considering various regular and variably graded grids over an idealized rectangular domain using a synthetic hurricane as the forcing function. Five grids over a rectangular domain of dimension 2500 $km \times$ 3000 km are constructed to correspond to the areal extent of the $WNAT$ domain. A land boundary representing the coastline lies along the left most 3000 km length of these rectangular domains and open ocean conditions apply at the remaining domain boundaries. Three uniform grids, designated as G01, G02, and G03, have regular nodal spacings of 50 km, 25 km, and 12.5 km, respectively. Two graded grids, VG01 and VG02 have nodal spacings ranging between 12.5 km and 50 km. A synthetic hurricane, H011, serves as the meteorological forcing for the hydrodynamic model. Hurricane H011 moves along a path perpendicular to the coastline and eventually makes landfall at 192 hours. Simulation of the storm surge generated by hurricane H011 is carried out over grids G01, G02, G03, VG01 and VG03. Storm surge predictions computed over the most finely uniformly discretized grid, G03, are considered the "truth" solution and are used to assess errors in the storm surge computations performed over the remaining grids.

Maximum overprediction (positive) and underprediction (negative) errors in the storm surge computed over uniform grids G01 and G02 are shown in Figures 3.6a and 3.6b respectively. The absolute error is calculated simply as the difference between computed storm surge values and the "true" storm surge elevations. The relative error normalizes the absolute error with respect to the highest surge elevation over the entire domain at the specific point in time being considered. Ocean bathymetries at the locations of the maximum over- and underprediction are also represented in Figures 3.6a and 3.6b. We note that up to about 174 hours, while the storm is predominately located in the deep ocean, the oscillating error pattern seen in Figures 3.6a and 3.6b is associated with insufficient resolution to capture the inverted barometer which forms in deep water. That is, the pressure forcing function is insufficiently resolved with our grids and results in an aliased response which fluctuates depending on the position of the eye of the hurricane relative to the nearest node. As the storm moves onto the continental shelf at about 180 hours, the absolute errors increase significantly. This error increase corresponds to the dramatic increase in storm surge as winds push water up onto the shelf against the coast. We also note a large increase in the relative error which coincides with a decreasing absolute error and a declining peak surge as the storm makes landfall. This behavior of the relative error indicates that processes which dissipate the peak surge are much more rapid than those which dissipate the errors in the storm surge. A comparison of the error magnitudes between Figures 3.6a and 3.6b demonstrates that halving the nodal spacing leads to a reduction of the error in deep waters by a factor of two and a threefold reduction of the errors generated on the continental shelf. The peak errors over grids G01 and G02 are concentrated on the continental shelf nearest the shoreline suggesting that a variably graded grid may provide an optimal grid structure for minimization of the storm surge prediction errors.

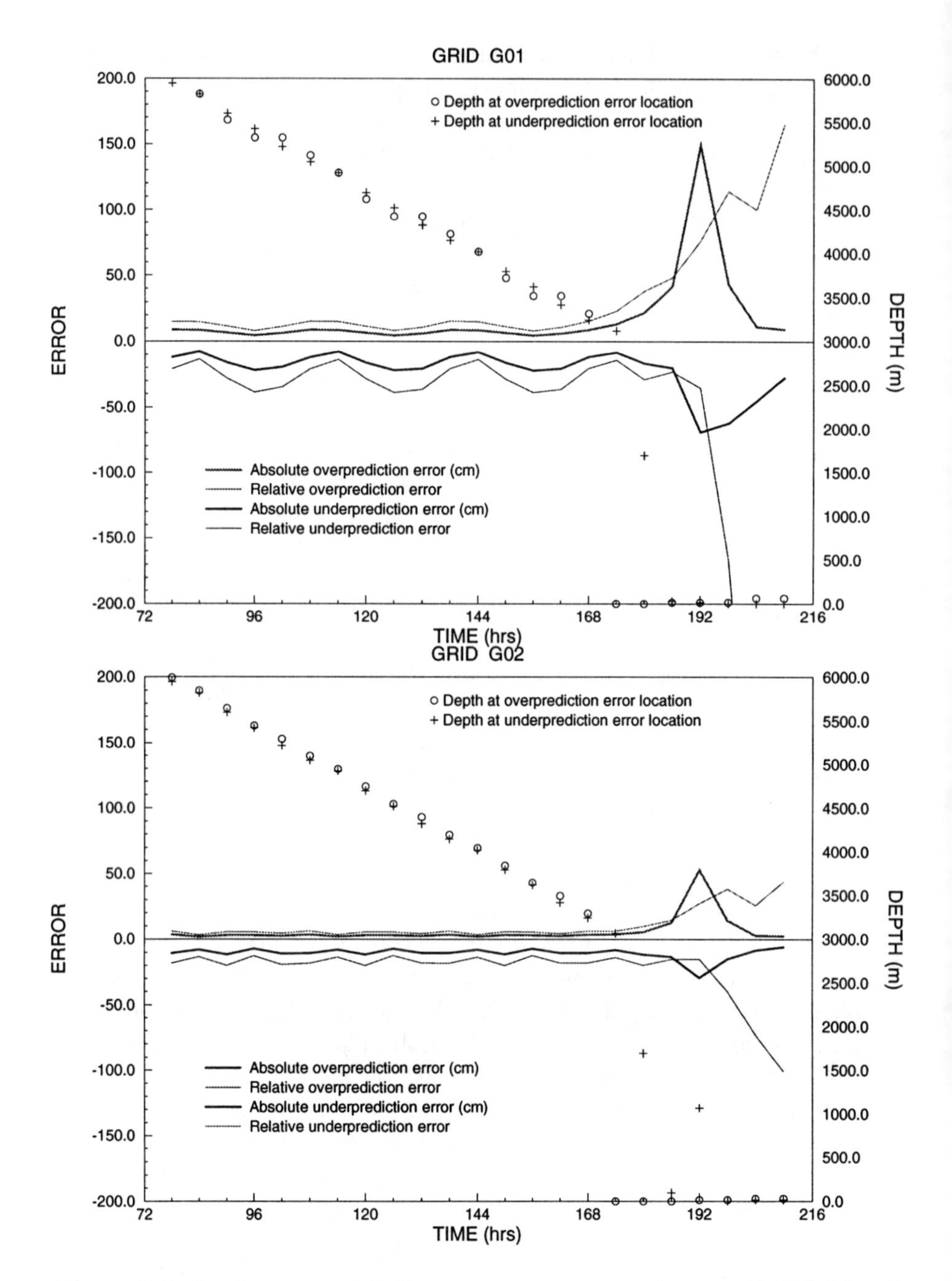

Figure 3.6: Maximum errors in the storm surge computed over grid G01 (a) and grid G02 (b)

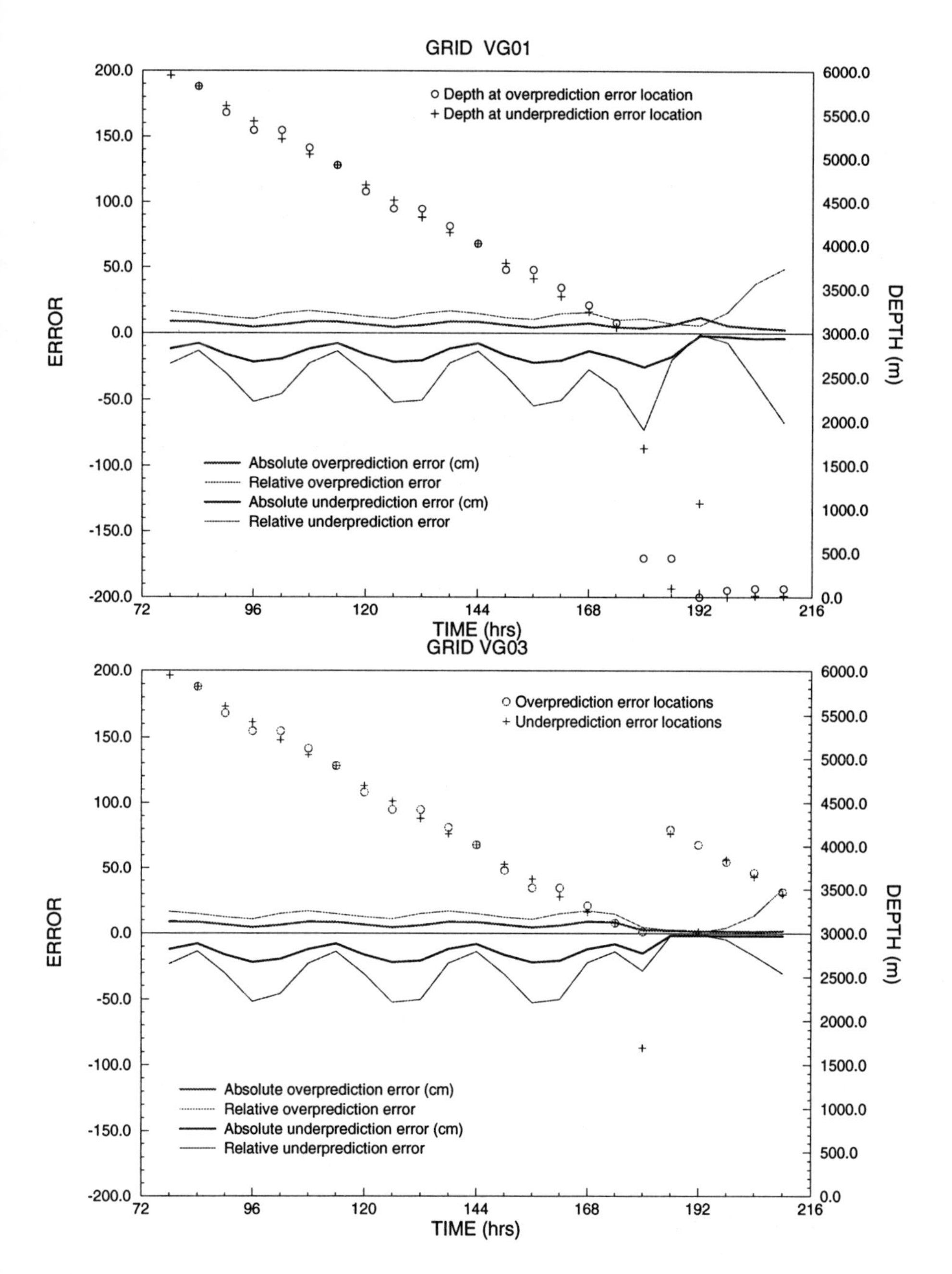

Figure 3.7: Maximum errors in the storm surge computed over grid VG01 and grid VG03.

A variably graded grid, VG01, is constructed such that the grid spacing ranges from a maximum of 50 km over the deep ocean and including regions up to 50 km offshore to a minimum of 12.5 km near the coastline. Maximum error values for grid VG01 are presented in Figure 3.7a. Errors in storm surge elevations over the deep ocean are the same as those computed over grid G01 due to an identical grid spacing of 50 km. However, errors over the continental shelf are less than those presented for grid G02 by a factor of four. In general, the maximum errors computed over grid VG01 are relatively uniform in time and space and are significantly less than those computed over either grids G01 or G02.

Finally, a variably graded grid, VG03, is constructed such that the grid spacing ranges from a maximum of 50 km over the deep ocean while the continental shelf and slope are resolved with a 12.5 km grid spacing. Maximum error values for grid VG03 are presented in Figure 3.7b. Error levels are not dramatically different than for grid VG01 indicating that the storm surge calculation does not require the same high level of grid resolution over the shelf break and slope as tidal computations.

3.5 Conclusions

Grids based on a minimum $\frac{\lambda}{\Delta x}$ criterion do not generally lead to a satisfactorily converged solution for tidal computations. This is related to the two-dimensional structure of the tides as well as the high local truncation errors caused by rapid changes in topography. The influence of changes in topography in the vicinity of the shelf break is clearly demonstrated using a truncation error analysis. Furthermore, numerical experiments using the *WNAT* domain with a coarse boundary clearly demonstrates the need for additional grid refinement over the shelf break. The unstructured graded grid T2, with a high level of refinement over the shelf break and slope, performs dramatically better than the unstructured graded grid T1, essentially the same as grid T2 except without the grid refinement over the shelf break/slope. Furthermore grid T2 performs better than the uniform grid SS4 which has approximately the same number of nodes as grid T2. Errors for grid T2 are predominantly in the vicinity of phase convergence zones such as the amphidromes situated in the Gulf and Caribbean. It appears that the positioning of these amphidromes is extremely sensitive to the details of the bathymetry. In fact, we believe that in order to achieve better convergence, we would have to provide additional resolution over the shelf break and slope.

The hurricane storm surge grid sensitivity studies demonstrate that both the size and discretization of the computational domain influence storm surge generation in the coastal region. A variable grid structure which has extensive refinement in the near-shore region leads to low uniform errors throughout the domain. Since discretization errors can never be completely eliminated, uniform errors over a grid are desired with the acceptable magnitude of these errors representing a balance between computational effort and required accuracy. The high level of resolution required over the continental shelf break and slope for tidal computations does not appear to be necessary for storm surge calculations. Rather, the highest level of resolution is only needed adjacent to the coast in the shallowest waters. The variable grid structure utilized here yields the smallest errors with the least computational effort particularly when implemented in conjunction with a large domain such as the *WNAT* domain.

Acknowledgments: This research was funded under contract DACW-39-90-K-0021 with the U.S. Army Engineers Waterways Experiment Station. We also thank Antonio Baptista of OGI for allowing us to use the Xmgredit software to generate the grids presented in this paper.

References

[1] Blain, C.A., J.J. Westerink, R.A. Luettich, "The influence of domain size on the response characteristics of a hurricane storm surge model," *J. Geophys. Res.*, 99,C9, 18467-18479, 1994.

[2] Foreman, M.G.G., "A comparison of tidal models for the southwest coast of Vancouver Island," *Proceedings of the VII International Conference on Computational Methods in Water Resources*, held in Cambridge, MA, Elsevier, 1988.

[3] Gray, W.G., "Some inadequacies of finite element models as simulators of two-dimensional circulation", *Adv. Water Resour.*, 5, 171-177, 1982.

[4] Gray, W.G., "A finite element study of tidal flow data for the North Sea and English Channel", *Adv. Water Resour.*, 12, 143-154, 1989.

[5] Hendershott, M.C., "Long Waves and Ocean Tides," *Evolution of Physical Oceanography*, 292-341, B.A. Warren and C. Wunsch, Eds., MIT Press, Cambridge, MA., 1981.

[6] Kinnmark, I.P.E., *The Shallow Water Wave Equations: Formulation, Analysis and Application*, Ph.D. Dissertation, Department of Civil Engineering, Princeton University, 1984.

[7] Kolar, R.L., W.G. Gray, J.J. Westerink and R.A. Luettich, "Shallow water modeling in spherical coordinates: equation formulation, numerical implementation, and application," *J. Hydraulic Res.*, 32, 1, 3-24, 1994.

[8] Kolar, R.L., J.J. Westerink, M.E. Cantekin and C.A. Blain, "Aspects of non-linear simulations using shallow water models based on the wave continuity equation," *Comp. Fluids*, 23, 3, 523-538, 1994.

[9] Le Provost, C. and P. Vincent, "Some tests of precision for a finite element model of ocean tides," *J. Comput. Phys.*, 65, 273-291, 1986.

[10] Luettich, R.A., J.J. Westerink and N.W. Scheffner, *ADCIRC: An Advanced Three-Dimensional Circulation Model for Shelves, Coasts and Estuaries, Report 1: Theory and Methodology of ADCIRC-2DDI and ADCIRC-3DL*, Dredging Research Program, Technical Report DRP-92-6, Department of the Army, Washington, D.C., 1992.

[11] Luettich, R.A. and J.J. Westerink, "Continental shelf scale convergence studies with a barotropic tidal model", *Quantitative Skill Assessment for Coastal Ocean Models*, D. Lynch and A. Davies, Eds., American Geophysical Union, Washington, D.C., 1994.

[12] Lynch, D.R. and W.G. Gray, "A wave equation model for finite element tidal computations," *Comput. Fluids*, 7, 207-228, 1979.

[13] Reid, R.O., "Waterlevel changes, tides and storm surges," *Handbook of Coastal and Ocean Engineering*, J. Herbich, Ed., Gulf Publishing, Houston, Texas, 1990.

[14] Schwiderski, E.W., "On charting global ocean tides," *Rev. Geophys. Space Phys.*, 18, 243-268, 1980.

[15] Turner, P.J. and A.M. Baptista, *ACE/Gredit Users Manual: Software for Semiautomatic Generation of Two-Dimensional Finite Element Grids,* CCALMR Software Report SDS2 (91-2), Oregon Graduate Institute, Beaverton, OR., 1991.

[16] Vincent, P. and C. Le Provost, "Semidiurnal tides in the northeast Atlantic from a finite element numerical model," *J. Geophys. Res.*, 93, C1, 543-555, 1988.

[17] Wahr, J.M., "Body tides on an elliptical, rotating, elastic and oceanless earth," *Geophys. J. Royal Astro. Soc.*, 64, 677-703, 1981.

[18] Walters, R.A. and F.E. Werner, "A comparison of two finite element models of tidal hydrodynamics using a North Sea data set," *Adv. Water Resour.*, 12, 4, 184-193, 1989.

[19] Werner, F.E. and D.R. Lynch, "Harmonic structure of English Channel/Southern Bight tides from a wave equation simulation," *Adv. Water Resour.*, 12, 121-142, 1989.

[20] Westerink, J.J. and W.G. Gray, "Progress in surface water modeling," *Rev. Geophys.*, 29, April Supplement, 210-217, 1991.

[21] Westerink, J.J., R.A. Luettich, A.M. Baptista, N.W. Scheffner and P. Farrar, "Tide and storm surge predictions using a finite element model," *J. Hydraul. Eng.*, 118, 1373-1390, 1992.

[22] Westerink, J.J., J.C. Muccino and R.A. Luettich, "Resolution requirements for a tidal model of the Western North Atlantic and Gulf of Mexico," *Proceedings of the IX International Conference on Computational Methods in Water Resources*, Denver, CO, T.F. Russell et al., Eds., Computational Mechanics Publications, Southampton, UK, 1992.

[23] Westerink, J.J., R.A. Luettich and J.C. Muccino, "Modeling tides in the Western North Atlantic using unstructured graded grids," *Tellus*, 46A, 178-199, 1994.

[24] Woodworth, P.L., "Summary of recommendations to the UK Earth Observation Data Centre (UK-EODC) by the Proudman Oceanographic Laboratory (POL) for tide model corrections on ERS-1 geophysical data records," Proudman Oceanographic Laboratory Communication, 1990.

Chapter 4
An Improved Finite Element Model for Shallow Water Problems

O.C. Zienkiewics[1] *and P. Ortiz*[2]

4.1 Introduction

The study of shallow water problems using numerical models has been widely developed in recent years for the fields of Coastal, Environmental and Hydraulic Engineering. The versatility of the finite element method for the analysis of complex domains makes it very attractive for the study of estuaries, harbors, lakes or hydraulic structures, characterized by complicated geometries and bathymetries (e.g. see [3,5,8,9,11,12,15]).

The present work introduces a new and general algorithm which can treat transient problems such as tidal flows, tsunami propagation, long wave amplification in harbors, as well as steady state flows of subcritical or supercritical kind, including "jump" or shock patterns in the latter. The algorithm can be extended in a straightforward manner to deal with transport equations and pollutant dispersion.

The basis of the present method is, essentially, the fractional step procedure [2], that has been successfully applied to incompressible flows [7] and

[1]Institute of Numerical Methods in Engineering, University College of Swansea, U.K.
[2]Centro de Estudios de Técnicas Aplicadas, Cedex, Madrid, Spain

more recently for compressible and incompressible flows in a unified formulation using the characteristic Galerkin procedure [16,18]. The fractional step methodology exploited here allows only a single characteristic velocity to be considered (the actual velocity) by means of the characteristic Galerkin method [17], thereby giving a rational definition of the balancing dissipation terms. The characteristic Galerkin approach also justifies the application of a Galerkin spatial discretization in the convective equation [16,17].

The semi-implicit form of the general formulation provides a critical time-step only dependent on the current velocity $\Delta t = h/u$, where h is the element size and u is the flow velocity. This can be compared with a critical time-step in terms of the wave celerity c, which places a severe constraint on fully explicit methods such as the Taylor-Galerkin approximation [13].

The present method computes, as in the fractional-step procedure, the pressure (or elevations of the free surface) by means of a Laplacian-type equation, whose self-adjointness makes the Galerkin space discretization optimal. Velocities are computed in two stages explicitly with the characteristic Galerkin method, first considering the momentum equations omitting the pressure gradient terms, and finally with a correction coming from the computed new pressure.

The first part of this work is devoted to a brief description of the governing shallow water equations derived, essentially, from the depth integration of the Navier-Stokes equations assuming hydrostatic pressure distribution and constant horizontal velocities with depth. In the following sections the numerical solution is described and some applications are studied.

4.2 Governing Equations

The shallow water equations, in their depth integrated form, can be written in a Cartesian system $x_i (i = 1, 2)$, using the summation convention, as

$$\frac{\partial h}{\partial t} + \frac{\partial U_i}{\partial x_i} = 0, \qquad i = 1,2 \tag{4.1}$$

$$\frac{\partial U_i}{\partial t} + \frac{\partial F_{ij}}{\partial x_j} + \frac{\partial p}{\partial x_i} + \frac{\partial G_{ij}}{\partial x_j} + Q_i = 0, \qquad i,j = 1,2 \tag{4.2}$$

where $h = H + \eta$ and $U_i = hu_i$ are the unknowns. Here h is the total height of water, $H = H(x_1, x_2)$ is the depth of water, η is the surface elevation with respect to the mean water level and u_i are the depth averaged components of the horizontal velocity. $F_{ij} = hu_iu_j = U_iu_j$ is the i component of the j flux vector and $\boldsymbol{G}$ represents the diffusive fluxes. The "pressure" p is defined as

$$p = \frac{1}{2}g(h^2 - H^2) \tag{4.3}$$

to maintain the analogy with the equations of compressible flow.

The vector $\mathbf{Q}$ contains the source terms, which in general can be specified in the form

$$Q_i = -g(h - H)\frac{\partial H}{\partial x_i} + g\frac{u_i \mid u \mid}{c^2 h} + r_i - \tau_i + \frac{h}{\rho}\frac{\partial p_a}{\partial x_i}, \qquad i = 1,2 \tag{4.4}$$

where the well-known Chezy-Manning formula is adopted for the bottom friction term, r_i is the i component of the Coriolis force $\mathbf{r}^T = [-fu_2,\ fu_1]$, with f the Coriolis parameter, τ_i are the surface wind tractions, ρ is the density of water and p_a is the atmospheric pressure.

Using (4.3), the variable h can be substituted in (4.1), defining c as

$$c^2 = \frac{dp}{dh} = gh \tag{4.5}$$

Now (4.1) - (4.2) can be rewritten, neglecting the diffusion terms, in the form

$$\frac{1}{c^2}\frac{\partial p}{\partial t} + \frac{\partial U_i}{\partial x_i} = 0, \qquad i = 1,2 \tag{4.6}$$

$$\frac{\partial U_i}{\partial t} + \frac{\partial F_{ij}}{\partial x_j} + \frac{\partial p}{\partial x_i} + Q_i = 0, \qquad i,j = 1,2 \tag{4.7}$$

where the unknowns are now U_i and p.

4.3 Numerical Procedure

A time discretization for (4.6) is introduced as

$$\frac{1}{c^2}\frac{\Delta p}{\Delta t} + \frac{\partial U_i^{n+\theta_1}}{\partial x_i} = 0 \tag{4.8}$$

and for (4.7) proceeding approximately along the characteristic paths

$$\frac{\Delta U_i}{\Delta t} = -\left[\frac{\partial F_{ij}}{\partial x_j} + Q_i\right]^n + \frac{\Delta t}{2}\left[u_k\frac{\partial}{\partial x_k}\left(\frac{\partial F_{ij}}{\partial x_j} + Q_i\right)\right]^n - \frac{\partial p^{n+\theta_2}}{\partial x_i} \tag{4.9}$$

for $i, j, k = 1,2$. Here the parameters θ_1, θ_2 can be chosen in the range 0 to 1. The first two terms of the right hand side of (4.9) come from the expansion along the characteristics. (The derivation of the equation is illustrated in the Appendix.)

By means of this explicit method, an appropriate treatment for the convective and source terms of equation (4.7) can be accomplished, justifying now the use of a Galerkin spatial discretization due to self adjointness.

Again, based on the discretization along the characteristics, the last term of (4.9) can be written as [16]

$$\frac{\partial p^{n+\theta_2}}{\partial x_i} = (1-\theta_2)\frac{\partial p^n}{\partial x_i} + \theta_2 \frac{\partial p^{n+1}}{\partial x_i} - (1-\theta_2)\frac{\Delta t}{2} u_k \frac{\partial}{\partial x_i}\left(\frac{\partial p^n}{\partial x_k}\right) \tag{4.10}$$

for $i, k = 1, 2$ where

$$U_i^{n+\theta_1} = U_i^n + \theta_1 \Delta U_i, \quad \Delta U_i = U_i^{n+1} - U_i^n \tag{4.11}$$

Inserting (4.11) into (4.8) and (4.10) into (4.9), introducing an auxiliary variable $\Delta U_i^*/\Delta t$ to replace the first two terms of the right hand side in (4.9) and observing that $\Delta p = p^{n+1} - p^n$, results in

$$\frac{1}{c^2}\frac{\Delta p}{\Delta t} + \frac{\partial U_i^n}{\partial x_i} + \theta_1 \frac{\partial(\Delta U_i)}{\partial x_i} = 0 \tag{4.12}$$

$$\frac{\Delta U_i}{\Delta t} = \frac{\Delta U_i^*}{\Delta t} - \frac{\partial p^n}{\partial x_i} - \theta_2 \frac{\partial(\Delta p)}{\partial x_i} + (1-\theta_2)\frac{\Delta t}{2} u_k \frac{\partial}{\partial x_k}\left(\frac{\partial p^n}{\partial x_i}\right) \tag{4.13}$$

Now ΔU_i can be eliminated, and the final expression for p is

$$\frac{1}{c^2}\frac{\Delta p}{\Delta t} - \Delta t \theta_1 \theta_2 \frac{\partial}{\partial x_i}\frac{\partial(\Delta p)}{\partial x_i} = -\frac{\partial}{\partial x_i}\left[U_i^n + \theta_1 \Delta U_i^*\right] + \theta_1 \Delta t \frac{\partial}{\partial x_i}\frac{\partial p^n}{\partial x_i} \tag{4.14}$$

where higher order terms have been neglected. A similar expression for the pressure can be reached if a splitting procedure is used [19]. The spatial operator (4.14) is self-adjoint in the variable Δp, and the standard Galerkin space discretization is appropriate.

The sequence of solution steps then follows as:

1. Computation of ΔU_i^*.

2. Computation of pressure by means of (4.14) and of the new surface elevation.

3. Computation of the final velocity $U_i^n + \Delta U_i$, using (4.13).

This procedure is similar to the fractional step method in [2]. However, in the process described above, a splitting of the variables has not been carried out explicitly.

Space discretization

For the sake of clarity, these three steps are now described in more detail:

1. Apply the standard Galerkin procedure and the Gauss theorem for the ΔU_i^* computation, assuming the discretization for the variables

$$U_i = \mathrm{N}\bar{U}_i, \quad \Delta U_i = \mathrm{N}\Delta\bar{U}_i, \tag{4.15}$$

$$\Delta U_i^* = \mathrm{N}\Delta\bar{U}_i^*, \quad Q_i = \mathrm{N}\bar{Q}_i \tag{4.16}$$

where the overbar denotes the appropriate nodal values. This results in

$$\left(\int_\Omega \mathrm{N}^T\mathrm{N}d\Omega\right)\frac{\overline{\Delta U_i^*}}{\Delta t} = \left[-\left(\int_\Omega \mathrm{N}^T u_j \frac{\partial \mathrm{N}}{\partial x_j} d\Omega\right)\bar{U}_i - \left(\int_\Omega \mathrm{N}^T\mathrm{N}d\Omega\right)\bar{Q}_i - \right.$$

$$\frac{\Delta t}{2}\left(\int_\Omega \frac{\partial}{\partial x_k}(\mathrm{N}^T u_k)\frac{\partial}{\partial x_j}(\mathrm{N}u_j)d\Omega\right)\bar{U}_i - \frac{\Delta t}{2}\left(\int_\Omega \frac{\partial}{\partial x_k}(\mathrm{N}^T u_k)\mathrm{N}d\Omega\right)\bar{Q}_i$$

$$\left. + \frac{\Delta t}{2}\int_\Gamma \mathrm{N}^T u_k \left(\frac{\partial F_{ij}}{\partial x_j} + Q_i\right) n_k d\Gamma\right]^n \tag{4.17}$$

where all terms in the right-hand-side are computed at time $n\Delta t = t_n$. Now, introducing the notation

$$\boldsymbol{M} = \int_\Omega \boldsymbol{N}^T\boldsymbol{N}d\Omega, \quad \boldsymbol{C} = \int_\Omega \boldsymbol{N}^T u_j \frac{\partial \boldsymbol{N}}{\partial x_j} d\Omega$$

$$\boldsymbol{K}_u = \int_\Omega \frac{\partial}{\partial x_k}(\boldsymbol{N}^T u_k)\frac{\partial}{\partial x_j}(\boldsymbol{N}u_j)d\Omega,$$

$$\boldsymbol{f}_Q = \left(\int_\Omega \frac{\partial}{\partial x_k}(\boldsymbol{N}^T u_k)\boldsymbol{N}d\Omega\right)\bar{Q}_i \tag{4.18}$$

the final matrix form of (4.17) is

$$\boldsymbol{M}\overline{\Delta\boldsymbol{U}}^* = -\Delta t[\boldsymbol{C}\bar{\boldsymbol{U}}^n + \boldsymbol{M}\bar{\boldsymbol{Q}}^n] - \frac{\Delta t}{2}[\boldsymbol{K}_u\bar{\boldsymbol{U}}^n + \boldsymbol{f}_Q] + \boldsymbol{b}_1 \tag{4.19}$$

where $\boldsymbol{b}_1$ represents the final (boundary) term of (4.17) and is identically zero on solid boundaries ($u_n = 0$).

This approximation is conditionally stable and its solution only involves the mass matrix inversion. For one-dimensional problems, neglecting the constraint coming from the source term, the stability condition for pure convection is

$$\Delta t \leq \Delta t_{crit} = \frac{h}{|\,u\,|} \tag{4.20}$$

where h is the element size. It can be observed that this limit is in terms of the current velocity, instead of the wave velocity as in the Taylor-Galerkin method [4,13].

When steady state solutions are studied, a local time-step defined by (4.20) is recommended for the right hand side of (4.17), giving identical diffusion to that included by the optimal streamline upwinding procedure [17].

2. In the standard Galerkin method with

$$p = \boldsymbol{N}\overline{p}_i \qquad , \qquad \Delta p = \boldsymbol{N}\overline{\Delta p}_i$$

and the expressions described above, (4.14) leads, after the use of Green's theorem, to

$$\left[\bar{\boldsymbol{M}} + \theta_1\theta_2\Delta t^2\boldsymbol{H}\right]\overline{\Delta \boldsymbol{p}} = -\Delta t\boldsymbol{Q}\left[\bar{\boldsymbol{U}}^n + \theta_1\overline{\Delta \boldsymbol{U}}^*\right] - \Delta t^2\boldsymbol{H}\boldsymbol{p}^n + \boldsymbol{b}_2 \tag{4.21}$$

where now

$$\bar{\boldsymbol{M}} = \int_\Omega \frac{1}{c^2}\boldsymbol{N}^T\boldsymbol{N}d\Omega, \quad \boldsymbol{H} = \int_\Omega \frac{\partial \boldsymbol{N}^T}{\partial x_i}\frac{\partial \boldsymbol{N}}{\partial x_i}d\Omega, \quad \boldsymbol{Q} = \int_\Omega \boldsymbol{N}^T\frac{\partial \boldsymbol{N}}{\partial x_i}d\Omega \tag{4.22}$$

and the boundary terms $\boldsymbol{b}_2$ are

$$\boldsymbol{b}_2 = \Delta t^2\int_\Gamma \boldsymbol{N}^T\left(\frac{\partial p^n}{\partial x_i}n_i\right)d\Gamma + \theta_1\theta_2\Delta t^2\int_\Gamma \boldsymbol{N}^T\left(\frac{\partial(\Delta p)}{\partial x_i}n_i\right)d\Gamma \tag{4.23}$$

For prescribed elevations in a portion of the boundary, both terms of $\boldsymbol{b}_2$ must be computed. For totally reflecting wall boundaries, the slip boundary condition is imposed for velocities, both in the solution of (4.19) and (4.13). The normal component of (4.13) to the wall boundary is

$$\frac{\Delta U_i}{\Delta t}n_i - \frac{\Delta U_i^*}{\Delta t}n_i = -\frac{\partial p^{n+\theta_2}}{\partial x_i}n_i = 0 \tag{4.24}$$

Therefore, the projection of the approximation to $\partial p^{n+\theta_2}/\partial x_i$ normal to the wall boundary is zero, and $\boldsymbol{b}_2$ vanishes.

3. The final velocity is obtained from the discretization of (4.13), which leads to

$$M\overline{\Delta U} = M\ \overline{\Delta U}^* - \Delta t Q[\bar{p}^n + \theta_2 \overline{\Delta p}] + \frac{\Delta t^2}{2} P\bar{p}^n \tag{4.25}$$

where

$$P = (1 - \theta_2) \int_\Omega \frac{\partial}{\partial x_k}(u_k N^T) \frac{\partial N}{\partial x_i} d\Omega \tag{4.26}$$

It can be observed that if other scalar transport equations are added to the model (e.g. pollutants, temperature), it is only necessary to carry out for them a computation totally analogous to step 1, having the same time-step limitation.

Semi-implicit and explicit forms

Assuming the values of the parameters θ_1, θ_2 such that

$$\theta_1 \geq 1/2 \quad , \quad \theta_2 \geq 1/2 \tag{4.27}$$

then (4.21) is solved implicitly and the stability limit of the whole solution is governed by the critical time step of (4.20). Taking $\theta_2 = 0$ (and θ_1 an arbitrary value $\theta_1 \geq 1/2$, e.g. $\theta_1 = 1/2$), then (4.21) becomes explicit and, denoting by $\Delta\tilde{t}$ the time-step on the right hand side,

$$M\frac{\overline{\Delta p}}{\Delta t} = -Q[\bar{U} + \theta_1 \overline{\Delta U}^*] - \Delta\tilde{t} H p^n + b_2 \tag{4.28}$$

For steady state solutions, the timestep Δt must fulfill the condition [19]

$$\Delta t \leq \frac{h^2}{2c^2\theta_1\Delta\tilde{t}} \tag{4.29}$$

We remark that $\Delta\tilde{t} = \Delta t$ is necessary for an accurate transient solution. (The global stability limit is nearly the same as that for the Taylor-Galerkin method.) Furthermore, $\Delta\tilde{t} = h/u$ gives the choice of local $\Delta\tilde{t}$ to yield optimal accuracy for steady state solutions as shown in [17].

4.4 Numerical Results

Some transient and steady state numerical results are now presented to illustrate the performance of the algorithm.

Transient Solutions

Case 1. Rectangular Channel

First, a simple case is studied. A rectangular channel is discretized using 40 triangular (linear) elements (total 33 nodes). A sinusoidal elevation of amplitude $A = 1$ is prescribed in the left boundary, and the rest of the boundary is considered to be totally reflecting. For a constant depth $H = 5\,m.$ and a channel length $L = 800\,m.$ the critical time-step for an explicit Taylor-Galerkin scheme [13] requires $\Delta t \leq 5\,sec.$ In the case of the limit governed by (4.20) the critical time-step is $\Delta t \leq 800\,sec.$, based on the maximum (theoretical) velocity of $u = 0.102\,m./sec.$ for a monochromatic wave with frequency $\omega = 0.005498\,rad./sec.$ (The wavelength is 10 times the channel length.)

To obtain at least 10 points per period, an upper limit of $\Delta t = 100\,sec.$ is needed. The results of the implicit version of the model ($\theta_1 = \theta_2 = 0.5$) for the elevations at the extreme right wall point and velocities at the left free boundary are compared with theory in Figure 4.1. These results were computed for $\Delta t = \Delta t(explicit) = 5\,sec.$, $\Delta t = 50\,sec.$ and $\Delta t = 100\,sec.$, with very good agreement. For instance, for $\Delta t = 50\,sec.$, the maximum error in amplitude is 0.8% and in velocity is 0.2%.

Case 2. Annular Section with Linearly Varying Depth

Here, computations for a problem with linear depth variation are compared with an existing analytical solution. This test is proposed in [10] and results are shown in Figure 4.2(a). Again, a sinusoidal elevation is prescribed on the circular open boundary for a frictionless flow. In Figures 4.2(b) and 4.2(c) we plot the numerical and theoretical results for maximum elevations and maximum velocities along a radial line ($\phi = \pi/4$), assuming the following data: $r_1 = 10\,m.$, $r_2 = 20\,m.$, $H(r_1) = 10\,m.$, $H(r) = H(r_1) + r$, $\omega = 0.9425\,rad./sec.$

The results were computed with a triangular mesh having 20 element rows in the radial direction and 10 in the circumferential direction. A bottom friction term, varying from the maximum at $t = 0$ to zero at $t = T$, is included to filter noise, and steady results are reached after two cycles. A more refined mesh is also used, giving an excellent agreement with the theory.

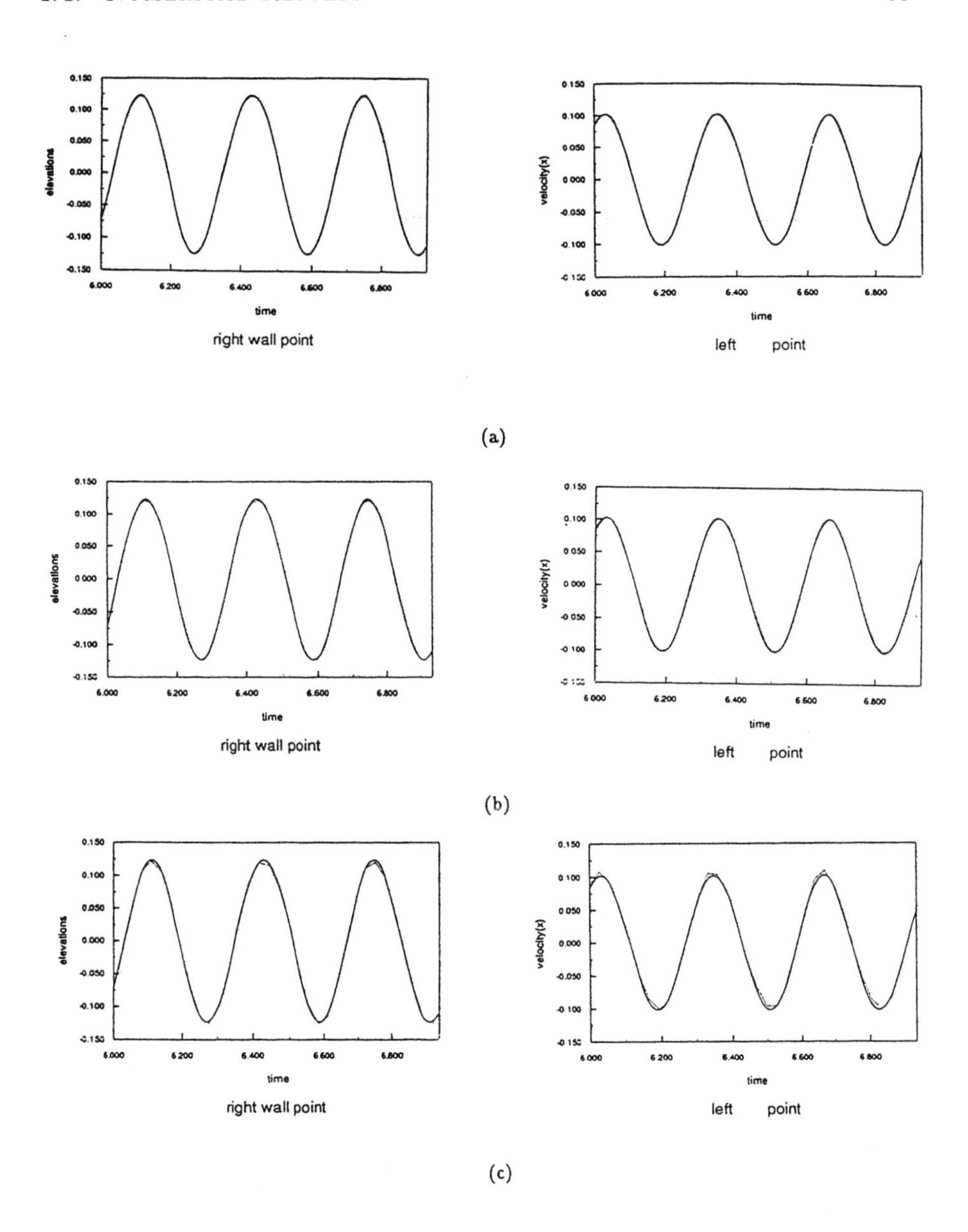

Figure 4.1: A rectangular channel with periodic excitation by prescribed input elevation. Numerical solution profiles (...) for numerical $dt = dt(\text{exptl})$, $10\ dt(\text{exptl})$ and $20dt(\text{exptl})$ in (a),(b),(c) resp. are close to theory(—).

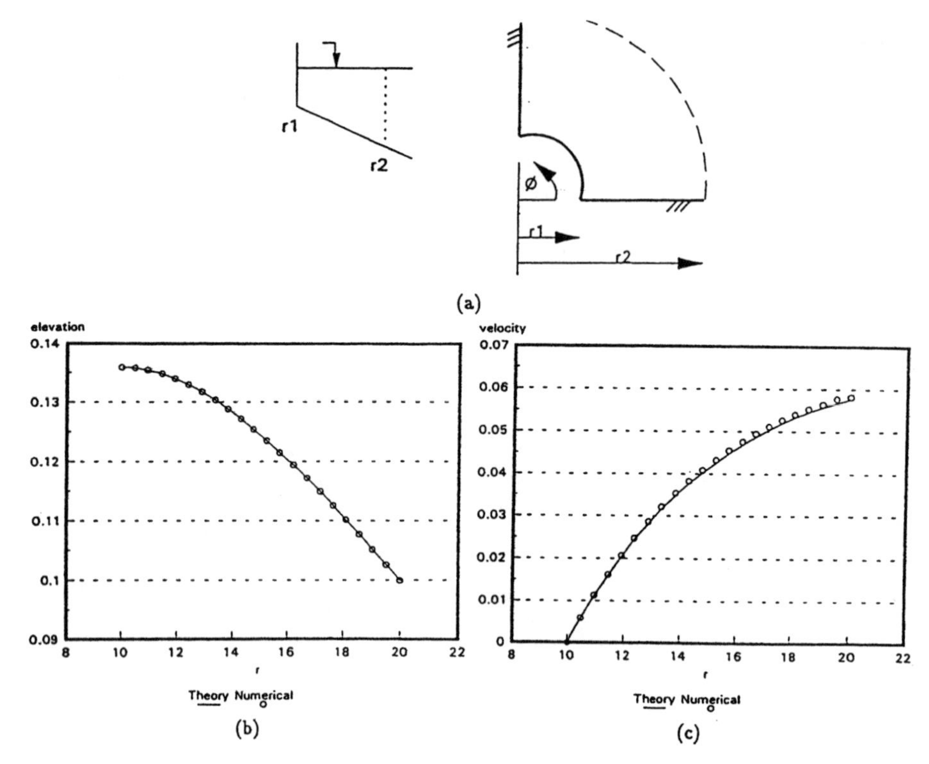

Figure 4.2: (a) Annular section with linear varying depth. (b) Maximum elevation $\eta(r)$ at $\phi = \pi/4$. (c) Maximum velocity $v(r)$ at $\phi = \pi/4$.

Case 3. Severn Estuary

Tidal propagation in the Severn estuary (Bristol Channel) (Figure 4.3(a)) is now studied with a relatively coarse mesh (Figure 4.3(b)) of 256 linear elements and 172 nodal points. A detailed bathymetry description is shown in Figure 4.3(c), where the depth (at mean water level) varies from 40 meters at the western limit to less than 10 meters at Avonmouth. The bottom friction is approximated using a Manning coefficient of $n = 0.04$.

Stable results are reached after two periods from a "cold" start, and are presented in Figures 4.4 and 4.5, for a period of $T = 12.5\,hours$. For Figure 4.4 four control points are chosen: Swansea bay, Weston Super Mare, Avonmouth and an exterior boundary point in the center of the western limit, where a sinusoidal input elevation is prescribed as 3.65 $m.$ of tidal amplitude.

The critical time-step for explicit computation is $\Delta t \approx 48\ sec.$ and the

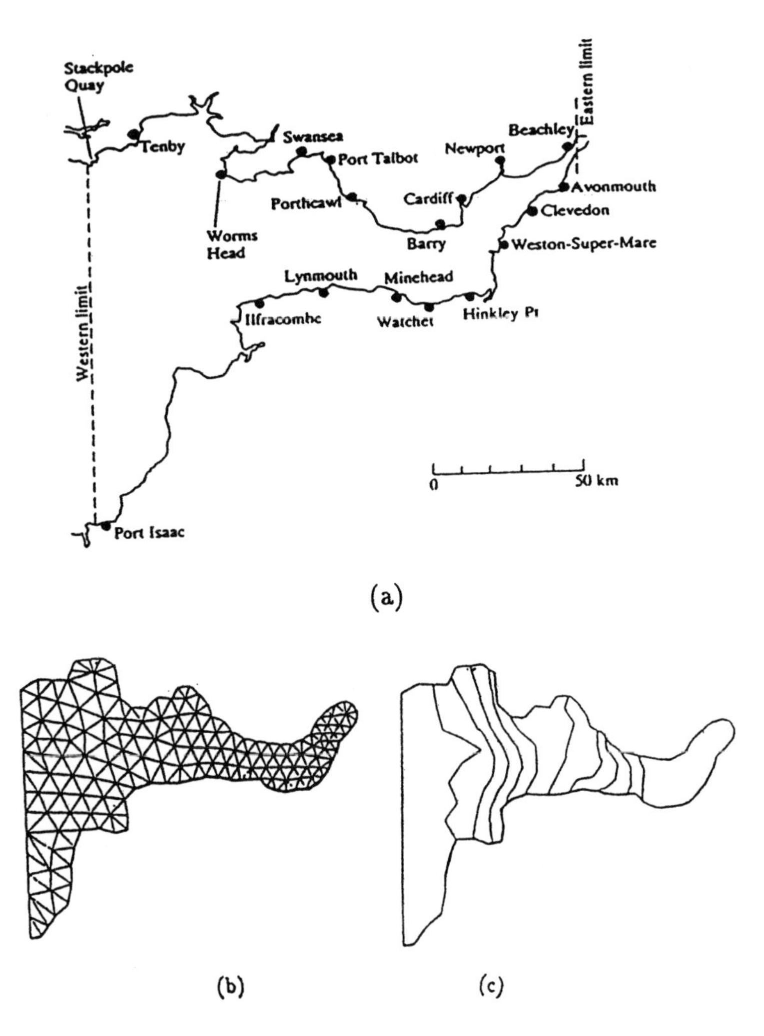

Figure 4.3: (a) Severn Estuary. (b) Mesh. (c) Bathymetry ($\Delta h = 3$ m).

semi-explicit model has been applied with $\Delta t = \Delta t(explicit) = 48\ sec.$, $\Delta t = 10\Delta t(explicit)$ and $\Delta t \approx 17\Delta t(explicit)$ with $\theta_1 = \theta_2 = 0.5$. In Figure 4.4(a) results for $\Delta t = 48\ sec.$ are compared at three points with measurements (elevations) [1], giving good results that show only slight differences, probably due to the simplified bathymetry. The good agreement is retained even when increasing the time-step up to $16\Delta t(explicit)$ as is shown in Figure 4.4(b) for the elevations of the Avonmouth control point.

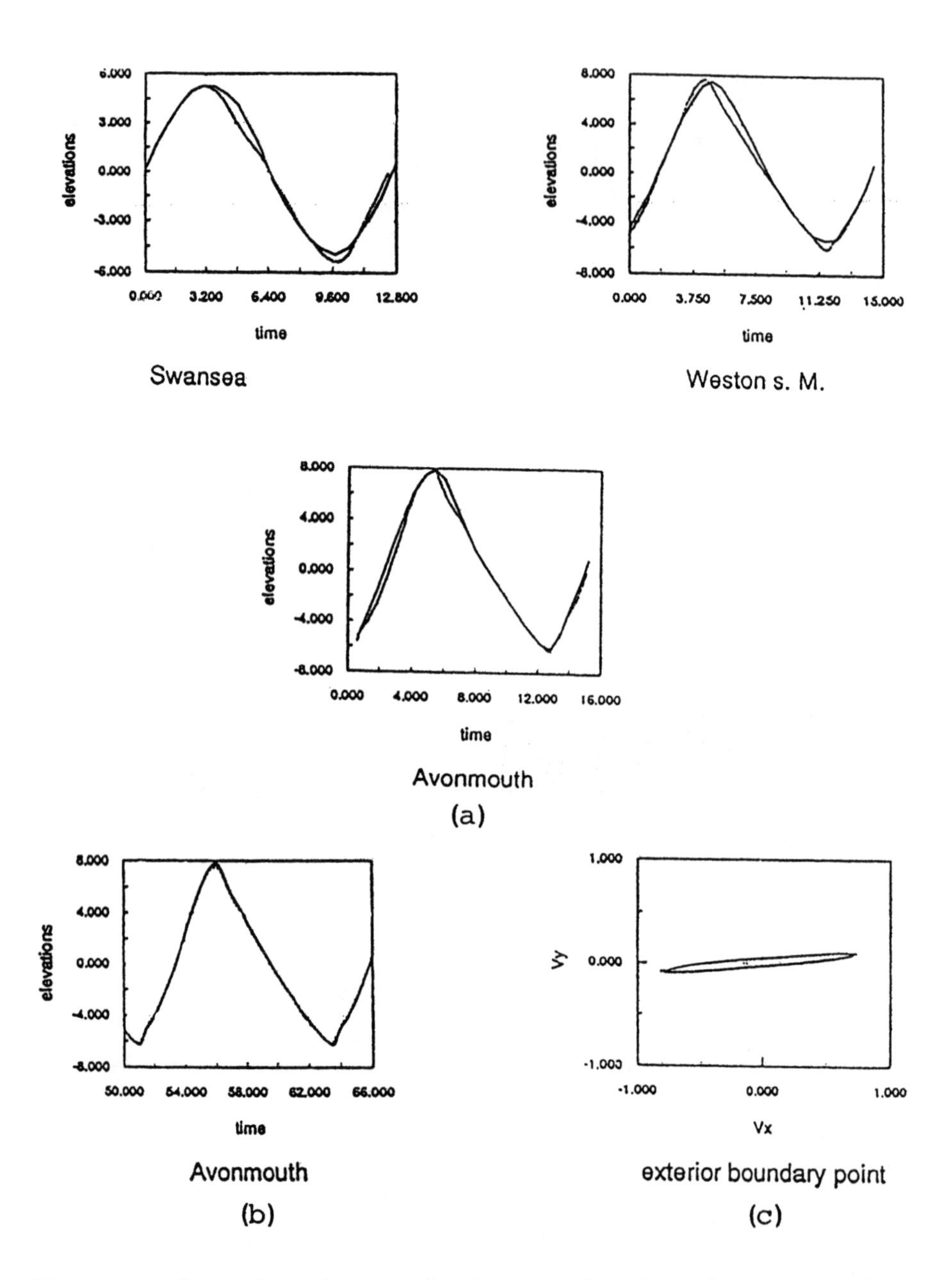

Figure 4.4: Comparison of measured and numerical results at four control points.

An exterior point is chosen to check velocity results, which are depicted in a phase diagram in Figure 4.4(c). Good phase and amplitude behavior is obtained for $\Delta t = 10\Delta t(explicit)$ and $\Delta t \approx 17\Delta t(explicit)$ as compared with the results obtained for $\Delta t = \Delta t(explicit)$. Finally, in Figure 4.5, a typical sequence of currents for a complete period is depicted. The prescribed elevations are imposed at the "Western limit" boundary (see Figure 4.3).

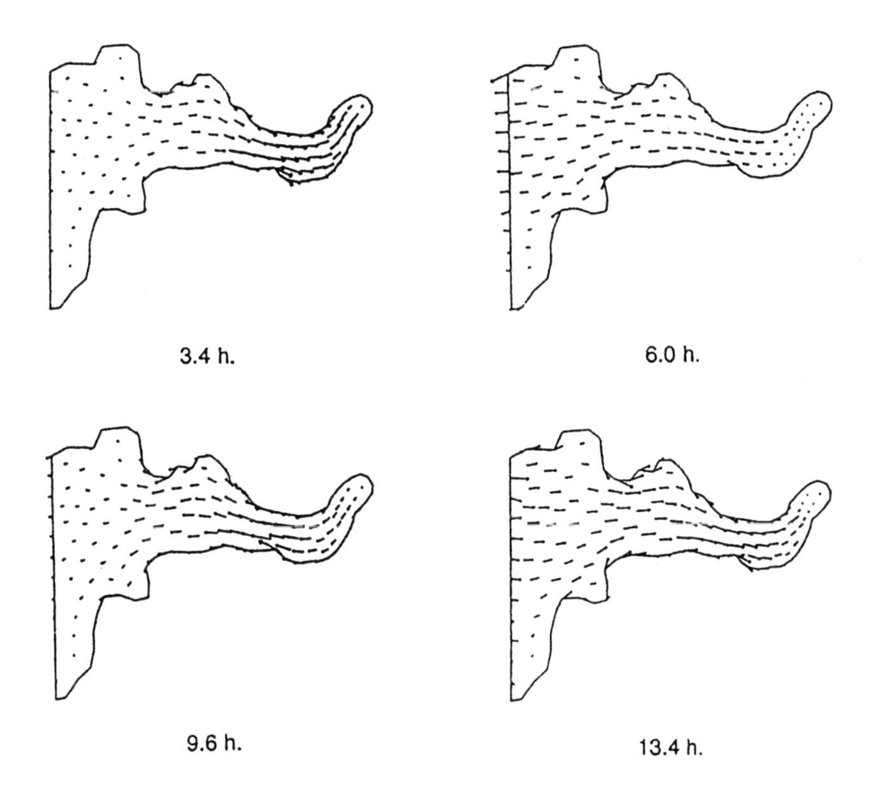

Figure 4.5: Velocity distributions at various times during the tidal cycle.

Steady State Solutions - Supercritical Flow

Obviously the algorithm can be used to obtain efficient steady state solutions for many shallow water problems. Here we illustrate this type of application on three problems involving shallow water flows at supercritical velocities

($F > 1$). These are: (1) a boundary wall constriction over one side of an otherwise unbounded flow [6]; (2) cross waves formed by a symmetrical wall constriction in a rectangular channel; and (3) a combination of cross waves and a "negative" jump in a variable width channel. These latter cases pose strong requirements on the performance of the model because of their "shock" type solution. We remark that in these problems, the vertical accelerations in the vicinity of the jumps are not considered (as a hydrostatic pressure distribution is assumed in the depth integration).

Case 1. Boundary-Wall Constriction

When the change in the alignment turns toward the center of a channel, as is represented in Figure 4.6a, a standing-wave pattern is created of the type shown in Figure 4.6b. A regular mesh of triangles was used (Figure 4.6c) with 2066 nodes and 3955 elements. A comparison is made here with the results obtained by the Taylor-Galerkin method, and the theoretical result in [6].

In Figure 4.7(a) we plot the elevation contours for the Taylor-Galerkin method and in Figure 4.7b for the new method, showing a smaller dispersion of the latter for F_1 (Froude number of the inflow)$= 2.5$. In both cases, the angle of the shock is close to theory: (Theory $\sim 39,58^o$; Taylor-Galerkin $\sim 39,7^o$ and the new method $\sim 39,7^o$) but more oscillations are observed for the Taylor-Galerkin scheme. This can be better seen in Figures 4.8a and 4.8b which compare the total height and Froude number along the bottom line with theory. No artificial or bottom friction were included but the interior damping was increased by using a limiting value of $\Delta\tilde{t}$ in the new algorithm.

The better behavior of the new algorithm is also confirmed when the problem is studied for $F(inflow) = 3$. Fewer oscillations are again observed (Figures 4.9a: Taylor-Galerkin versus Figure 4.9(b): New). See also Figures 4.10a and 4.10b. Now, the angles of the shock are: (Theory $\sim 34,36^o$; Taylor-Galerkin $\sim 34,5^o$ and new method $\sim 34,4^o$). Again, no artificial diffusion is required.

Height and velocities are prescribed at the inflow boundary (left boundary), a slip boundary condition at the wall, free variables at the outflow boundary (right boundary) and a symmetric boundary condition on the upper boundary. An explicit ($\theta_1 = 0.5$, $\theta_2 = 0$) solution scheme was adopted, and local time-stepping with $\Delta\tilde{t} = h/u$ was used.

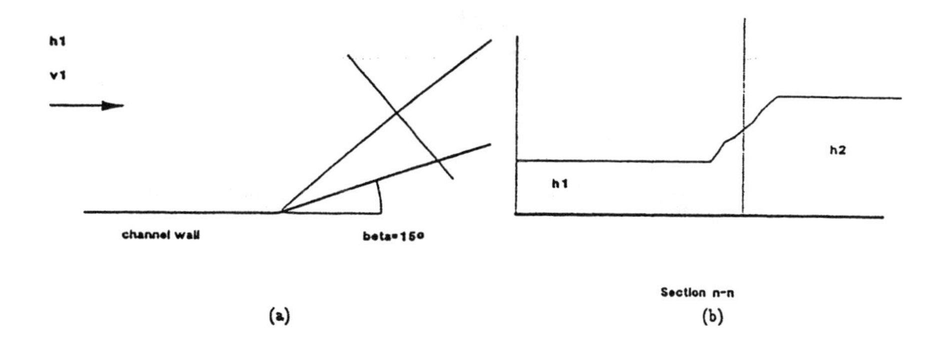

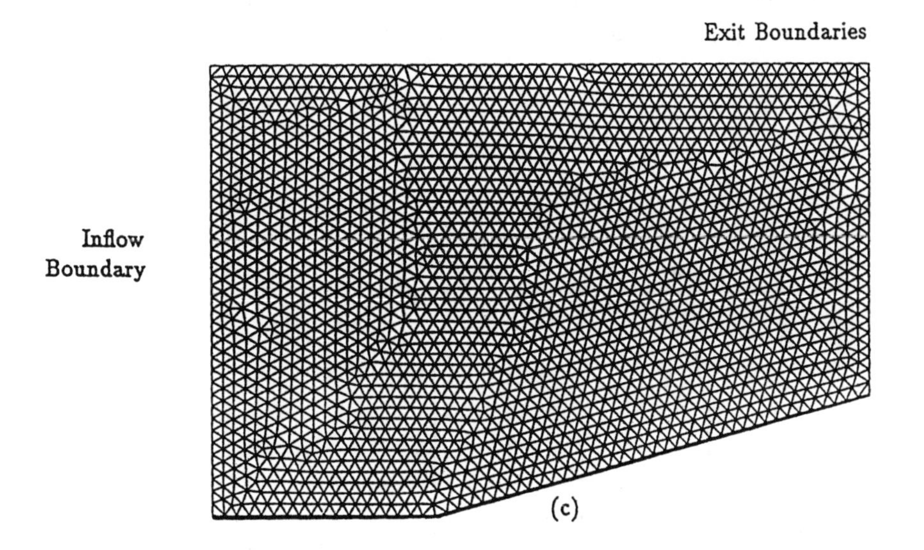

Figure 4.6: (a) Problem of Boundary wall constriction in supercritical flow; (b) Section of the jump formed $(n - n)$; (c) Mesh.

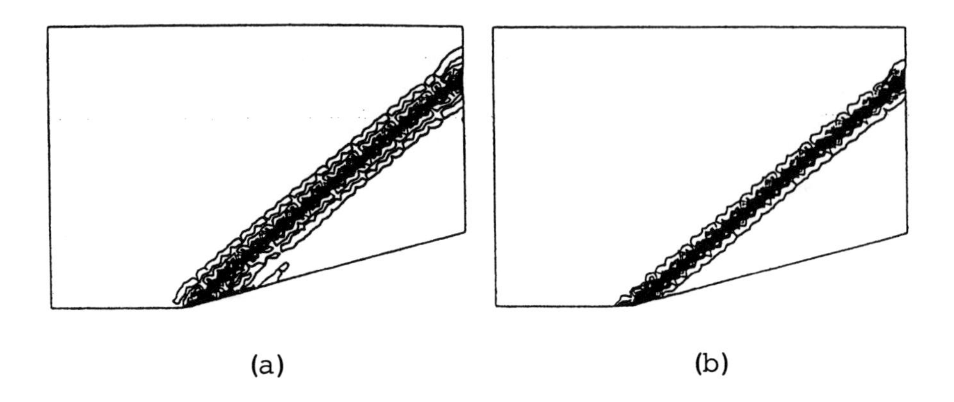

Figure 4.7: Boundary wall constriction. Contours of h for inflow $F = 2.5$: (a) Taylor-Galerkin; (b) New algorithm.

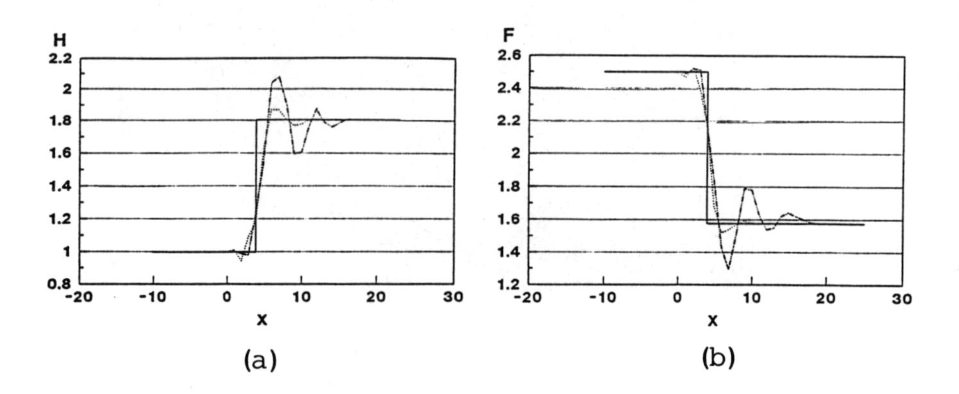

Figure 4.8: Comparison of methods and theory for problem with boundary wall constriction. (a) Total height along bottom line; (b) Froude number along bottom line. Inflow $F = 2.5$.

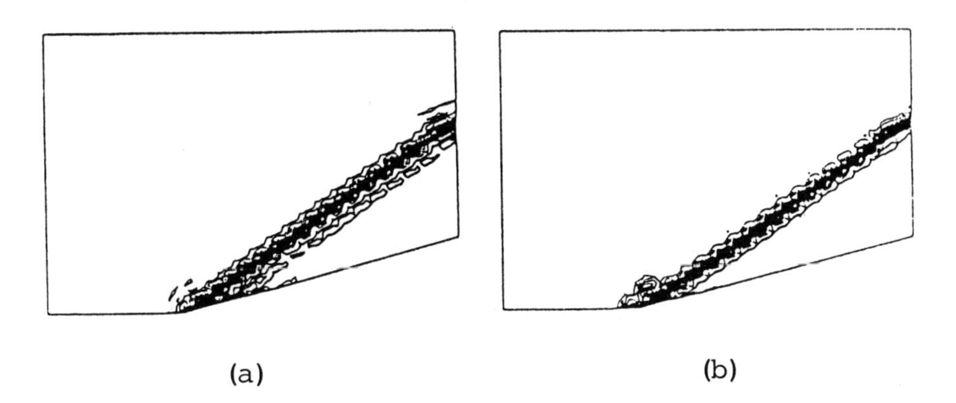

Figure 4.9: Boundary wall constriction. Inflow $F = 3$. Contours of h: (a) Taylor-Galerkin; (b) New algorithm.

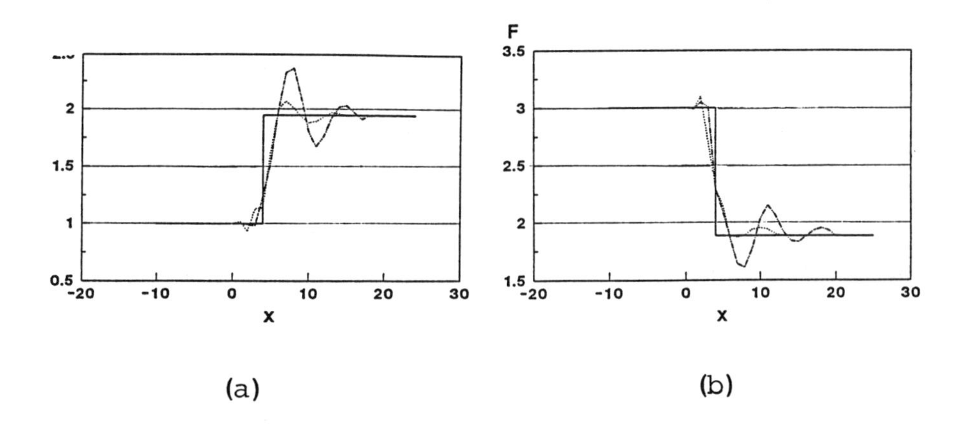

Figure 4.10: Comparison of methods and theory for problem with boundary wall constriction (Inflow $F = 3$). (a) Total height along bottom line; (b) Froude number along bottom line.

Case 2. Symmetric Channel Constriction

If the change of the alignment described above appears on both sides of the channel, a "cross wave" pattern is developed. A fine regular mesh (Figure 4.11a) of 9181 nodes and 17975 elements was generated for a symmetrical constriction with $\beta = 5^o$. The boundary conditions for this problem are: Supercritical inflow at the left boundary (h, u_i , prescribed). Free variables at the outflow (right boundary) and slip wall boundary on the sides. The inflow Froude number was taken as 2.5. The explicit version ($\theta_1 = 0.5$; $\theta_2 = 0$) with lumped mass matrix and local timestepping (in terms of current velocity) was again used. The results in Figure 4.11b are for $h = 1\ m.$ in region 1. In region 2, the theoretical result is $h = 1.254$ m and the numerical result compares well at 1.25 m. Similarly, in region 3 theory predicts $h = 1.55$ m and the present numerical scheme yields $h = 1.54$ m. These results could be further improved by an adaptive remeshing methodology [14].

Case 3. Symmetrical Channel of Variable Width

For supercritical flow in a rectangular channel with a transition of the form described in Figure 4.12a, a combination of a "positive" jump, as in the previous example, and "negative" waves, causing a decrease of depth, occurs. An approximate solution can be obtained assuming no energy losses and that the flow near the wall turns without separation. Here we study a constriction and enlargement of 15°, using a mesh with 9790 nodes and 19151 elements. Applying the same boundary conditions and version of the model as in the previous example, the final results containing "cross" waves and "negative" waves are represented in Figure 4.12b (heights). Note the "gradual" change in the behavior of the negative wave, created at the origin of the wall enlargement. Finally, in Figure 4.13 heights at the wall boundary are plotted, showing that the negative jump at the origin of the enlargement still has a "shock" character.

4.5 Conclusions

The new algorithm presented here is shown to perform well throughout the range of flows encountered in conventional shallow water problems. It represents an improvement (and indeed a simplification) of the procedure introduced in [19] and its accuracy is demonstrated for several examples. In the formulation here we used the simplest linear triangular element but it appears from preliminary studies that use of quadratic triangles may be

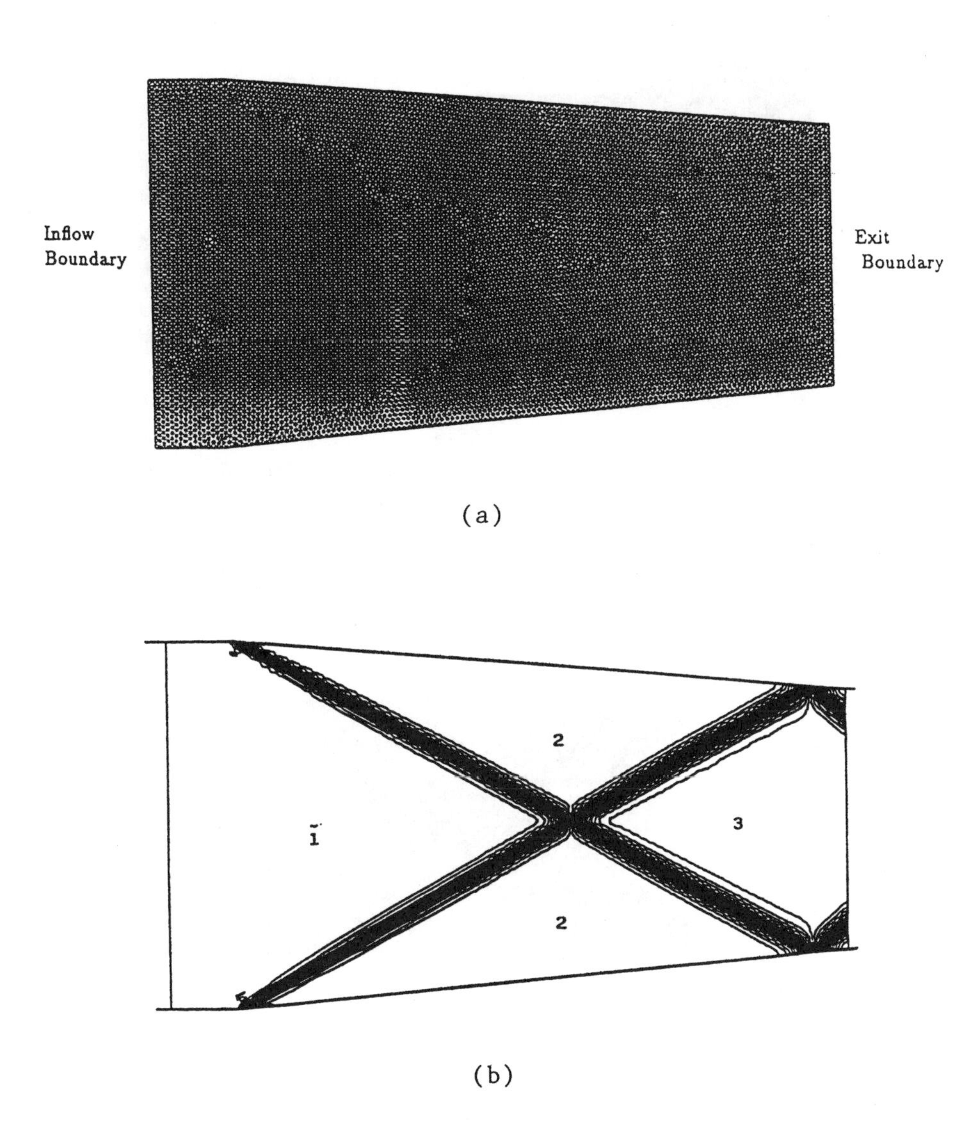

Figure 4.11: Symmetric channel constriction: (a) mesh; (b) h Contours.

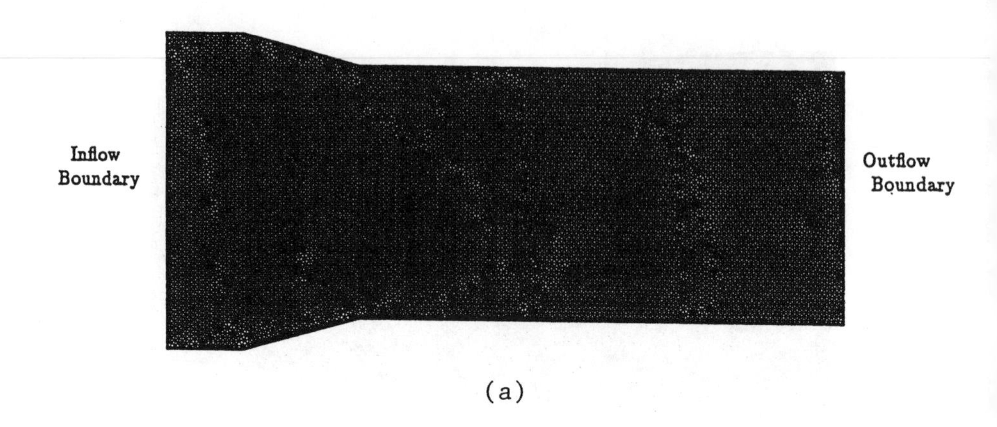

(a)

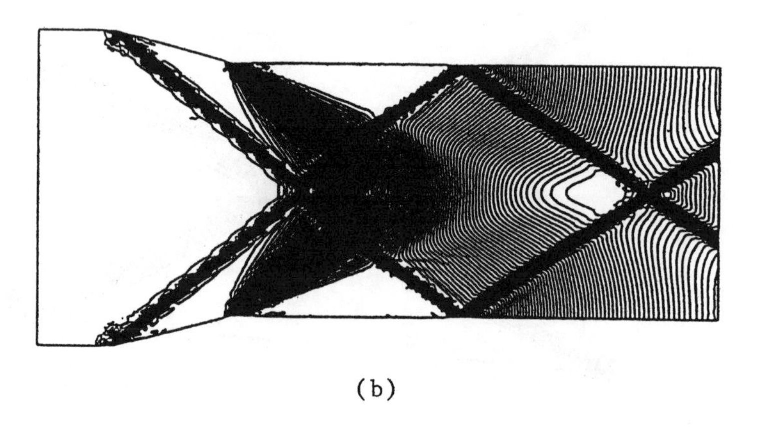

(b)

Figure 4.12: Symmetric channel of Variable width: (a) Mesh; (b) h Contours.

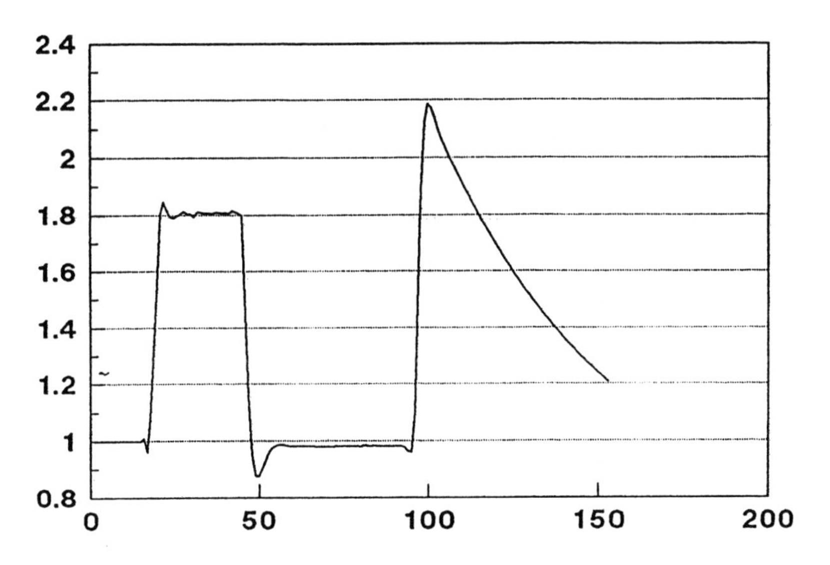

Figure 4.13: Height at wall boundary for Case 3.

advantageous. This issue will be addressed in subsequent work together with the problem of "drying" areas, where the geometry of the estuarine flow changes due to tidal and wind driven circulation.

Acknowledgments: Dr. Pablo Ortiz is grateful to DGYCIT (Ministerio de Educación y Ciencia, Spain) for the fellowship which permitted his visit to University College of Swansea (Institute of Numerical Methods in Engineering) in 1993.

Appendix

Consider, for simplicity, convective transport of a scalar variable ϕ

$$\frac{\partial \phi}{\partial t} + \frac{\partial F_j}{\partial x_j} + Q = 0 \tag{4.30}$$

where $F_j = u_j \phi$ and Q is a source term. This can be written as

$$\frac{\partial \phi}{\partial t} + u_j \frac{\partial \phi}{\partial x_j} + \phi \frac{\partial u_j}{\partial x_j} + Q = 0 \tag{4.31}$$

Let $\bar{\phi}^n$ denote the value at $n\Delta t$ and position $(x_i - \delta_i)$ (origin of the characteristic, as seen in Figure 4.14 for the space coordinate x_i). Along the

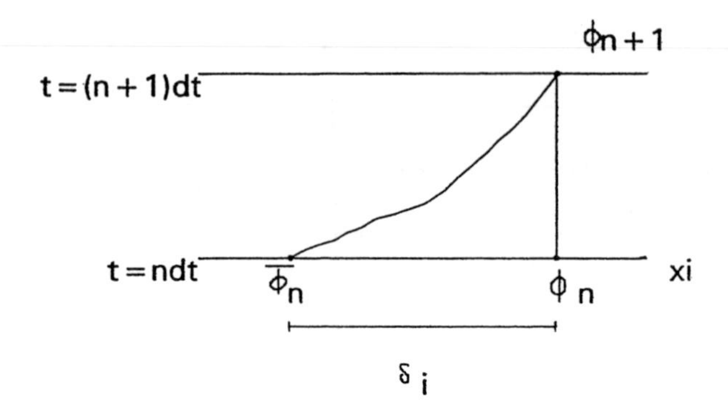

Figure 4.14: Local Approximation.

characteristic

$$\phi^{n+1} - \bar{\phi}^n = -\Delta t \left[\bar{Q} + \left(\overline{\phi \frac{\partial u_j}{\partial x_j}} \right) \right]^n \tag{4.32}$$

where the overbar denotes values which are some average along the characteristic at x_i. The distance δ_i can be approximated as

$$\delta_i \approx \Delta t \left(u_i - \frac{\Delta t}{2} u_j \frac{\partial u_i}{\partial x_j} \right)^n \tag{4.33}$$

Further, approximating $\bar{\phi}^n$ as

$$\bar{\phi}^n = \phi^n - \delta_i \frac{\partial \phi}{\partial x_i} |^n + \frac{\delta_i \delta_j}{2} \frac{\partial^2 \phi}{\partial x_i \partial x_j} |^n \tag{4.34}$$

and inserting the approximation for δ_i we can write (omitting terms of higher order than Δt^2)

$$\bar{\phi}^n = \phi^n - u_i \Delta t \frac{\partial \phi}{\partial x_i} |^n + \frac{\Delta t^2}{2} u_j \frac{\partial u_i}{\partial x_j} \frac{\partial \phi}{\partial x_i} |^n + \frac{\Delta t^2}{2} u_i u_j \frac{\partial^2 \phi}{\partial x_i \partial x_j} |^n \tag{4.35}$$

The evaluation of $\bar{Q}$ and $\left(\overline{\phi \frac{\partial u_j}{\partial x_j}} \right)$ in (4.32) gives

$$\bar{Q} = Q^n - u_i \frac{\Delta t}{2} \frac{\partial Q}{\partial x_i} |^n \tag{4.36}$$

$$\overline{\phi \frac{\partial u_j}{\partial x_j}} = \left(\phi \frac{\partial u_j}{\partial x_j} \right)^n - u_i \frac{\Delta t}{2} \frac{\partial}{\partial x_i} \left(\phi \frac{\partial u_j}{\partial x_j} \right)^n \tag{4.37}$$

Now, setting $\Delta\phi = \phi^{n+1} - \phi^n$ and reordering,

$$\Delta\phi = -\Delta t \left[\frac{\partial F_i}{\partial x_i} + Q\right]^n + \frac{\Delta t^2}{2}\left[u_i \frac{\partial}{\partial x_i}\left(\frac{\partial F_j}{\partial x_j} + Q\right)\right]^n \tag{4.38}$$

Note that the final expression does not now include the term $\phi \frac{\partial u_j}{\partial x_j}$. (For a more detailed derivation see [16].)

References

[1] Allen, B., Private Communication, Univ. Coll. of Swansea, 1993.

[2] Chorin, A.J., "Numerical solution of the Navier-Stokes equations", *Math. Comput.*, 22, 745-762, 1968.

[3] Daubert, O., J. Hervouet and A. Jami, "Description of some numerical tools for solving incompressible turbulent and free surface flows", *IJNME*, 27, 3-20, 1989.

[4] Donea, J.,"A Taylor-Galerkin method for convective transport problems", *IJNME*, 20, 101-119, 1984.

[5] Foreman, M.G., "An analysis of the 'wave equation' model for finite element tidal computation", *J.Comp. Phys.*, 52, 290-312, 1983.

[6] Ippen, A.T., "High velocity flow in open channels: Symposium", *Trans. ASCE*, V. 116, 1951.

[7] Jiang, C.B. and M. Kawahara, "A three-step finite element method for unsteady incompressible flows". *Comp.Mech.*, 11, 355-370, 1993.

[8] Kawahara, M., H. Hirano, K. Tsuhora and K. Iwagaki, "Selective lumping finite element method for shallow water flow", *IJNME*, 2, 99-112, 1982.

[9] Kinnmark, I. and W. Gray, "An implicit wave equation model for the shallow water equations", *Adv. in Water Res.*, 7, 168-171, 1984.

[10] Lynch, D.R. and W. Gray, "Analytic solutions for computer flow model testing", *J. of Hydr. Div., ASCE*, 104, HY10, 1409-1428, 1978.

[11] Lynch, D.R. and W. Gray, "A wave equation model for finite element tidal computations", *Computers and Fluids*, 7, 207-228, 1979.

[12] Navon, I.M., "A review of finite element methods for solving the shallow water equations", *Computer Modelling in Ocean Eng.*, B. Schreffler and O.C.Zienkiewicz (eds), Balkema, 273-278, 1988.

[13] Peraire, J.,"A finite element method for convection dominated flows", *PhD thesis*, Univ.Coll. of Swansea, 1986.

[14] Peraire, J., M. Vahadati, K. Morgan and O.C. Zienkiewicz, "Adaptive remeshing for compressible flow computations", *J.Comp. Phys.*, 72, 449-466, 1987.

[15] Peraire, J., O.C. Zienkiewicz and K. Morgan, "Shallow water problems. A general explicit formulation", *IJNME*, 22, 547-574, 1986.

[16] Zienkiewicz, O.C. and R. Codina, "A general algorithm for compressible and incompressible flows. Part I : The split characteristic based scheme", to be published, *IJNME*, 1995.

[17] Zienkiewicz, O.C. and R. Taylor, *The Finite Element Method*, Vol. 2, Mc Graw-Hill, 1991.

[18] Zienkiewicz, O.C. and J. Wu, "A general explicit or semi-explicit algorithm for compressible and incompressible flows", *IJNME*, 35, 457-479, 1992.

[19] Zienkiewicz, O.C., J. Wu and J. Peraire, "A new semi-implicit or explicit algorithm for Shallow Water equations", *Math. Mod. and Scient. Comp.*, 1, 31-49, 1993.

Chapter 5

An Entropy Variable Formulation and Petrov-Galerkin Methods for the Shallow Water Equations

S.W. Bova[1] *and G.F. Carey*[1]

5.1 Introduction

Various forms of the shallow water equations and finite element models based on these are in common use, particularly in such areas as coastal and estuarine hydrodynamics (*e.g.* see [10, 19, 31]). These equations apply when the free surface wavelength is very long compared to the depth, and are often coupled with other transport models for chemical and biological species. Under certain standard assumptions, the shallow water equations constitute a hyperbolic system of partial differential equations that describe the elevation and velocity of the free surface; *e.g.*, in response to storm surges or for the design of spillways and other channels [3].

Finite element methods are particularly appealing for these applications because of the ease with which irregular boundaries, obstructions, islands, and strongly varying bathymetry can be handled. However, many of the well-known difficulties associated with solving convective transport problems still must be accommodated [28, 29]. These difficulties, common to other problems involving convection-dominated transport, have led to the

[1]The Computational Fluid Dynamics Laboratory, The University of Texas at Austin

development of special finite element methods. The present work concerns a theoretical framework for developing such a class of methods for the shallow water equations.

The outline of this chapter is as follows: we first define the first-order, hyperbolic, shallow water equation system to be analyzed. Then, using the total energy of the water column, a transformation of variables is introduced and a new, symmetric form of the system is constructed. This form becomes the foundation for a SUPG finite element formulation. The one-dimensional case is then considered and a time-accurate Runge-Kutta scheme introduced to solve the semi-discrete system. Numerical results follow. Finally, a lumped mass, time-iterative strategy is developed for the solution of two-dimensional, steady-state problems and applied to representative channel flow problems.

5.2 Symmetric Shallow Water System

Consider a body of water with mean free surface level in the (x_1, x_2) plane. The bottom depth is given by the positive function $h(x_1, x_2)$, and the unknown surface elevation, $\eta(x_1, x_2, t)$ is measured from the mean free surface. Thus, the total height is $H = h + \eta$. A body force due to gravity is present and is assumed to act in the $-x_3$ direction. The problem is to determine the free surface $\eta(x_1, x_2, t)$ and surface velocity $\boldsymbol{u}(x_1, x_2, t) = (u_1, u_2)^t$ for a given bottom shape and known initial disturbance.

The shallow water equations may be derived by depth-integration of the incompressible Navier-Stokes equations, and the assumption of a hydrostatic pressure distribution (*e.g.* see [26]). If the additional assumptions of negligible friction and Coriolis effects are made, these equations may be written in the divergence form,

$$\boldsymbol{U}_{,t} + \boldsymbol{\nabla} \cdot \boldsymbol{F}(\boldsymbol{U}) = \boldsymbol{S}(\boldsymbol{U}) \tag{5.1}$$

for the state vector $\boldsymbol{U} = (H, Hu_1, Hu_2)^t$. In (5.1), the subscripted comma denotes differentiation with respect to the indicated variable. The divergence term may be written as

$$\boldsymbol{\nabla} \cdot \boldsymbol{F}(\boldsymbol{U}) = \boldsymbol{F}_{1,x_1} + \boldsymbol{F}_{2,x_2} \tag{5.2}$$

where

$$\boldsymbol{F}_1(\boldsymbol{U}) = \begin{pmatrix} Hu_1, & Hu_1^2 + gH^2/2, & Hu_1u_2 \end{pmatrix}^t \tag{5.3}$$

$$\boldsymbol{F}_2(\boldsymbol{U}) = \begin{pmatrix} Hu_2, & Hu_1u_2, & Hu_2^2 + gH^2/2 \end{pmatrix}^t \tag{5.4}$$

and g is the acceleration due to gravity. Finally, the source term is

$$\boldsymbol{S}(\boldsymbol{U}) = \begin{pmatrix} 0, & gHh_{,x_1} - b_1, & gHh_{,x_2} - b_2 \end{pmatrix}^t \tag{5.5}$$

where the bottom friction stresses are given by the multidimensional extension of the Manning-Chézy formula [7,30],

$$b_i = \frac{gn^2 u_i \sqrt{u_1^2 + u_2^2}}{\alpha^2 \sqrt[3]{H}} \quad \text{for } i = 1, 2. \tag{5.6}$$

In (5.6), n is Manning's roughness coefficient, the conversion factor $\alpha = 1.0$ for metric units and 2.21 for British units, and the assumption of a wide channel has been made so that hydraulic radius is given by the depth [7].

Rather than the divergence form (5.1), the shallow water equations may also be written using the chain rule as the quasilinear system

$$\boldsymbol{U}_{,t} + \boldsymbol{A}_1(\boldsymbol{U})\boldsymbol{U}_{,x_1} + \boldsymbol{A}_2(\boldsymbol{U})\boldsymbol{U}_{,x_2} = \boldsymbol{S}(\boldsymbol{U}) \tag{5.7}$$

where the flux Jacobian matrices, $\boldsymbol{A}_1(\boldsymbol{U})$ and $\boldsymbol{A}_2(\boldsymbol{U})$, are given by

$$\boldsymbol{A}_1(\boldsymbol{U}) = \begin{pmatrix} 0 & 1 & 0 \\ gH - u_1^2 & 2u_1 & 0 \\ -u_1 u_2 & u_2 & u_1 \end{pmatrix}, \quad \boldsymbol{A}_2(\boldsymbol{U}) = \begin{pmatrix} 0 & 0 & 1 \\ -u_1 u_2 & u_2 & u_1 \\ gH - u_2^2 & 0 & 2u_2 \end{pmatrix} \tag{5.8}$$

It is well known that the system (5.7) (or, equivalently (5.1)) is hyperbolic: *i.e.* that the Jacobians $\boldsymbol{A}_i$ have real eigenvalues and a complete set of eigenvectors (*e.g.* see [26]). For most systems of this form that arise from convective transport models, the flux Jacobian matrices, $\boldsymbol{A}_1$ and $\boldsymbol{A}_2$, are not symmetric. Nor can they usually be simultaneously diagonalized [32]. Indeed, this is the case for the shallow water system presently under consideration. Similar types of hyperbolic systems arise in gas dynamics and other applications. These systems may be symmetrized if an appropriate generalized entropy function exists as shown in [11,13,27]. However, the physical entropy is independent of the flow variables in the present context, since the assumptions of incompressible, adiabatic, isothermal flow are implicit in the shallow water equations. Nevertheless, the analogy between shallow water theory and gas dynamics can be used to provide guidance in constructing an appropriate transformation. For example, the hydraulic jumps (or bores) that appear in supercritical shallow water flow are analogous to the shock waves that arise in the gas dynamics case. It is a matter of observation that bores form in such a way that the water always moves from a region of lower depth to one of higher depth, although the opposite case is also admissible mathematically. As in the gas dynamics (*e.g.* transonics) problem, this nonuniqueness in the mathematical solution may be reconciled by appealing to the physics. This situation is discussed at length by Stoker [26], and resolved by requiring that the total energy of the water column always decrease across a bore. For this reason, the analog to the entropy is defined to be the total energy of the water column.

Finite element methods based on entropy variable formulations have been developed for gas dynamics applications [14, 15, 24]. These methods are based on the analysis of Harten [11], who investigated the symmetrization of conservation laws that have associated, generalized entropy functions. In the case of the compressible Navier-Stokes equations with heat conduction, symmetrization occurs only if the generalized entropy function is at most trivially different from the physical entropy [13]. With respect to the shallow water equations, Tadmor [27] used the total energy as a generalized entropy function and derived a skew-self-adjoint form of the shallow water equations in terms of the resulting variables. A theorem for hyperbolic conservation systems is presented in the same paper, which establishes an equivalence among the various properties of symmetrizability, having an entropy function, and having a skew-self-adjoint form. In the present study, we derive the associated symmetric form of the system (5.7). (In [6] we show that the same generalized entropy function (5.16) symmetrizes the incompletely parabolic system of shallow water equations that results when viscous stresses are included in the formulation.)

To proceed, let us define a change of dependent variables, $\boldsymbol{U} = \boldsymbol{U}(\boldsymbol{V})$, under which the new flux Jacobians of the resulting system are symmetric. If the chain rule is applied to (5.7), the result may be written as

$$\boldsymbol{A}_0(\boldsymbol{V})\boldsymbol{V}_{,t} + \tilde{\boldsymbol{A}}_1(\boldsymbol{V})\boldsymbol{V}_{,x_1} + \tilde{\boldsymbol{A}}_2(\boldsymbol{V})\boldsymbol{V}_{,x_2} = \tilde{\boldsymbol{S}}(\boldsymbol{V}) \tag{5.9}$$

where

$$\boldsymbol{A}_0 = \frac{\partial \boldsymbol{U}}{\partial \boldsymbol{V}} \tag{5.10}$$

and $\tilde{\boldsymbol{A}}_i = \boldsymbol{A}_i\boldsymbol{A}_0$ for $i = 1, 2$. The change of variables is to be chosen so that $\boldsymbol{A}_0$ is symmetric and positive-definite, and the new flux Jacobians $\tilde{\boldsymbol{A}}_1$ and $\tilde{\boldsymbol{A}}_2$ are symmetric. Under these conditions, (5.9) is, by definition, a symmetric hyperbolic system. The symmetric form of the transport equations is interesting because weighted residual formulations based upon the symmetric form automatically possess certain stability properties associated with exact solutions of the governing equations [13, 27].

Let us assume that an appropriate (convex), generalized entropy function $\mathcal{F} = \mathcal{F}(\boldsymbol{U})$ can be identified. Then the change of variables may be obtained by setting $\boldsymbol{V}^t = \frac{\partial \mathcal{F}}{\partial \boldsymbol{U}}$. Convexity is necessary but not sufficient to guarantee symmetrization [13]. There must also exist scalar-valued functions σ_i associated with $\mathcal{F}$ such that

$$\frac{\partial \sigma_i}{\partial \boldsymbol{U}} = \boldsymbol{V}^t \boldsymbol{A}_i \tag{5.11}$$

These functions are called entropy fluxes. If $\mathcal{F}$ is convex and corresponding entropy fluxes exist, then the system can be symmetrized.

It is convenient to nondimensionalize the governing equations before deriving the symmetric form for the shallow water system. This normalization is accomplished by introducing a length scale L and a velocity scale $u_\infty = \sqrt{gL}$. The time is nondimensionalized by the ratio L/u_∞. The resulting nondimensional system is identical to (5.1), with the definition of $\boldsymbol{U}$ unchanged. However, the nondimensional flux and source vectors are given by

$$\boldsymbol{F}_1(\boldsymbol{U}) = \Big(Hu_1, \ \ Hu_1^2 + H^2/2, \ \ Hu_1u_2 \Big)^t, \tag{5.12}$$

$$\boldsymbol{F}_2(\boldsymbol{U}) = \Big(Hu_2, \ \ Hu_1u_2, \ \ Hu_2^2 + H^2/2 \Big)^t, \tag{5.13}$$

and

$$\boldsymbol{S}(\boldsymbol{U}) = \Big(0, \ \ Hh_{,x_1} - b_1, \ \ Hh_{,x_2} - b_2 \Big)^t, \tag{5.14}$$

respectively, with

$$b_i = \frac{n_*^2 u_i \sqrt{u_1^2 + u_2^2}}{\alpha \sqrt[3]{H}} \quad \text{for } i = 1, 2, \tag{5.15}$$

and $n_*^2 = n^2 g / \sqrt[3]{L}$. The nondimensional flux Jacobians are given by (5.8) with g replaced by 1. It should be understood that all subsequent equations are to be regarded as nondimensional.

The total energy at a point (x_1, x_2, x_3) in the fluid at time t is the sum of the potential and the kinetic energies: $E = x_3 + V^2(x_1, x_2, x_3, t)/2$, where $V^2 = u_1^2 + u_2^2$. Integrating over the depth, we obtain $\frac{H^2 + HV^2}{2}$. Since this quantity must always decrease across a bore, let

$$\mathcal{F} = \frac{H^2 + HV^2}{2} \tag{5.16}$$

be the nondimensional, generalized entropy function. This choice leads to the change of variables

$$\boldsymbol{V}^t = \frac{\partial \mathcal{F}}{\partial \boldsymbol{U}} = \left(\begin{array}{ccc} H - V^2/2, & u_1, & u_2 \end{array} \right) \tag{5.17}$$

It may be noted that the new variables are similar to the primitive variable form of the equations that is usually considered (*e.g.* see [30]), in the sense that V_2 and V_3 are simply the cartesian velocity components. However, the proposed variables differ in that instead of the surface elevation, V_1 may be interpreted as the difference in the potential and kinetic energies.

Differentiation of (5.17) with respect to $\boldsymbol{U}$ results in the symmetric matrix

$$\boldsymbol{A}_0^{-1} = \frac{\partial \boldsymbol{V}}{\partial \boldsymbol{U}} = \frac{\partial^2 \mathcal{F}}{\partial \boldsymbol{U}^2} = \left(\begin{array}{ccc} 1 + V^2/H & -u_1/H & -u_2/H \\ & 1/H & 0 \\ \text{symm.} & & 1/H \end{array} \right) \tag{5.18}$$

It may be readily verified that $\boldsymbol{A}_0^{-1}$ never becomes singular for physical values of $\boldsymbol{V}$: the determinant of $\boldsymbol{A}_0^{-1}$ is $1/H^2$. We remark that the matrix $\boldsymbol{A}_0^{-1}$ is the Hessian of $\mathcal{F}$, and is therefore symmetric and positive definite (SPD) for any convex $\mathcal{F}$. Positive definiteness may be verified by considering the eigenvalues of $\boldsymbol{A}_0^{-1}$. Since $\boldsymbol{A}_0^{-1}$ is real and symmetric, its eigenvalues are real. These eigenvalues may be written as

$$\mu_1 = 1/H, \quad \mu_{2,3} = \frac{\alpha \pm \sqrt{\alpha^2 - 4H}}{2H} \tag{5.19}$$

where $\alpha = V^2 + 1 + H$. The depth H is always positive, so $\mu_1 > 0$, and since the eigenvalues are real, $\alpha^2 > 4H$. Therefore, μ_2 and μ_3 are also positive. Hence, $\boldsymbol{A}_0^{-1}$ is SPD. The symmetrizer may be found by inverting (5.18) to obtain

$$\boldsymbol{A}_0 = \left(\begin{array}{ccc} 1 & u_1 & u_2 \\ & H + u_1^2 & u_1 u_2 \\ \text{symm.} & & H + u_2^2 \end{array} \right) \tag{5.20}$$

The corresponding entropy fluxes then follow on integrating (5.11) and may be written as

$$\sigma_i = u_i \left(H^2 + \frac{HV^2}{2} \right), \text{ for } i = 1, 2. \tag{5.21}$$

Finally, postmultiplying the flux Jacobian matrices $\boldsymbol{A}_i, i = 1,2$, by $\boldsymbol{A}_0$ yields the symmetrized flux Jacobian matrices,

$$\tilde{\boldsymbol{A}}_1 = \begin{pmatrix} u_1 & H + u_1^2 & u_1 u_2 \\ & u_1(3H + u_1^2) & u_2(H + u_1^2) \\ \text{symm.} & & u_1(H + u_2^2) \end{pmatrix} \tag{5.22}$$

and

$$\tilde{\boldsymbol{A}}_2 = \begin{pmatrix} u_2 & u_1 u_2 & H + u_2^2 \\ & u_2(H + u_1^2) & u_1(H + u_2^2) \\ \text{symm.} & & u_2(3H + u_2^2) \end{pmatrix} \tag{5.23}$$

associated with the new dependent variables $\boldsymbol{V}$.

We remark that the flux vectors are not homogeneous functions of $\boldsymbol{V}$. If this were the case, then by Euler's theorem on homogeneous functions, they would satisfy $\gamma \tilde{\boldsymbol{F}}_i(\boldsymbol{V}) = \tilde{\boldsymbol{A}}_i(\boldsymbol{V})\boldsymbol{V}$, where γ is the degree of homogeneity. For example, the Euler equations of gas dynamics written in the form (5.7) have homogeneous flux functions of degree one: they satisfy $\boldsymbol{F}_i(\boldsymbol{U}) = \boldsymbol{A}_i(\boldsymbol{U})\boldsymbol{U}$. This property may be useful for performing stability analyses or implementing flux split algorithms. The lack of homogeneous flux functions is not a major disadvantage since the equations for the proposed symmetric formulation of the shallow water equations do not have this property even when written in the more familiar conservation form (5.7).

This completes the transformation to the symmetric form of the shallow water equations (5.9), with the definitions (5.17) and (5.20)-(5.23). In the next section, we develop a streamline upwind Petrov-Galerkin finite element formulation based on this symmetric system.

5.3 Petrov-Galerkin Formulation

The finite element method used in the present study is based on the approach in [14,15] for the Euler equations of gas dynamics. The variational formulation is obtained by taking a duality pairing of the transport equations with test functions and integrating over the domain. Thus, (5.9) becomes

$$\int_{\Omega} \hat{\boldsymbol{W}}^t \left(\boldsymbol{A}_0 \boldsymbol{V}_{,t} + \tilde{\boldsymbol{A}}^t \boldsymbol{\nabla} \boldsymbol{V} - \tilde{\boldsymbol{S}} \right) d\Omega = \boldsymbol{0} \tag{5.24}$$

where $\tilde{\boldsymbol{A}}^t = (\tilde{\boldsymbol{A}}_1, \tilde{\boldsymbol{A}}_2)$ and

$$\boldsymbol{\nabla V} = \begin{pmatrix} \boldsymbol{V}_{,x_1} \\ \boldsymbol{V}_{,x_2} \end{pmatrix}.$$

For the Petrov-Galerkin formulation, the test functions are defined as the standard Galerkin test functions plus a perturbation [17]. (For the class of methods considered here, this perturbation may be regarded as a directional derivative.) More specifically, we set

$$\hat{\boldsymbol{W}} = \boldsymbol{W} + \tilde{\boldsymbol{\tau}}\tilde{\boldsymbol{A}}^t\boldsymbol{\nabla}\boldsymbol{W} + \boldsymbol{\mathcal{D}}^t\boldsymbol{\nabla}\boldsymbol{W} \tag{5.25}$$

where $\tilde{\boldsymbol{\tau}}$ denotes a symmetric, positive semidefinite matrix of intrinsic time scales which we discuss in detail later. Briefly, it acts to normalize the directional derivative $\tilde{\boldsymbol{A}}^t\boldsymbol{\nabla}$. A discontinuity capturing operator, $\boldsymbol{\mathcal{D}}^t = (\boldsymbol{\mathcal{D}}_1^t, \boldsymbol{\mathcal{D}}_2^t)$, is useful for eliminating spurious oscillations in the vicinity of steep, local gradients. Note that if $\tilde{\boldsymbol{\tau}} = \boldsymbol{\mathcal{D}}_i = \boldsymbol{0}$ the Galerkin method is obtained.

If (5.25) is substituted into (5.24), and Gauss' divergence theorem is applied, the result may be written as

$$\begin{aligned}\int_\Omega \hat{\boldsymbol{W}}^t\boldsymbol{A}_0\boldsymbol{V}_{,t}d\Omega &= -\int_{\partial\Omega}\boldsymbol{W}^t\tilde{\boldsymbol{F}}_n d\Gamma + \int_\Omega \boldsymbol{\nabla}\boldsymbol{W}^t\tilde{\boldsymbol{F}}d\Omega \\ &- \int_\Omega \boldsymbol{\nabla}\boldsymbol{W}^t(\tilde{\boldsymbol{A}}\tilde{\boldsymbol{\tau}} + \boldsymbol{\mathcal{D}})\tilde{\boldsymbol{A}}^t\boldsymbol{\nabla}\boldsymbol{V}d\Omega + \int_\Omega \hat{\boldsymbol{W}}^t\tilde{\boldsymbol{S}}d\Omega,\end{aligned} \tag{5.26}$$

where $\tilde{\boldsymbol{F}}_n = \tilde{\boldsymbol{F}}_1 n_1 + \tilde{\boldsymbol{F}}_2 n_2$, and n_1 and n_2 are the components of the local, outward, unit, normal vector. The term on the left side of (5.26) leads to a mass matrix. The first two terms on the right arise from the application of the divergence theorem to the convective flux vector. The third term on the right is due to the upwinding on the flux term, and the final term accounts for the effects of the source function.

The discretization proceeds by introducing the usual semidiscrete finite element expansion

$$\boldsymbol{V}_h = \sum_{j=1}^{N}\boldsymbol{V}_j(t)\psi_j(x_1, x_2), \tag{5.27}$$

where N is the number of nodes in the finite element mesh and $\psi_j(x_1, x_2)$ are the basis functions. (Linear Lagrange basis functions are used exclusively in the present work.) Substituting $\boldsymbol{V}_h$ for $\boldsymbol{V}$ in (5.26), and setting the components of $\boldsymbol{W}_h$ successively to ψ_i, the semidiscrete system of ODE's has the form

$$\sum_{j=1}^{N}\boldsymbol{M}_{ij}\boldsymbol{V}_j'(t) = \boldsymbol{f}_i(\boldsymbol{V}(t)), \quad i = 1, \ldots, N \tag{5.28}$$

with the nonlinear mass matrix

$$\boldsymbol{M}_{ij} = \int_\Omega \left\{\left[\boldsymbol{I}\psi_i + \psi_{i,x_1}\left(\tilde{\boldsymbol{A}}_1\tilde{\boldsymbol{\tau}} + \boldsymbol{\mathcal{D}}_1\right) + \psi_{i,x_2}\left(\tilde{\boldsymbol{A}}_2\tilde{\boldsymbol{\tau}} + \boldsymbol{\mathcal{D}}_2\right)\right]\boldsymbol{A}_0\right\}\psi_j d\Omega \tag{5.29}$$

and the forcing function

$$
\begin{aligned}
\boldsymbol{f}_i &= -\int_{\partial\Omega} \psi_i \tilde{\boldsymbol{F}}_n d\Gamma + \int_{\Omega} \left(\psi_{i,x_1} \tilde{\boldsymbol{F}}_1 + \psi_{i,x_2} \tilde{\boldsymbol{F}}_2\right) d\Omega \\
&\quad - \int_{\Omega} \left[\psi_{i,x_1} \left(\tilde{\boldsymbol{A}}_1 \tilde{\boldsymbol{\tau}} + \boldsymbol{\mathcal{D}}_1\right) + \psi_{i,x_2} \left(\tilde{\boldsymbol{A}}_2 \tilde{\boldsymbol{\tau}} + \boldsymbol{\mathcal{D}}_2\right)\right] \tilde{\boldsymbol{A}}^t \nabla \boldsymbol{V} d\Omega + \\
&\quad \int_{\Omega} \left[\boldsymbol{I}\psi_i + \psi_{i,x_1} \left(\tilde{\boldsymbol{A}}_1 \tilde{\boldsymbol{\tau}} + \boldsymbol{\mathcal{D}}_1\right) + \psi_{i,x_2} \left(\tilde{\boldsymbol{A}}_2 \tilde{\boldsymbol{\tau}} + \boldsymbol{\mathcal{D}}_2\right)\right] \tilde{\boldsymbol{S}} d\Omega \qquad (5.30)
\end{aligned}
$$

Except for the specification of the operators $\tilde{\boldsymbol{\tau}}$ and $\boldsymbol{\mathcal{D}}$, the spatial discretization is complete. The selection of the matrix of intrinsic time scales, $\tilde{\boldsymbol{\tau}}$, is an open problem: linear error estimates, convergence proofs, and dimensional analysis provide design conditions to be satisfied, but are insufficient to provide a unique definition [24]. Moreover, the choice of $\tilde{\boldsymbol{\tau}}$ is somewhat dependent on the problem being solved. In [14], a formula that works well in practice for the compressible Navier-Stokes equations is presented, while in [24] a more general form is proposed. In general, $\tilde{\boldsymbol{\tau}}$, is a symmetric, positive semidefinite matrix that, loosely speaking, acts to normalize the magnitude of the test function perturbation. For example, consider the one-dimensional, scalar convection equation. In this case, $\tilde{\boldsymbol{\tau}}$ reduces to a scalar quantity, and it may be shown that an optimal choice is the ratio of the local element size to the magnitude of the convective velocity [14]. For systems of equations, the situation is complicated by the presence of multiple wave modes and in general, an eigenproblem must be solved to obtain a formula for $\tilde{\boldsymbol{\tau}}$. This situation is discussed in [24], from which we obtain the expression

$$\tilde{\boldsymbol{\tau}} = \frac{\ell}{2} \boldsymbol{A}_0^{-1} \left[\boldsymbol{A}_1^2 + \boldsymbol{A}_2^2\right]^{-1/2} \qquad (5.31)$$

where ℓ denotes the element length, and the $\boldsymbol{A}_i$ are the flux Jacobians (5.8). Note that the inverse square root is taken on a 3×3 matrix. Furthermore, $\tilde{\boldsymbol{\tau}}$ may be interpreted as an inverse norm of the rectangular, convective operator $\boldsymbol{A}$ with respect to $\boldsymbol{A}_0$. To compute $\tilde{\boldsymbol{\tau}}$, we write (5.31) as

$$\tilde{\boldsymbol{\tau}} = \boldsymbol{A}_0^{-1} \boldsymbol{\mathcal{B}}^{-1/2} \qquad (5.32)$$

where

$$\boldsymbol{\mathcal{B}} = \frac{4}{\ell^2} \left(\boldsymbol{A}_1^2 + \boldsymbol{A}_2^2\right) \qquad (5.33)$$

Then we solve the associated eigenproblem to factor $\boldsymbol{\mathcal{B}}$ using the similarity transformation

$$\boldsymbol{\mathcal{B}} = \boldsymbol{M} \operatorname{diag}(\mathcal{B}_k) \boldsymbol{M}^{-1} \qquad (5.34)$$

where $\boldsymbol{M}$ is a modal matrix, the columns of which are the eigenvectors $\boldsymbol{e}_k$, $k = 1, 2, 3$, and the $\mathcal{B}_k$ are the corresponding eigenvalues. It is important to note that the $\boldsymbol{e}_k$ are scaled so that $\boldsymbol{M}\boldsymbol{M}^t = \boldsymbol{A}_0$ [14]. In practice, since $\tilde{\boldsymbol{\tau}}$ is symmetric, it may be computed from the expansion

$$\tilde{\boldsymbol{\tau}} = \sum_{k=1}^{3} \tau_k \boldsymbol{\phi}_k \boldsymbol{\phi}_k^t \tag{5.35}$$

where $\tau_k = 1/\sqrt{\mathcal{B}_k}$ and each eigenvector $\boldsymbol{\phi}_k = \boldsymbol{A}_0^{-1}\boldsymbol{e}_k$. Formulas for the eigenvalues τ_k and the eigenvectors $\boldsymbol{\phi}_k$ are given for the one and two-dimensional cases in Sections 5.4 and 5.6, respectively.

The motivation and derivation of the discontinuity-capturing operator (DCO) used in the present study is given in [15] for the Euler equations of gas dynamics. For completeness, we present below the final formulas. The DCO is computed as the product

$$\boldsymbol{\mathcal{D}} = \boldsymbol{A}_{||}\tilde{\boldsymbol{\tau}}_2 \tag{5.36}$$

The matrix $\boldsymbol{A}_{||}$ may be interpreted as the projection of $\tilde{\boldsymbol{A}}$ onto the direction $\boldsymbol{\nabla V}$. It is defined by

$$\boldsymbol{A}_{||} = \frac{\text{diag}_2(\boldsymbol{A}_0)\boldsymbol{\nabla V}(\boldsymbol{\nabla V})^t\tilde{\boldsymbol{A}}}{|\boldsymbol{\nabla V}|^2_{\boldsymbol{A}_0}} \tag{5.37}$$

where $\text{diag}_2(\boldsymbol{A}_0)$ denotes a 6×6 operator with two copies of $\boldsymbol{A}_0$ on the diagonal and zeroes elsewhere. In (5.37), we have also introduced the norm

$$|\boldsymbol{\nabla V}|_{\boldsymbol{A}_0} = \sqrt{(\boldsymbol{\nabla V})^t \text{diag}_2(\boldsymbol{A}_0)\boldsymbol{\nabla V}} \tag{5.38}$$

Note that $\boldsymbol{A}_{||}$ is not symmetric. The operator $\tilde{\boldsymbol{\tau}}_2$ in (5.36) is a symmetric, positive semidefinite, rank one matrix that contains the time scales associated with the gradient information. It is computed according to

$$\tilde{\boldsymbol{\tau}}_2 = \max(0, \tau_{||} - \tau)\boldsymbol{\phi}_{||}\boldsymbol{\phi}_{||}^t \tag{5.39}$$

with the eigenvector

$$\boldsymbol{\phi}_{||} = \frac{\boldsymbol{A}_0^{-1}\tilde{\boldsymbol{A}}^t\boldsymbol{\nabla V}}{\left|\tilde{\boldsymbol{A}}^t\boldsymbol{\nabla V}\right|_{\boldsymbol{A}_0^{-1}}} \tag{5.40}$$

Computing the eigenvalue for (5.39) first requires the evaluation of the scalar

$$\tau_{||} = \frac{|\boldsymbol{\nabla V}|^2_{\boldsymbol{A}_0}}{\left|\tilde{\boldsymbol{A}}^t\boldsymbol{\nabla V}\right|_{\boldsymbol{A}_0^{-1}} |\boldsymbol{DV}|_{\boldsymbol{A}_0}} \tag{5.41}$$

where the metric derivative

$$\boldsymbol{DV} = \begin{pmatrix} \boldsymbol{\nabla V}^t \boldsymbol{\nabla} \xi_1 \\ \boldsymbol{\nabla V}^t \boldsymbol{\nabla} \xi_2 \end{pmatrix} \tag{5.42}$$

has been introduced, with ξ_1 and ξ_2 the computational coordinates. Then the definition of the eigenvalue in (5.39) is completed by constructing

$$\tau = \left| \boldsymbol{A}_0 \boldsymbol{\phi}_{||} \right|^2_{\tilde{\boldsymbol{T}}} \tag{5.43}$$

In closing, we remark that the nonlinear set of ordinary differential equations (5.28) may be time-integrated using standard ODE system integrators. We apply Runge-Kutta strategies for this problem in Section 5.4.

5.4 The One-Dimensional Shallow Water System

In one dimension, the shallow water equations model the propagation of planar free surface waves. The symmetric form (5.9) reduces to

$$\boldsymbol{A}_0 \boldsymbol{V}_{,t} + \tilde{\boldsymbol{A}}(\boldsymbol{V}) \boldsymbol{V}_{,x} = \tilde{\boldsymbol{S}}(\boldsymbol{V}) \tag{5.44}$$

for $x_l \leq x \leq x_r$, $t > 0$, where, from (5.5), (5.17), (5.20), and (5.22)

$$\boldsymbol{V} = \begin{pmatrix} H - u^2/2, & u \end{pmatrix}^t \tag{5.45}$$

$$\boldsymbol{A}_0 = \begin{pmatrix} 1 & u \\ u & H + u^2 \end{pmatrix} \tag{5.46}$$

$$\tilde{\boldsymbol{A}} = \begin{pmatrix} u & H + u^2 \\ H + u^2 & u(3H + u^2) \end{pmatrix} \tag{5.47}$$

and

$$\tilde{\boldsymbol{S}}(\boldsymbol{V}) = \begin{pmatrix} 0, & Hh' - \frac{u|u|n_*^2}{\alpha^2 \sqrt[3]{H}} \end{pmatrix}^t \tag{5.48}$$

with the slope of the bottom given by h'.

The semidiscrete equations have the form (5.28) with (5.29) and (5.30) simplifying to

$$\boldsymbol{M}_{ij} = \int_{x_l}^{x_r} \left\{ \left[\boldsymbol{I} \psi_i + \psi_i' \left(\tilde{\boldsymbol{A}} \tilde{\boldsymbol{\tau}} + \boldsymbol{\mathcal{D}} \right) \right] \boldsymbol{A}_0 \right\} \psi_j dx \tag{5.49}$$

and

$$\begin{aligned} \boldsymbol{f}_i &= -\psi_i \tilde{\boldsymbol{F}} \Big|_{x_l}^{x_r} + \int_{x_l}^{x_r} \psi_i' \tilde{\boldsymbol{F}} dx + \int_{x_l}^{x_r} \left[\boldsymbol{I}\psi_i + \psi_i' \left(\tilde{\boldsymbol{A}}\tilde{\boldsymbol{\tau}} + \boldsymbol{\mathcal{D}} \right) \right] \tilde{\boldsymbol{S}} dx \\ & - \int_{x_l}^{x_r} \psi_i' \left(\tilde{\boldsymbol{A}}\tilde{\boldsymbol{\tau}} + \boldsymbol{\mathcal{D}} \right) \tilde{\boldsymbol{A}} \boldsymbol{V}_{,x} dx \end{aligned} \tag{5.50}$$

respectively, where

$$\tilde{\boldsymbol{F}} = \left(\; Hu, \quad Hu^2 + H^2/2 \; \right)^t \tag{5.51}$$

The integrals in (5.49) and (5.50) are approximated with a two point, Gauss quadrature rule. The resulting system is block tridiagonal and non-symmetric. All terms in (5.50) are evaluated as a function of the most recent solution 'iterate' in the integration/solution algorithm given below by (5.58) and (5.59).

In one dimension, (5.31) reduces to

$$\tilde{\boldsymbol{\tau}} = \frac{\ell}{2} \boldsymbol{A}_0^{-1} \left| \boldsymbol{A} \right|^{-1} \tag{5.52}$$

where

$$\boldsymbol{A} = \begin{pmatrix} 0 & 1 \\ H - u^2 & 2u \end{pmatrix} \tag{5.53}$$

The matrix $\tilde{\boldsymbol{\tau}}$ is computed according to the expansion (5.35) and has eigenvectors

$$\boldsymbol{\phi}_1 = \left(\; \frac{\lambda_2}{\sqrt{2H}}, \; \frac{-1}{\sqrt{2H}} \; \right)^t, \quad \boldsymbol{\phi}_2 = \left(\; \frac{-\lambda_1}{\sqrt{2H}}, \; \frac{1}{\sqrt{2H}} \; \right)^t \tag{5.54}$$

where $\lambda_1 = u - \sqrt{H}$ and $\lambda_2 = u + \sqrt{H}$. The eigenvalues τ_1 and τ_2 are given by

$$\tau_1 = \frac{\ell}{2\left|\lambda_1\right|}, \quad \tau_2 = \frac{\ell}{2\left|\lambda_2\right|} \tag{5.55}$$

Finally, the DCO (5.36) reduces to

$$\boldsymbol{\mathcal{D}} = \frac{\max(0, \tau_{\|} - \tau)}{\left|\boldsymbol{V}_{,x}\right|_{\boldsymbol{A}_0}^2} \boldsymbol{A}_0 \boldsymbol{V}_{,x} \left(\boldsymbol{A}_0^{-1} \boldsymbol{F}_{,x} \right)^t \tag{5.56}$$

where

$$\tau_{\|} = \frac{\ell \left|\boldsymbol{V}_{,x}\right|_{\boldsymbol{A}_0}}{2\left|\boldsymbol{F}_{,x}\right|_{\boldsymbol{A}_0^{-1}}}, \quad \tau = \frac{\left|\boldsymbol{F}_{,x}\right|_{\tilde{\boldsymbol{\tau}}}^2}{\left|\boldsymbol{F}_{,x}\right|_{\boldsymbol{A}_0^{-1}}^2} \tag{5.57}$$

This completes the derivation of the one-dimensional SUPG formulation.

Time Discretization

A second-order accurate, two step Runge-Kutta (RK) scheme [23] is used to integrate the system of ordinary differential equations (5.28). This reduces to solving the pair of systems

$$\boldsymbol{M}(\boldsymbol{V}^n)\left(\boldsymbol{V}^{n+1/2} - \boldsymbol{V}^n\right) = \gamma_1 \Delta t \boldsymbol{f}(\boldsymbol{V}^n) \tag{5.58}$$

$$\boldsymbol{M}(\boldsymbol{V}^{n+1/2})\left(\boldsymbol{V}^{n+1} - \boldsymbol{V}^n\right) = \gamma_2 \Delta t \boldsymbol{f}(\boldsymbol{V}^{n+1/2}) \tag{5.59}$$

with RK parameters $\gamma_1 = 1/2$ and $\gamma_2 = 1$. Given a solution $\boldsymbol{V}^n$ from the previous time step, the solution at the half step, $\boldsymbol{V}^{n+1/2}$, is computed according to (5.58). The forcing function and mass matrix are then reevaluated at the half step, and the solution at time level $n+1$ is computed according to (5.59). We remark that, since the systems (5.58) and (5.59) must be solved at each time step, an implicit strategy would be as efficient, and possibly offer greater stability.

Boundary Conditions

The number of boundary conditions enforced after each RK step is determined via the method of characteristic projection [9,12,20]: For supercritical inflow, Dirichlet data are applied on both equations at the associated node; for supercritical outflow, no boundary conditions are applied. The remaining subcritical cases have both incoming and outgoing modes and are treated as follows: First, note that the flux Jacobian $\boldsymbol{A}$ in conservation variables is diagonalized by the similarity transformation

$$\boldsymbol{A} = \boldsymbol{T}\begin{pmatrix} \lambda_1 & 0 \\ 0 & \lambda_2 \end{pmatrix}\boldsymbol{T}^{-1}, \quad \boldsymbol{T} = \frac{1}{\sqrt{2}}\begin{pmatrix} 1 & 1 \\ \lambda_1 & \lambda_2 \end{pmatrix} \tag{5.60}$$

where the eigenvalues λ_1 and λ_2 are given above. Note also that these eigenvectors have been scaled so that $\boldsymbol{T}\boldsymbol{T}^t = \boldsymbol{A}_0$. While this is not necessary for the projection strategy, it simplifies the relationship between the characteristic variables, $\hat{\boldsymbol{U}}$, and the entropy variables, $\boldsymbol{V}$. The left eigenvectors are used to separate the flow into incoming and outgoing modes at the boundaries. This projection may be written as

$$\Delta\hat{\boldsymbol{U}} = \boldsymbol{T}^t \Delta \boldsymbol{V} \tag{5.61}$$

or, more explicitly in terms of changes in the depth and velocity as

$$\Delta\hat{\boldsymbol{U}} = \frac{1}{\sqrt{2}}\begin{pmatrix} \Delta H - \sqrt{H}\Delta u \\ \Delta H + \sqrt{H}\Delta u \end{pmatrix} \tag{5.62}$$

For example, at a subcritical inflow boundary, the mode corresponding to λ_1 is outgoing and the other mode that corresponds to λ_2 is incoming. Accordingly, we specify one boundary condition and leave the other variable unconstrained. If we wish to specify the depth, V_1 is determined by the most recent value of the velocity and the prescribed depth. We do not alter the predicted value of V_2. Similarly, at a subcritical outflow, λ_1 is incoming and λ_2 is outgoing. Again, only one boundary condition is required.

5.5 Numerical Results in 1D

The performance of the proposed method is evaluated by considering a few representative test cases. All of the results presented in this section were obtained by solving the consistent mass matrix (5.49) at each RK step using the LAPACK bandsolver SGBSV [1]. We remark that attempts were made to solve the given examples using biconjugate gradient (BCG), an iterative method appropriate for nonsymmetric systems [18]. The results so obtained were generally poor, even with Jacobi preconditioning, and were often divergent.

Hydraulic Jump

The first case is a steady-state calculation that demonstrates the ability of the proposed formulation to capture hydraulic jumps. The channel is rectangular and has zero slope, bottom friction is neglected so that $n = 0.0$, and the incoming flow has depth $H_i = 1$ and Froude number $F_r = 6$. Theory states that if the depth at the exit boundary, H_e, is specified according to

$$\frac{H_e}{H_i} = \frac{1}{2}\left(\sqrt{1 + 8F_r^2} - 1\right) \tag{5.63}$$

then the analytic solution to this problem is given by two regions of uniform depth and velocity, separated by a hydraulic jump [7]. The initial condition for the depth is a linear distribution between the two boundaries, and the velocity is set equal to the local ratio F_r/H. The numerical solution obtained on representative uniform meshes is shown in Figures 5.1 and 5.2.

From the figures, it is clear that the jump is captured within four elements on each mesh. Furthermore, as the mesh is refined, the slope of the jump increases considerably. This decrease in numerical viscosity associated with the mesh refinement is also evident in the increasing magnitude of the overshoot on the downstream side of the jump. Also note that the location of the jump, as defined by the location of the first grid point downstream of

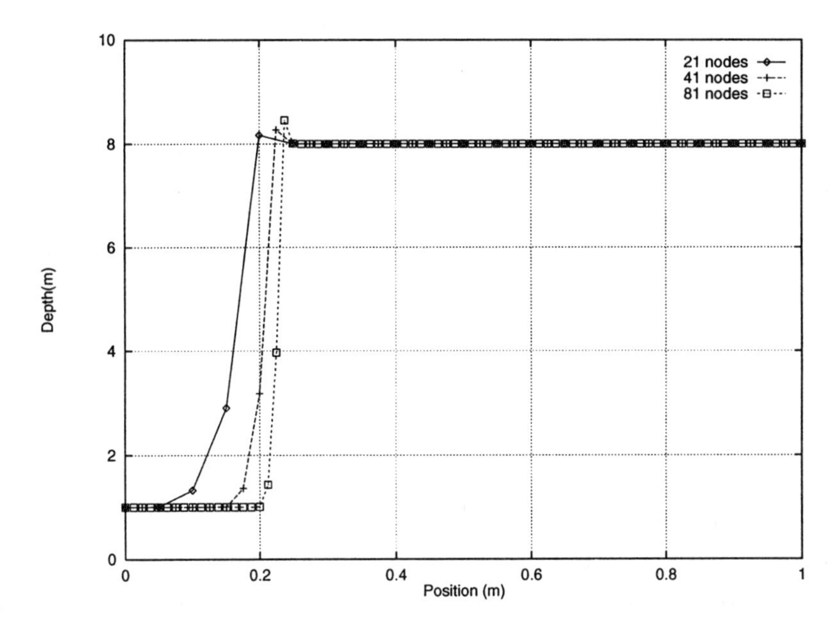

Figure 5.1: Hydraulic jump: effect of uniform refinement on depth. $CFL = 0.1$

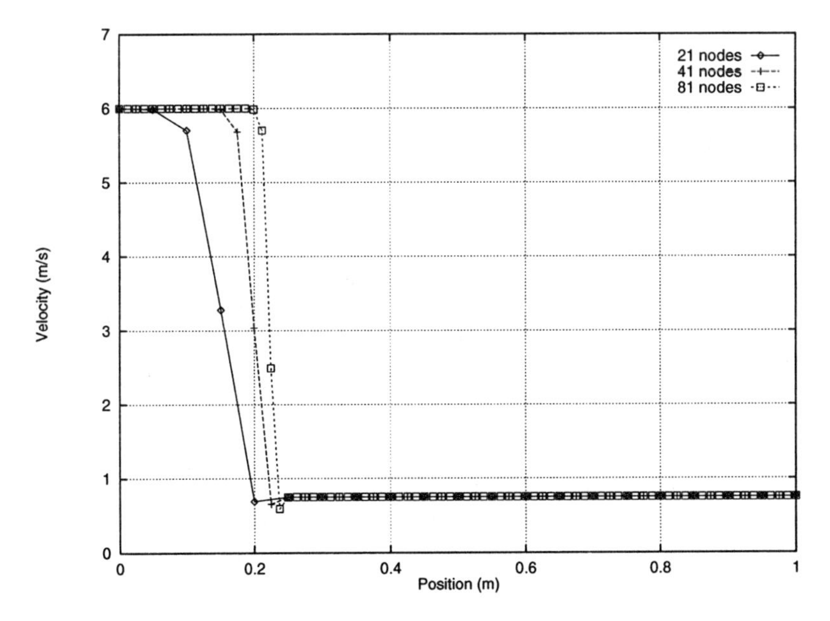

Figure 5.2: Hydraulic jump: effect of uniform refinement on velocity. $CFL = 0.1$.

the jump, does not change with refinement. Finally, we remark that no solution could be obtained for this problem unless the DCO (5.36) was included in the formulation.

Shoaling of a Wave

The next example is the propagation of a wave along a channel having uniform bottom slope. The wave begins in deep water and progresses to the right as the depth decreases. The initial conditions are

$$\eta = \frac{a}{\cosh^2\left[\sqrt{3a}(x - 1/\alpha)/2\right]}, \qquad u = \frac{\eta(1 + a/2)}{\alpha x + \eta} \tag{5.64}$$

where $\alpha = 1/30$ and $a = 1/10$. The bed is given by

$$h = -(x - 60)/30, \quad 10 \le x \le 58 \tag{5.65}$$

and bottom friction is again neglected, so that $n = 0$.

This problem has been considered by many authors (*e.g.* see [19, 21]). The discretization and time steps chosen for the present study are identical to those in [19]. The problem is first solved with 40 equally spaced elements and $\Delta t = 0.3125$. The initial condition and computed surface elevation are shown in Figure 5.3 at various times. The wave does increase in amplitude and steepen somewhat before collapsing at the shallow boundary. Some smaller reflected waves are evident for $15 < x < 30$. The analytic solution to the linearized problem [19] indicates that the maximum elevation should be approximately 0.12, significantly larger than the predicted value, which implies that the scheme is excessively dissipative on this coarse mesh. For $\Delta t = 0.3125$ and 40 elements, no steepening or increase in amplitude is reported in [19], which suggests that their results are more dissipative than those of the method proposed herein. As the mesh is uniformly refined, the behavior predicted by the analytic solution is captured numerically. For further comparison, the proposed SUPG method was applied to a uniform mesh of 160 elements with $\Delta t = 0.06667$. These results are shown in Figure 5.4. For this case, the value 0.12 is attained and the wave becomes fairly steep. Localized oscillations are still evident at the foot of the wave, but they are smaller in amplitude compared to the coarse mesh solution.

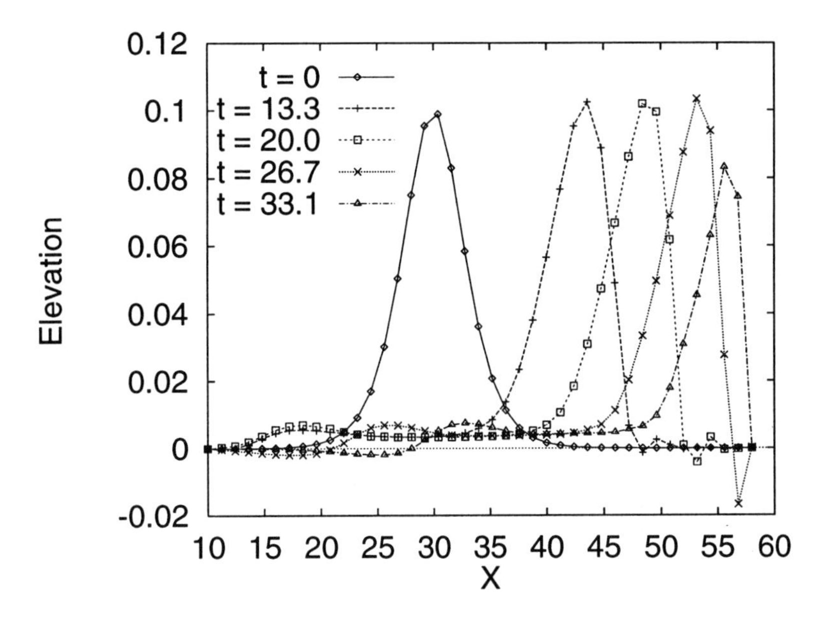

Figure 5.3: Shoaling of a wave: surface elevation, $\Delta t = 0.3125$, $h = 1/40$.

Spillway Flow

The final one-dimensional example models the flow over a spillway and stilling basin. The bed profile is given by

$$0.2 \exp\left[-\frac{1}{2}(x/0.24)^2\right] \tag{5.66}$$

where x is the horizontal distance in meters from the center of the crest. This geometry was used for numerical and experimental studies in [25] and numerical studies in [3]. We consider the conditions described in [3]. A discharge $Q = 0.03599$ m^3/s is specified at the inflow boundary, and a depth of 0.1m is specified at the outflow boundary. The bottom is fairly smooth, so that $n = 0.016$. For this discharge and tailwater depth, the steady-state solution contains a hydraulic jump. The solution is marched to steady-state using a mesh of 59 nodes which is graded towards the spillway crest.

It should be noted that the shallow water equations described in the present study are the Saint-Venant, or mild slope equations: the pressure is assumed hydrostatic, and the velocity is constant on a cross section. More general models exist in which the equations are written in bed-fitted coordinates, and the assumptions of mild slope and zero curvature are not made [8]. As shown in [3], these assumptions are clearly violated in this test case. In

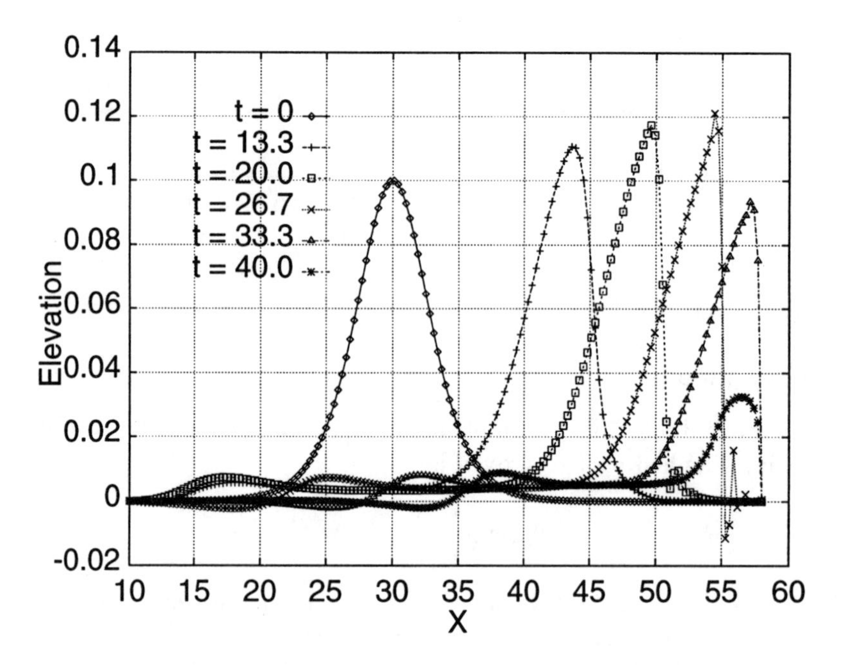

Figure 5.4: Shoaling of a wave: surface elevation, $\Delta t = 0.06667$, $h = 1/160$.

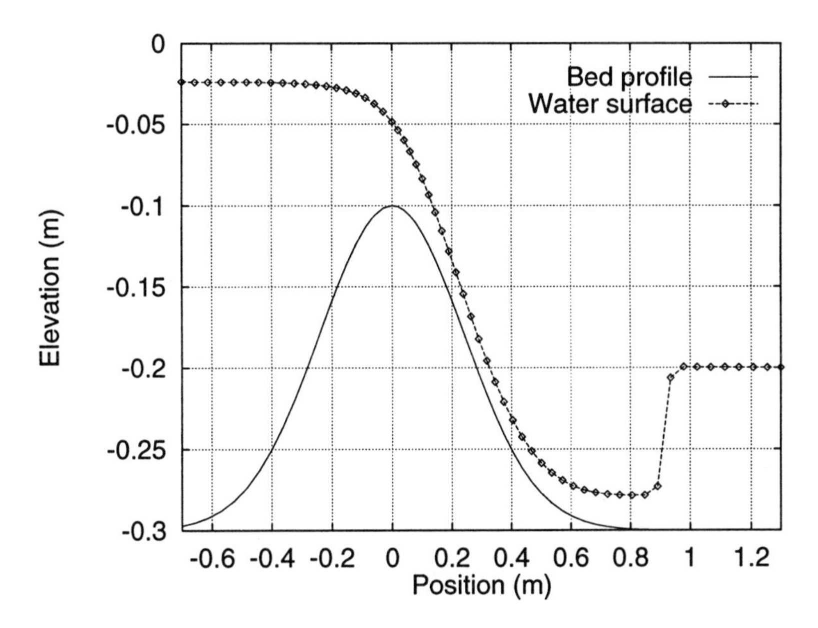

Figure 5.5: Spillway bed elevation and water surface profile. The y coordinate has been stretched for clarity.

particular, downstream of the crest, the velocity should be parallel to the spillway, and not in the $x-$direction as assumed by the Saint-Venant equations. Furthermore, the pressure distribution deviates substantially from being hydrostatic. Nevertheless, the problem is fairly difficult numerically and serves as a useful test of the numerical scheme.

The resulting surface elevation is shown in Figure 5.5. Note that the jump occurs across two elements and that there is some deepening of the water just upstream of the jump which is due to the effect of the bottom friction. The velocity and Froude number distributions are shown in Figure 5.6. The peak velocity is 1.905m/s, and the maximum Froude number is about 4.430. The Froude number just upstream of the jump is 3.1097. Finally, we remark that the jump occurs at about $x = 0.9$m. This position is slightly upstream of the that which results when the effects of bed curvature are included [3].

5.6 The Two-Dimensional Shallow Water System

In the present two-dimensional studies, we restrict our attention to the time-iterative solution of steady-state, open channel problems. Consequently,

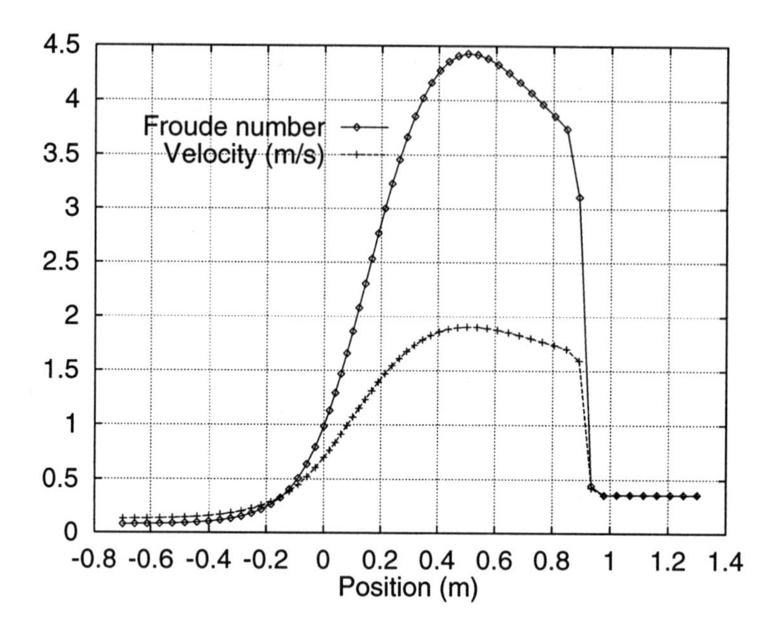

Figure 5.6: Froude number and velocity distribution along spillway.

the complete mass matrix (5.29) is inconvenient. If the asymmetric terms arising from upwinding are neglected in the mass matrix, we are led to the approximation

$$\boldsymbol{M}_{ij} \simeq \int_{\Omega} \psi_i \psi_j \boldsymbol{A}_0 d\Omega. \tag{5.67}$$

This integral may be conveniently treated using a nodal quadrature rule. For linear triangles, the element contribution to node i is simply $1/3$ times the product of the triangle area and $\boldsymbol{A}_0(\boldsymbol{V}_i)$. Then (5.28) becomes decoupled for each mesh node i and may be written as

$$\boldsymbol{V}_i'(t) = (\boldsymbol{A}_0^{-1})_i \boldsymbol{L}_i^{-1} \boldsymbol{f}_i(t), \tag{5.68}$$

where $(\boldsymbol{A}_0^{-1})_i$ denotes (5.18) evaluated at mesh point i, and $\boldsymbol{L}_i$ is the ith diagonal entry of the standard Galerkin lumped mass matrix. All of the integrals in (5.30) are evaluated using a one point Gauss quadrature rule. We emphasize that (5.68) is no longer time accurate and should be regarded as a point iterative method for solving the stationary equations. By discarding the upwind part of the mass matrix, it is no longer possible to recover the variational form (5.24), except at the steady-state.

All that remains to fully define the proposed SUPG method is to specify $\tilde{\tau}$ and $\mathcal{D}$. As in the one-dimensional case, we again compute $\tilde{\tau}$ according to the formula

$$\tilde{\tau} = \sum_{k=1}^{3} \tau_k \phi_k \phi_k^t \tag{5.69}$$

with the eigenvalues

$$\tau_{1,2} = \frac{\sqrt{2}\ell}{\sqrt{3H+2V^2 \mp \beta}}, \qquad \tau_3 = \frac{\ell}{\sqrt{V^2+H}} \tag{5.70}$$

where ℓ is the local element length scale and $\beta = \sqrt{H^2 + 16HV^2}$. The eigenvectors are given by

$$\phi_1 = \varphi_1 \left(\frac{-4V^2+H-\beta}{H-\beta}, \quad \frac{4u_1}{H-\beta}, \quad \frac{-u_2(H+\beta)}{4HV^2} \right)^t \tag{5.71}$$

$$\phi_2 = \varphi_2 \left(\frac{-4V^2+H+\beta}{H+\beta}, \quad \frac{4u_1}{H+\beta}, \quad \frac{-u_2(H-\beta)}{4HV^2} \right)^t \tag{5.72}$$

$$\phi_3 = \varphi_3 \left(0, \quad u_2, \quad -u_1 \right)^t \tag{5.73}$$

where the scale factors

$$\varphi_{1,2} = \sqrt{\frac{\beta \mp H}{2\beta}} \quad , \quad \varphi_3 = \frac{1}{V\sqrt{H}} \tag{5.74}$$

are chosen as described in Section 5.3. The two-dimensional SUPG formulation is completed by implementing the DCO as defined by equations (5.36)-(5.43).

Time Discretization

Since only the steady-state solution is of interest, forward Euler time-stepping is used. This strategy can be regarded as a one step RK method. Obviously, a multistep RK integrator could also be implemented and the parameters γ_i that appear in (5.59) could be chosen to maximize the damping properties of the integration scheme at the expense of time accuracy [22]. This approach, when combined with the local time step and lumped mass matrix, can dramatically accelerate convergence. For the forward Euler, lumped mass scheme, the iteration amounts to computing the correction

$$\Delta \boldsymbol{V}_i = \Delta t (\boldsymbol{A}_0^{-1})_i \boldsymbol{L}_i^{-1} \boldsymbol{f}_i(t) \tag{5.75}$$

It is well known that explicit schemes are subject to a stability condition on the allowable size of the time step, Δt. For linear systems, a Fourier analysis can be performed to obtain an estimate of the allowable time step [2]. For the linear hyperbolic case, this corresponds to specifying a Courant-Friedrichs-Lewy (CFL) parameter less than unity. However, the nonlinearity of the present problem and the shock-capturing operator complicate the situation. It is then standard practice to perform a stability analysis of the linearized equations and this result is then used to guide the selection of a suitable time step. Accordingly, in the present study, the time step is computed by simply using the relation

$$\Delta t = CFL \frac{\sqrt{2A_e}}{V + \sqrt{H}} \tag{5.76}$$

where A_e is the local element area. In practice, the time integration can be computed using a constant value of CFL or Δt. For time-accurate simulations, a global, minimum value of Δt is required. For time-iterative, steady-state simulations, it is beneficial to use a constant CFL and compute the time step locally using (5.76), particularly when the element size varies significantly throughout the mesh. Since only the steady-state is of interest, it is not necessary to synchronize the computations before carrying out the next iteration. Hence the scheme is analogous to a pointwise iterative method for the stationary equations using local relaxation. In practice, the time step is computed as follows: at every iteration, Δt for each node is initialized to a very large value. Next, a loop over the elements is performed and Δt for node i is computed on element e according to

$$(\Delta t)_i = \min \left[(\Delta t)_i, CFL \frac{\sqrt{2A_e}}{V_n + \sqrt{H}} \right] \tag{5.77}$$

where V_n is the magnitude of the velocity component normal to the edge opposite node i.

Boundary Conditions

We consider two basic types of boundary conditions: The first type satisfies an inflow/outflow condition in which the normal component of the velocity is nonzero; the second type is a zero mass flux boundary which implies that the normal velocity component vanishes and the flow is tangential to the boundary. In each case, we evaluate the normal flux, $\tilde{\boldsymbol{F}}_n$, in the boundary integral that appears in (5.30) using the current iterate for $\boldsymbol{V}$. Then we again use the method of characteristic projections [9, 12, 20] to determine the number of boundary conditions to apply.

The classification of the boundary type in two dimensions depends on the velocity component normal to the boundary, u_n. Accordingly, the projections are defined by the eigenvalues and left eigenvectors of the matrix

$$\boldsymbol{A}_n = \boldsymbol{A}_1 n_1 + \boldsymbol{A}_2 n_2 = \begin{pmatrix} 0 & n_1 & n_2 \\ Hn_1 - u_1 u_n & u_1 n_1 + u_n & u_1 n_2 \\ Hn_2 - u_2 u_n & u_2 n_1 & u_2 n_2 + u_n \end{pmatrix}. \tag{5.78}$$

The matrix (5.78) has eigenvalues

$$\lambda_{1,2} = u_n \mp \sqrt{H}, \quad \lambda_3 = u_n \tag{5.79}$$

with eigenvectors given by the columns of

$$\boldsymbol{T} = \frac{1}{\sqrt{2}} \begin{pmatrix} 1 & 1 & 0 \\ (u_1 - n_1\sqrt{H}) & (u_1 + n_1\sqrt{H}) & \sqrt{2H}n_2 \\ (u_2 - n_2\sqrt{H}) & (u_2 + n_2\sqrt{H}) & -\sqrt{2H}n_1 \end{pmatrix}. \tag{5.80}$$

We remark that the eigenvectors in (5.80) have again been scaled so that $\boldsymbol{T}\boldsymbol{T}^t = \boldsymbol{A}_0$. The number of incoming modes is given by the number of negative eigenvalues in (5.79). Equation (5.61) can be written more explicitly for the two dimensional case in terms of changes in the depth and velocity components as

$$\Delta \hat{U} = \begin{pmatrix} \left(\Delta H - \sqrt{H}\Delta u_n\right)/\sqrt{2} \\ \left(\Delta H + \sqrt{H}\Delta u_n\right)/\sqrt{2} \\ -\sqrt{H}\Delta u_s \end{pmatrix}, \tag{5.81}$$

where the tangential velocity component $u_s = -u_1 n_2 + u_2 n_1$.

In the present work, we only consider problems with supercritical inflow/outflow boundaries. Zero mass flux boundaries correspond to boundaries along which the flow is locally tangent. In this case, only the first mode is incoming, so only one boundary condition need be applied. That condition is, of course, $u_n = 0$, or, equivalently, $\Delta u_n = 0$. This condition can be enforced weakly, by specifying $u_n = 0$ in the boundary integral that appears in (5.30). Weak implementations, while convenient, satisfy the boundary constraints only in an average sense along the boundary. In our experience, substantial nonzero normal velocity components can arise, particularly on coarse meshes, which can lead to instabilities. For this reason, we project the velocities at each iteration so that the zero mass flux condition is satisfied strongly at the boundary nodes. We do this by projecting the second and third components of $\Delta \boldsymbol{V}$, and leave the first component unconstrained.

5.7 Numerical Results in 2D

In order to demonstrate the proposed method, we simulate the flow through two rectangular channels. For each of the following test cases, a Froude number, F_r, depth profile, and a flow angle, θ, must be specified. Then the starting iterate at each node in the mesh is computed according to

$$V = \begin{pmatrix} H(1 - F_r^2/2) \\ F_r \cos\theta\sqrt{H} \\ F_r \sin\theta\sqrt{H} \end{pmatrix} \tag{5.82}$$

except for nodes on the zero mass flux boundaries, in which case the tangential component of (5.82) is taken.

Step Transition

The first example is that of supercritical flow in a channel whose width is suddenly reduced from 9 units to 7 units. The contraction occurs 9.5 units downstream of the inlet boundary. The inlet Froude number (based on the incoming unit depth) is 2.5, the channel is 28.5 units long, has zero bed slope and no bottom friction. This example is obviously not a practical design, but has features that make it an interesting test case. The mesh of linear triangular elements used for this calculation has 1,120 nodes and 2,073 triangles and is shown in Figure 5.7. The computed depth contours are presented in Figure 5.8 and range from 0.8694 to 3.441. The Froude numbers are not shown, but they range from 2.4×10^{-3} to 2.5. After passing through the hydraulic jump, the flow becomes subcritical, then becomes supercritical again as it rounds the corner. The velocity vectors as they are computed at the nodes are shown in Figure 5.9. Note that the flow is essentially one-dimensional through the jump, which is captured within a band of about three elements. Then the velocity vectors begin to turn well upstream of the corner, as is appropriate for a subcritical flow. Near the concave portion of the corner, the flow is nearly stagnant, then accelerates rapidly around the corner. Finally, the direction becomes uniform again as the exit boundary is approached. We remark that the apparent irregularity in the upper right corner of the domain is due to the nonuniformity of the grid in that region. Careful observation reveals that the direction and magnitude of these vectors are the same as those of their neighbors; the grid points at which the vectors are drawn are simply shifted with respect to their neighbors.

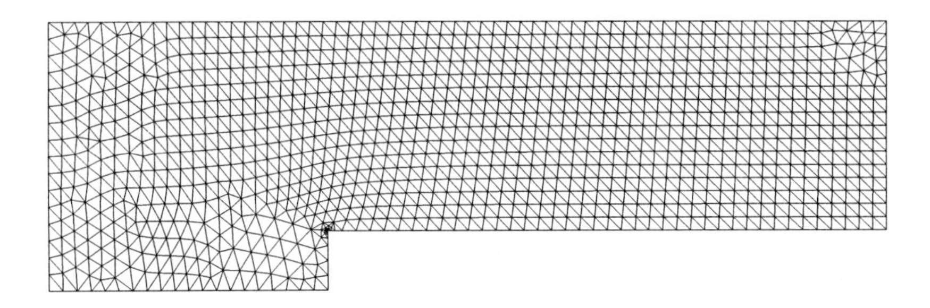

Figure 5.7: Mesh for step transition. 1,120 nodes and 2,073 triangles.

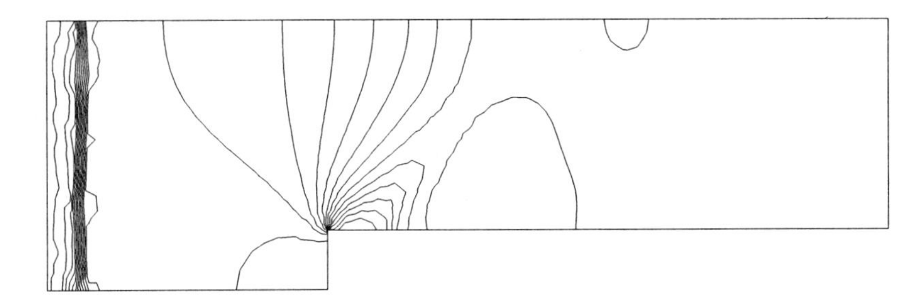

Figure 5.8: Depth contours for step transition problem. $0.8694 \leq H \leq 3.441$

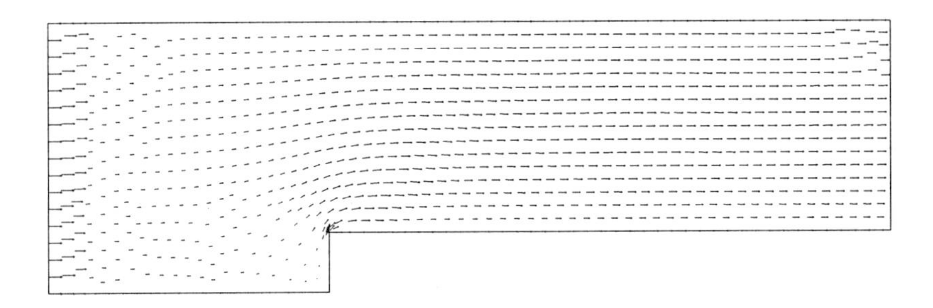

Figure 5.9: Nodal velocity vectors for step transition problem.

Curved Wall Transition

This problem involves a supercritical transition from a rectangular flume model of width two feet to one foot. The reduction in area is accomplished using two circular arcs of radius 75in and a transition length of 41.375in. The flume is smooth and has a constant bed slope so that a uniform flow is achieved upstream of the transition. We follow [4] and use a bed slope of 0.0125. The Froude number is 4.0, based on the incoming depth of 0.1ft. Consequently, for a uniform flow to exist upstream of the transition we must have Manning's friction factor $n = 0.005$.

The mesh used in the present study has 4,585 nodes and 8,628 linear triangles (Figure 5.10.) Computed depth contours are shown in Figure 5.11. Waves of the same family intersect and merge into an oblique hydraulic jump. This occurs at both walls so that the two jumps intersect and reflect off the opposite wall. This interaction continues downstream, becoming progressively weaker. In the numerical solution, the depth ranges from 0.3606ft to 0.09908ft. These extrema compare well with the experimentally observed values of 0.40ft and 0.10ft, respectively [16]. The peak depth as predicted by the numerical model occurs just downstream of the first intersection of the two oblique jumps. This location is about 0.5ft upstream of the experimentally observed location [16]. The computed Froude numbers (not shown) vary from 4.020 to 1.691.

Also note that there is some mesh-related asymmetry in the computed solution. Although the distribution of nodes is symmetric about the centerline of the flume, the orientation of triangles is not. This local grid orientation effect also explains why the waves emanating from the lower wall are resolved more sharply relative to those which emanate from the upper wall. Also, there is a local area of increased resolution of the upper wave in the vicinity of the inflection point of the boundary curve. Referring to the mesh detail, this area coincides with a band of elements whose edges are aligned roughly parallel with the wave. Indeed, this rough alignment is present throughout most of the domain with respect to the waves that are created by the lower boundary. In contrast, the waves generated by the upper boundary are roughly orthogonal to the triangle edges throughout most of the domain. Consequently, hydraulic jumps may be more highly resolved if the element edges are aligned with the front. This effect could be ameliorated through the use of an adaptive refinement algorithm which locally reduces the scale of the triangles (and therefore the numerical dissipation of the scheme) in the vicinity of the jumps [5].

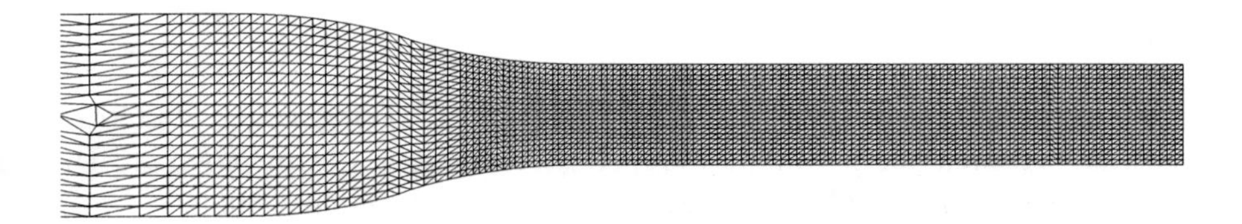

Figure 5.10: Mesh detail for curved wall transition. (4,585 nodes , 8,628 triangles.)

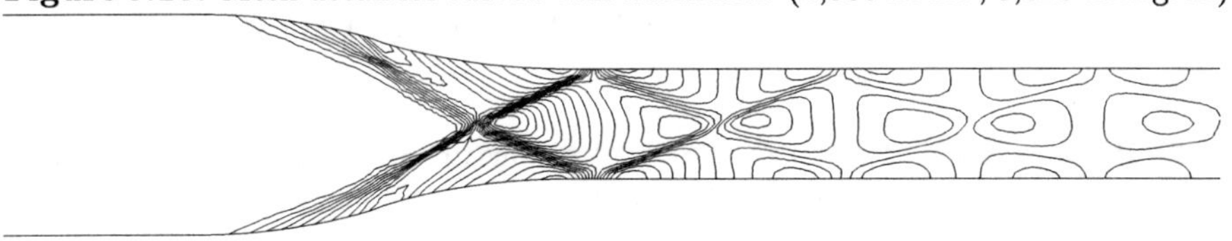

Figure 5.11: Detail of depth contours for curved wall transition problem. $0.09908\text{ft} \leq H \leq 0.3606\text{ft}$

5.8 Concluding Remarks

A new, symmetric form of the shallow water equations has been developed using variables derived from the total energy of the water column. The symmetric equations are then used as a starting point for an SUPG finite element method. Explicit Runge-Kutta methods are introduced for the time integration, although other time integration strategies may be used. Test cases in one and two dimensions were considered, and the proposed algorithms were evaluated for both smooth and nonsmooth solutions. The results agree well with those published by other authors, and demonstrate that the new method can accurately solve the shallow water equations.

The performance of the method for steady-state problems is quite good, although the method could benefit from a more sophisticated time-stepping strategy. In particular, the second-order accurate, two-step RK scheme does not allow significantly larger time steps over forward Euler time stepping, although its transient performance for a given time step is much improved. For stationary problems, RK strategies that offer increased stability in exchange for time accuracy might provide more efficient solution [22]. We plan to explore this issue further in future studies.

Acknowledgements: We are grateful to Dr. Charlie Berger at the U.S. Army Corps of Engineers' Waterways Experiment Station for helpful discussions and for providing the nodal coordinate set for the curved wall transition mesh. This work was supported in part by the National Science Foundation.

References

[1] Anderson, E., Z. Bai, C. Bischof, J. Demmel, J. Dongarra, J. Du Croz, A. Greenbaum, S. Hammarling, A. McKenney, S. Ostrouchov, and D. Sorenson, *LAPACK User's Guide*, SIAM, Philadelphia, 1992.

[2] Anderson, D.A., J.C. Tannehill, and R.H. Pletcher, *Computational fluid mechanics and heat transfer*, Hemisphere Publishing, NY, 281-283, 1984.

[3] Berger, R.C., and G.F. Carey, "A perturbation analysis and finite element approximate model for free surface flow over curved beds", *IJNME*,31, 493-507, 1991.

[4] Berger, R.C., *Free-Surface Flow Over Curved Surfaces*, Ph.D. thesis, The University of Texas at Austin, May, 1992.

[5] Bova, S.W., *Finite Element Solution of Hyperbolic Transport Systems using Adaptive Methods*, PhD thesis, The University of Texas at Austin, May, 1994.

[6] Bova, S.W. and G.F. Carey, "An entropy variable formulation and applications for the two-dimensional shallow water equations", submitted to *Int. J. Num. Meth. Fluids*, 1995.

[7] Chow, V.T., *Open Channel Hydraulics*, McGraw-Hill, NY, 1959.

[8] Dressler, R.F., "New nonlinear shallow flow equations with curvature", *Journal of Hydraulic Research*, 16, 205-272, 1978.

[9] Engquist, B. and A. Majda, "Absorbing boundary conditions for the numerical simulation of waves", *Mathematics of Computation*, 31, 629-651, 1977.

[10] Gray, W.G., and D.R. Lynch, "Time-stepping schemes for finite element tidal model computations", *Adv. Water Resources*, 2, 83-95, 1977.

[11] Harten, A., "On the symmetric form of systems of conservation laws with entropy", *J. Comp. Phys.*, 49, 151-164, 1983.

[12] Higdon, R. L. "Initial boundary value problems for linear hyperbolic systems", *SIAM Review*, 28, 177-217, 1986.

[13] Hughes, T.J.R., L.P. Franca, and M. Mallet, "A New Finite Element Formulation for Computational Fluid Dynamics: I. Symmetric Forms of the Compressible Euler and Navier-Stokes Equations and the Second Law of Thermodynamics", *Computer Methods in Applied Mechanics and Engineering*, 54, 223-234, 1986.

[14] Hughes, T.J.R., and M. Mallet, "A New Finite Element Formulation for Computational Fluid Dynamics: III. The Generalized Streamline Operator for Multidimensional Advective-Diffusive Systems", *Computer Methods in Applied Mechanics and Engineering*, 58, 305-328, 1986.

[15] Hughes, T.J.R., and M. Mallet, "A New Finite Element Formulation for Computational Fluid Dynamics: IV. A Discontinuity-Capturing Operator for Multidimensional Advective-Diffusive Systems", *Computer Methods in Applied Mechanics and Engineering*, 58, 329-336, 1986.

[16] Ippen, A.T., and J.H. Dawson, "Design of channel contractions", *Transactions of the A.S.C.E.*, 116, 326-346, 1951.

[17] Johnson, C., "Streamline diffusion methods for problems in fluid mechanics", *Finite Elements in Fluids–Volume 6*, R.H. Gallagher *et al* (eds), Wiley, 251-261, 1985.

[18] Joubert, W., and A. Manteuffel, "Iterative methods for nonsymmetric linear systems", *Iterative Methods for Large Linear Systems*, D. Kincaid, and L. Hayes, eds., Academic Press, Boston, 149-171, 1990.

[19] Kawahara, M., H. Hirano, K. Tsubota, and K. Inagaki, "Selective lumping finite element method for shallow water flow", *Int. J. Num. Meth. Fluids*, 2, 89-112, 1982.

[20] Kreiss, H. O., "Initial boundary value problems for hyperbolic systems", *Communications in Pure and Applied Mathematics*, 23, 277-298, 1970.

[21] Löhner, R., K. Morgan, and O.C. Zienkiewicz, "The solution of nonlinear hyperbolic equation systems by the finite element method", *Int. J. Num. Meth. Fluids*, 4, 1043-1063, 1984.

[22] Lorber, A.A., G. F. Carey, and W.D. Joubert, "ODE recursions and iterative solvers for linear equations", *SIAM Journal of Scientific Computing*, in press, 1994.

[23] Ralston, A. and P. Rabinowitz, *A First Course in Numerical Analysis*, McGraw-Hill, NY, 209-225, 1987.

[24] Shakib, F., T.J.R. Hughes, and Z. Johan, "A new finite element formulation for computational fluid dynamics: X. The compressible Euler and Navier-Stokes equations", *Computer Methods in Applied Mechanics and Engineering*, 89, 4, 141-219, 1991.

[25] Sivakumaran, N.S., T. Tingsanchali, and R.J. Hosking, "Steady shallow flow over curved beds", *Journal of Fluid Mechanics*, 128, 469-487, 1983.

[26] Stoker, J.J., *Water Waves*, Interscience Publishers, Inc., NY, 1957.

[27] Tadmor, E. , "Skew-selfadjoint form for systems of conservation laws", *Journal for Mathematical Analysis and Applications*, 103, 428-442, 1984.

[28] Walters, R. and G.F. Carey, "Analysis of spurious oscillation modes for the shallow water and Navier-Stokes equations", *Computers and Fluids*, 11, 51-68, 1983.

[29] Walters, R., "Numerically induced oscillations in finite element approximations to the shallow water equations", *Int. J. Num. Meth. Fluids*, 3, 591-604, 1983.

[30] Walters, R. and R. Cheng, "A two-dimensional hydrodynamic model of a tidal estuary", *Adv. Water Resources*, 2, 177-184, 1979.

[31] Walters, R., "A model for tides and currents in the English Channel and Southern North Sea", *Adv. Water Resources*, 10, 138-148, 1987.

[32] Warming, R.F., R.M. Beam, and B.J. Hyett, "Diagonalization and simultaneous symmetrization of the gas-dynamic matrices", *Mathematics of Computation*, 29, 1037-1045, 1975.

Chapter 6

Tidal Simulation using Conjugate Gradient Methods

E. Barragy[1], *R. Walters*[2] *and G.F. Carey*[3]

6.1 Introduction

The shallow water equations are a nonlinear, coupled set of equations for surface elevation and velocity that have wide application in hydrology, oceanography, and meteorology. As computers have become more powerful and grid resolution greater, there is a need to find more efficient numerical methods, particularly matrix solvers. For small problems of the order of a few thousand nodes, direct solvers such as frontal solvers have been popular [21]. For large problems, iterative methods have considerable appeal, particularly for problems in 3D.

In this chapter we explore the relative merits of an optimized frontal solver as compared to a variety of iterative solution methods for this class of problems. As it turns out, this comparison depends greatly on geometric irregularities, the presence of the Coriolis term that generates asymmetries in the matrices, the form of the equations and whether the problem is 2D or 3D. In the present study, the equations arise from a harmonic decomposition

[1]Intel Corporation, 15201 NW Greenbrier Pkwy., MS CO1-05, Beaverton, OR, 97006
[2]U.S. Geological Survey, 1201 Pacific Ave., Suite 901, Tacoma, WA, 98402
[3]TICAM,Univ. of Texas at Austin, Austin, TX, 78712

of the primitive shallow water equations and the 3D advection diffusion equation and are complex-valued.

In the next section we briefly state the harmonic formulation and the finite element model. Following this, we describe the test problems used in this analysis and present performance studies for several solution schemes applied to the associated complex linear systems. As the results indicate, in 2D, one test problem proves to be favorable for the frontal solver, and the other for the iterative solver. For the 3D transport problem, the frontal solver proves inferior for all test cases considered.

6.2 Harmonic Formulation

The shallow water equations are derived from a vertical integration of the Navier-Stokes equations for a problem with a free surface. In many cases, such as astronomical tides, radiational tides, and residual circulation, it is computationally much more efficient to express the dependent variables in terms of harmonics

$$\begin{aligned} \eta(\boldsymbol{x},t) &= \tfrac{1}{2} \sum_{k=-N}^{k=N} \eta_k(\boldsymbol{x}) \exp^{-i\omega_k t} \\ \boldsymbol{u}(\boldsymbol{x},t) &= \tfrac{1}{2} \sum_{k=-N}^{k=N} \boldsymbol{u}_k(\boldsymbol{x}) \exp^{-i\omega_k t} \\ s(\boldsymbol{x},t) &= \tfrac{1}{2} \sum_{k=-N}^{k=N} s_k(\boldsymbol{x}) \exp^{-i\omega_k t} \end{aligned} \tag{6.1}$$

where η is sea level, $\boldsymbol{u}$ is depth averaged velocity, and s is the salinity concentration. Subscripted quantity η_k is the amplitude of sea level,$\boldsymbol{u}_k$ is the amplitude of horizontal velocity with components (u_k, v_k) and s_k is the amplitude of salinity, each for the kth harmonic with frequency ω_k. The horizontal coordinate (x, y) is also given as $\boldsymbol{x}$, where x is positive eastward and y positive northward; t is time, and N is the number of harmonic constituents. (Note N is small for the applications of interest here.)

These expressions are substituted into the shallow water equations and the 3D advection diffusion equation followed by an application of harmonic decomposition (a general case of Fourier decomposition where the frequencies are not necessarily integer multiples of each other). The details of this derivation can be found in [20,21]. The result is that the governing equation for sea level becomes a set of complex-valued equations of Helmholtz type for $\eta \equiv \eta_k$ of the form

$$-i\omega\eta - \nabla \cdot [a\nabla\eta - \boldsymbol{f} \times \nabla\eta] = \boldsymbol{b}(\eta) \text{ in } \Omega \tag{6.2}$$

where $\omega \equiv \omega_k$ is a known real forcing frequency, $\boldsymbol{f} = \boldsymbol{f}(\eta)$ represents the Coriolis parameter multiplied by a complex function of velocity, $a = a(\eta)$ includes the effects of depth, bottom friction and Coriolis force, and $b = b(\eta)$ includes various nonlinear coupling terms and forcing terms. The variables η, a, b are all complex-valued functions. Similarly, the advection-diffusion equation for component $s \equiv s_k$ becomes

$$-i\omega s + \boldsymbol{u}_0 \cdot \nabla s - \nabla \cdot [A\nabla s] = c(\boldsymbol{u}, s) \text{ in } \Omega \tag{6.3}$$

where A is the diffusivity tensor, and $\boldsymbol{u}_0$ may be either a 2D or 3D velocity field. Suitable boundary conditions must be supplied for both open ocean boundaries and coastal boundaries. Usually, the complex amplitude for sea level and salinity are specified at open boundaries. Velocities are computed as a post-processing step once η is found. The salinity distribution is computed once the velocity field is known. The salinity distribution can then be applied as a density forcing term for the sea level solution on successive substitution sweeps.

Applying the method of weighted residuals to the equation for sea level and integrating by parts produces a weak form of equation (6.2): find $\eta \in H$ such that

$$\begin{aligned} -i\omega \int_\Omega \eta\bar{\eta}dx + \int_\Omega a\nabla\eta \cdot \nabla\bar{\eta}dx + \int_\Omega \boldsymbol{f} \times \nabla\eta \cdot \nabla\bar{\eta}dx \\ = \int_\Omega b\bar{\eta}dx + \int_{\partial\Omega} \bar{\eta}\boldsymbol{Q} \cdot \hat{\boldsymbol{n}}ds \qquad \forall\bar{\eta} \in G \end{aligned} \tag{6.4}$$

where $\boldsymbol{Q}$ is the depth integrated velocity (discharge), $\hat{\boldsymbol{n}}$ is the outward unit normal at the boundary and H, G are appropriate trial and test spaces. Expanding in terms of a finite element basis yields a set of complex, nonsymmetric algebraic equations for the sea surface. Note that the asymmetry scales with the Coriolis parameter, included in $\boldsymbol{f}$. In the algorithm applied here, a, $\boldsymbol{f}$ and b are evaluated at the current estimate of η corresponding to an outer successive approximation iteration. The formulation of the transport equation (6.3) follows a similar methodology. However, the elements are now of prismatic type for this 3D case: piecewise-linear triangles in the horizontal plane with a piecewise-linear 1D approximation in the vertical direction. The number of piecewise-linear segments in the vertical direction is referred to as the number of layers. Thus each prismatic element belongs to a layer in the vertical direction. Integration in the vertical direction produces an element matrix which is block tridiagonal. Each diagonal block corresponds to the degrees of freedom in a given horizontal layer. Off-diagonal blocks correspond to couplings to the immediately adjacent layers above and below the vertical.

Test Problems

Our initial test problem is taken from a previous study of the English Channel [20]. A coarse mesh triangulation (unstructured) is shown in Figure 6.1. Three problem sizes are considered corresponding to the coarse mesh of Figure 1 supplemented with two steps of uniform refinement. This produces an order of magnitude variation in the problem sizes investigated: 990, 3741, and 14529 nodes for the three meshes respectively. This baseline test problem assumes a quadratic bottom friction term and has variable depth. Only ω corresponding to the M2 tidal component is considered with no coupling to other tidal modes. Modifications of this baseline problem include a linear bottom friction model, constant depth, and a variable Coriolis term (e.g. the latitude is varied) to control the degree of asymmetry in the resulting linear systems. The 3D advection diffusion problem uses the English Channel mesh set in the horizontal plane coupled with five layers in the vertical direction.

A second, simpler, test problem is also briefly considered. It consists of a square basin of fixed depth with open ocean conditions on one side and a linear bottom friction model. The basin is located at a latitude of 75 degrees for the purposes of setting the Coriolis parameter. The solution is computed on uniform, structured meshes of 126, 461, 1796, and 6881 nodes. All computations are carried out on a Sun SparcStation 2 in double precision. All timings are given in CPU seconds.

Frontal Solver

Before examining iterative methods we first consider the performance of a direct method on the complex system for the Channel test problem. The relatively high aspect ratio of the domain for this problem is quite favorable for direct solution methods, at least for the 2D sea level problem, as the front width is small compared with the total number of degrees of freedom. Hence, this will provide a more demanding test for the iterative methods. The 3D transport problem, even with only 5 depth layers, will handicap the frontal solver due to the large frontwidths encountered.

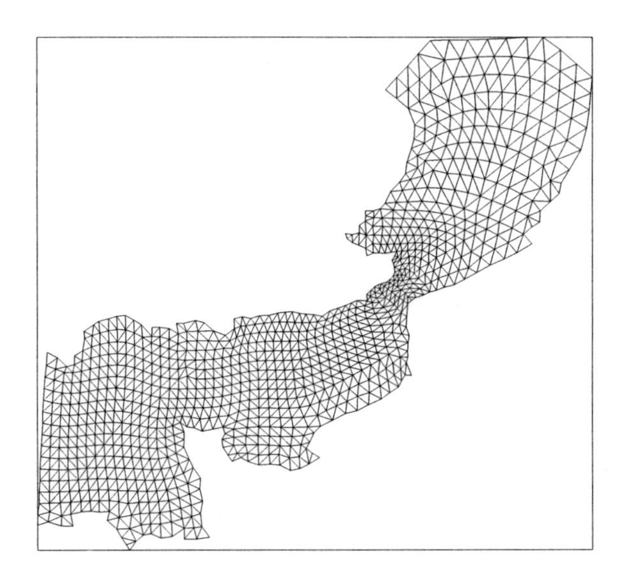

Figure 6.1: Finite element mesh for the Channel problem; coarse grid with 990 nodes.

The frontal solver has been extensively optimized, and although it does incorporate pivoting, this was found to be unnecessary for the current class of applications. The run time for a frontal solver can be bounded by an expression such as cNR^2, where c is a constant, N is the number of degrees of freedom in the problem, and R is the RMS frontwidth. Noting that for rectangular domains with quasiuniform refinement $R \approx O(N^{1/2})$, then the run time should be bounded by cN^2. Table 6.1 shows observed (t_{exp}) and predicted (t_{prd}) run times in seconds for both the two-dimensional sea level problem and the three-dimensional transport problem with 5 vertical layers. A value of $c = 0.845 \times 10^{-6}$ was estimated for the predicted run times. The timings in Table 6.1 do not include the work of forming the element stiffness matrices, only the factorization and back substitution process are considered. The agreement is reasonably close and the result provides a qualitative estimate of how computational effort scales with the number of nodes N.

Table 6.1: Performance of frontal solver on the Channel problem.

N	R	t_{exp}	t_{prd}
990	23.6	1.13	0.83
3741	46.8	11.83	11.83
14529	94.4	166.3	178.4
990 × 5	120.0	101.9	99.8
3741 × 5	235.0	1351.5	1446.0

6.3 Iterative Methods

As an initial step, the complex algebraic system corresponding to equation (6.5) was solved using a battery of iterative methods applied to an equivalent system of coupled real equations. The performance of this approach was poor as could be anticipated from theoretical results in [6]. Next, complex versions of the Biconjugate Gradient (BCG) [11] and Biconjugate Gradient Squared (BCGS) [17, 18] methods were investigated, coupled with Jacobi and ILUT(k) preconditioners, nodal reordering strategies and some other specific optimizations.

System of Real Equations

The complex algebraic system resulting from equation (6.5) can be broken into corresponding coupled systems of real equations. The real systems remain highly nonsymmetric and indefinite. Several iterative methods from the NSPCG package [12] were applied to these systems. These included both BCG and BCGS as well as the Generalized Minimum Residual Method (GMRES) [14] and the Orthomin method (OMIN). Both Jacobi and Modified Incomplete LU (MILU) preconditioners were applied. All of the methods either diverged or failed to converge in a reasonable number of iterations (stagnated). Selected results are summarized in Table 6.2. The Table gives the method and preconditioner, the result of the iteration (stagnation or divergence), and the number of iterations at which a given residual norm (relative) was achieved (at stagnation). Evaluating when an iterative method has stagnated can be somewhat subjective. For the cases given in Table 6.2, the method was assumed to stagnate when the relative residual would not decrease appreciably below 5×10^{-3} with further iteration. The iteration counts given in the Table refer to the number of iterations required for the relative residual to reach 5×10^{-3}. Further evidence to support the divergence is given by the fact that the outer successive approximation iteration for the nonlinear process also did not converge, indicating that the final residual on the inner iteration was not small enough. To give an idea of the degree of stagnation, GMRES(18) required 626 iterations to reach a relative residual of 5×10^{-3}; by 1069 iterations the relative residual was 2.5×10^{-3}. The results observed here are found to be in agreement with theoretical arguments put forward in [6] which indicate that the transformation of a complex system to a coupled real system always has detrimental effects on convergence. Here it is observed that for the tidal system the degradation is very significant.

Table 6.2: Performance of NSPCG on Channel problem

Method	Preconditioner	Result	Iterations	$\|\|r\|\|$
BCG	Jacobi	diverged	1600	4.5
BCGS	Jacobi	diverged	600	1.e6
GMRES(6)	Jacobi	stagnates	983	5.e-3
GMRES(18)	Jacobi	stagnates	626	5.e-3
OMIN(18)	Jacobi	stagnates	527	5.e-3
GMRES(6)	MILU	stagnates	10	0.1
OMIN(18)	MILU	stagnates	10	0.1

Complex Iterative Methods

The next step in the investigation involved working with the complex equations directly. This necessitates reformulating the iterative methods in a complex form. The complex forms of the BCG and BCGS methods were selected as representative schemes. More advanced methods, such as complex BiCGStab and BiCGStab2, could be developed similarly. A complex version of the BCG iteration (unpreconditioned) for the system $Ax = b$ can be stated as:

$$\lambda_0 = 0,\ \alpha_0 = 0,\ p_0 = 0,\ \tilde{p}_0 = 0,\ v_0 = 0,\ \tilde{v}_0 = 0$$
$$r_0 = b - Ax_0,\quad \tilde{r}_0 = r_0$$

$$for\ i = 1, maxits$$
$$x_i = x_{i-1} + \lambda_{i-1} p_{i-1}$$
$$\tilde{r}_i = \tilde{r}_{i-1} - \lambda_{i-1}{}^* \tilde{v}_{i-1}$$
$$r_i = r_{i-1} - \lambda_{i-1} v_{i-1}$$

$$\theta_i = < \tilde{r}_i, r_i > /(||\tilde{r}_i||\,||r_i||)$$
$$if(||r_i|| < tol) done;$$
$$if(|\theta_i| < tol2) breakdown;$$

$$
\begin{aligned}
&if(i = 1)\\
&\ \alpha_i = 0\\
&else\\
&\ \alpha_i = <\tilde{r}_i, r_i> / <\tilde{r}_{i-1}, r_{i-1}>\\
&\tilde{p}_i = \tilde{r}_i + \alpha_i{}^*\tilde{p}_{i-1}\\
&p_i = r_i + \alpha_i p_{i-1}\\
&v_i = Ap_i\\
&\tilde{v}_i = A^*\tilde{p}_i\\
&\lambda_i = <\tilde{r}_i, r_i> / <\tilde{p}_i, v_i>
\end{aligned}
$$

where, x_0 is the initial solution iterate, x_i is the approximate solution, r_i represents the residual, p_i represents the search direction and λ_i is the step length. Note that $<,>$ denotes the Hermitian inner product, $*$ denotes the complex conjugate for scalars and the conjugate transpose for matrices and $\tilde{.}$ denotes the "left" vectors in the BCG iteration. A complex version of the BCGS iteration (unpreconditioned) can be given similarly as:

$$
\begin{aligned}
&p_0 = 0,\ q_0 = 0,\ v_0 = 0,\ u_0 = 0\\
&r_0 = b - Ax_0,\ \tilde{r}_0 = r_0\\
&\rho_0 = <\tilde{r}_0, r_0>,\ \beta_0 = 0\\
&\\
&for\quad i = 1, maxits\\
&\quad u_i = r_{i-1} + \beta_{i-1}q_{i-1}\\
&\quad v_i = u_i + \beta_{i-1}(q_{i-1} + \beta_{i-1}v_{i-1})\\
&\quad p_i = Av_i \qquad \lambda_i = \rho_{i-1}/ <\tilde{r}_0, p_i> \qquad q_i = u_i - \lambda_i p_i\\
&\quad x_i = x_{i-1} + \lambda_i(u_i + q_i)\\
&\quad r_i = r_{i-1} - \lambda_i A(u_i + q_i)\\
&\quad \rho_i = <\tilde{r}_0, r_i>\\
&\quad \beta_i = \rho_i/\rho_{i-1}\\
&\quad if(||r_i|| < tol)done
\end{aligned}
$$

Additional discussion of these and related methods can be found in [3, 7, 10]. These iterative methods must be supplemented with appropriate preconditioners to enhance their convergence rates. As in the previous calculations, we focus on Jacobi and ILU preconditioners.

6.4 Preconditioning

The objective of preconditioning is to find some nonsingular matrix, M, which is easy to invert and which approximates the coefficient matrix A

in the system $Ax = b$. One can then iteratively solve the equivalent system $M^{-1}Ax = M^{-1}b$. If M^{-1} is a good approximation to A^{-1}, then the conditioning of $M^{-1}A$ should be much better than the conditioning of the original system, thus leading to accelerated convergence of gradient type iterative methods.

The simplest of all preconditionings is the Jacobi preconditioner, where M is just taken as the diagonal of A. While this preconditioning is easy to implement it is often insufficient in terms of accelerating convergence. A more sophisticated class of preconditioners is based on the incomplete LU factorization of A. In their simplest form ILU preconditioners can be formed by simply performing a row-wise Gaussian elimination on A discarding any fill-in which may occur at any stage. Thus one obtains factors L, U with a sparsity pattern identical to that of A. In actual implementation, there are several modifications to forming the ILU preconditioner which prove important. These include thresholding and allowing some additional fill-in, or variable fill-in in each row of the matrix. As pointed out in [13], for nonsymmetric problems it is also important to vary the sparsity pattern itself. Combining all of these modifications leads to the ILUT(k) methods [7,13,19] which we have considered.

Construction of ILUT(k) Preconditioners

Before describing the two specific methods explored here, we first consider the ILU Threshold, or ILUT, method. Recall that the ILU preconditioner can be formed row-wise. That is, entries in a specific row of L and U, denoted l^T and u^T respectively, are formed by considering the corresponding row a^T of A. Diagrammatically we have:

$$\begin{bmatrix} \cdots & \cdots & \cdots \\ -- & a^T & -- \\ \cdots & \cdots & \cdots \end{bmatrix} = \begin{bmatrix} \ddots & 0 & 0 \\ -- & l^T & 0 \\ \cdots & \cdots & \ddots \end{bmatrix} \begin{bmatrix} \ddots & \cdots & \cdots \\ 0 & u^T & -- \\ 0 & 0 & \ddots \end{bmatrix} \tag{6.5}$$

One moves across the entries in a^T successively eliminating entries to the left of the diagonal. After each entry is eliminated, a^T must be updated with a multiple of a previously computed row of U. After all of the entries in l^T have been computed, what remains in a^T is just u^T. Details can be found in [1,2,7,8,13,15]. It is important to note that at this stage l^T and u^T hold the respective row contributions of the complete LU factorization of A. One expects that for A banded, then L and U will exhibit complete fill-in within the band. The incomplete factorization strategy seeks to approximate the complete factorization by dropping much of the fill within the band. Thus most of the fill-in in l^T and u^T is dropped. Similarly, the thresholding strategy simply notes that if an entry in a^T is small relative to the initial norm of a^T, then it need not be eliminated at all, but can simply be dropped. Varying the threshold tolerance produces a family of approximate factorizations. In the present work, a tolerance of 10^{-10} is used, resulting in very little thresholding.

An important consequence of dropping excessive fill-in and incorporating a thresholding strategy is that the incomplete factorization becomes very inexpensive to compute (relative to the full factorization). As entries in a^T to the left of the diagonal are eliminated, multiples of previously computed rows of U are added in. However, these previously computed rows are very sparse, typically with a sparsity comparable to the original sparsity of A. Hence, there is very little arithmetic associated with these update computations. The full factorization must work with the full bandwidth of A to compute these updates. Thus while the full factorization requires $O(NR^2)$ operations, the incomplete factorization requires only $O(NR)$ operations, where N is the number of degrees of freedom and R is the RMS frontwidth. Depending on the effectiveness of the thresholding strategy, the factor of R can be considerably reduced.

Once the elimination process for a^T has been completed, a strategy must be developed for dropping some or most of the entries in l^T and u^T. Retaining all of the entries would result in a full factorization of the thresholded problem. Strategies for retaining selected fill elements while dropping the rest are typically characterized by a parameter k. This parameter represents the number of extra nonzeros to be added to the sparsity patterns for l^T and u^T beyond the sparsity pattern implied by the original a^T. For example, if there are initially seven nonzeros in a^T (the diagonal entry plus three left and three right of the diagonal) then an ILUT(k) method with k=4 would allow two additional nonzeros to the left and right of the diagonal. In other words, a total of five nonzeros are stored for l^T and five nonzeros plus the diagonal are stored for u^T. With this prescription only even values of k are allowed. Having selected k, one knows how many nonzeros are to be allowed in l^T and u^T after scanning the original sparsity pattern of a^T. One strategy, proposed in [13], sorts the entries in l^T and u^T and keeps only those of largest magnitude. We will denote this as strategy 1. Another approach is to retain the original sparsity pattern of a^T plus the k largest fill entries in both l^T and u^T. We will denote this as strategy 2. Note that strategy 1 is somewhat more expensive in that it requires larger sorts.

The above procedure for computing the ILUT(k) preconditioner is quite straightforward for 2D problems. Recall however, that the 3D transport problem is based on prismatic elements. The number of vertical layers is constant throughout the mesh and a data structure for a 2D mesh is used everywhere. This considerably reduces the complexity of the 3D code in terms of the required data structures and element computations. However, it makes the implementation of a fully 3D incomplete factorization quite complex. In addition, it is expected that the computational expense of forming this factorization will be quite high, roughly proportional to the number of vertical layers. Given an expectation of at most 10 - 20 layers in the vertical, and a data structure developed for 2D problems and 2D incomplete factorizations, we have chosen to develop a layer-based preconditioner. Thus the 3D preconditioner simply consists of considering each horizontal layer in turn and forming its ILUT(k) preconditioner. All coupling between layers in the vertical is neglected in constructing this preconditioner. Within each horizontal layer, the existing 2D data structure is used directly. This preconditioner is consistent in some sense with the 2D shallow water approximation. It also allows one to use a frontal solver within each horizontal layer as a more sophisticated form of preconditioning.

6.5 Test Results

Effect of Ordering

In finite element applications one often encounters refined or adaptively refined meshes. Such refinement techniques can produce coefficient matrices of unusually large bandwidth or frontwidth unless a nodal reordering scheme is first applied. As is well known, the effectiveness of sparse elimination methods depends critically upon the sparsity pattern of the matrix. It is not surprising then that similar issues arise in the implementation of incomplete factorization preconditioners for sparse matrices. One study of such effects can be found in [4], where a battery of node reordering schemes are applied before constructing the ILU preconditioner. In the present work we explore the effect of only one reordering, a variant of the Reverse Cuthill-McKee scheme proposed by Sloan [16]. No variation in iteration count with ordering is observed for Jacobi preconditioning as expected. However, comparing a "refinement" type ordering and that of Sloan's scheme for ILUT(k) preconditioners based on strategy 2 for retaining fill produces remarkably different results. These results are summarized in Table 6.3 in terms of iteration counts for the BCG method applied to the Channel problem. A convergence tolerance of 10^{-4} on the relative residual was applied. The threshold tolerance was set to 10^{-10}. The first number listed for each combination of node number N and fill k refers to the refinement ordering iteration count while the second refers to that for the "Sloan ordering".

Table 6.3: Effect of node ordering on iteration count for fill k.

N	$k = 0$	$k = 4$	$k = 8$
990	60 / 50	35 / 28	25 / 20
3741	159 / 102	83 / 56	55 / 40
14529	355 / 209	201 / 118	124 / 88

As can be seen, the effect of ordering is most pronounced for low levels of fill and the most highly refined meshes. For $k = 0, N = 14529$ the iteration count for the refinement ordering is 70% greater than that for the "Sloan ordering". In terms of run times, the difference is even more pronounced. Computing the "Sloan ordering" requires negligible CPU time, (typically much less than 1% of the total run time for the problem sizes considered). However, the cost of computing the ILUT(k) preconditioner depends strongly on the bandwidth. Experimentally, it is observed that for the refinement ordering the construction of the ILUT(k) preconditioner varies as N^2. For Sloan's method it varies as $N^{3/2}$. The code executed for each ordering is identical. The difference appears to lie in how the ILUT(k) complexity analysis changes as the bandwidth is changed. It must be emphasized that for the refinement ordering the bandwidths, R, are roughly $O(N)$. In this case, terms of $O(NR)$ in the complexity analysis, such as thresholding, elimination, and sorting, become dominant, yielding an overall $O(N^2)$ operation count. On the other hand, for the "Sloan ordering" the bandwidths are roughly $O(N^{1/2})$ (similar to the frontal solver RMS frontwidths). Note that terms $O(NRS)$, where S represents the number of nonzeroes retained in each row, are not dominant here. For a bandsolver, where $S = R$, this term is dominant and yields the familiar $O(NR^2)$ operation count. The sensitive dependence of cost on bandwidth is reflected in run-time breakdowns where the construction of the ILUT(k) preconditioner was observed to account for roughly 1/2 of the overall run time for the Channel problem if a refinement ordering is used. If Sloan's ordering scheme is used, the fraction drops to 10 – 15% , depending on k.

The Effect of Fill

Assuming that the nodes have been ordered according to Sloan's scheme, we now proceed to investigate the effect of the level of fill, k, on the iteration count. Table 6.4 presents iteration count results for BCG combined with ILUT(k) for the Channel problem for various combinations of N, k. A BCG convergence tolerance of 10^{-4} was applied to the relative residual. The threshold tolerance was set to 10^{-10}. Results for both strategies 1 and 2 for retaining fill are presented. The first number in each category refers to strategy 1, while the second refers to strategy 2. Similar results for the overall run time in CPU seconds are presented in Table 6.5.

Table 6.4: Effect of fill level and retention strategy on iteration count.

N	$k = 0$	$k = 4$	$k = 8$
990	49 / 50	27 / 28	19 / 20
3741	95 / 102	56 / 56	40 / 40
14529	201 / 209	118 / 118	84 / 88

Table 6.5: Effect of fill level and retention strategy on run time.

N	$k = 0$	$k = 4$	$k = 8$
990	8.0s / 8.4s	6.2s / 6.2s	5.8s / 5.4s
3741	53.8s / 56.1s	41.8s / 40.6s	37.6s / 35.6s
14529	422.3s / 425.9s	313.0s / 290.8s	264.4s / 260.1s

Table 6.4 indicates that strategy 1 for retaining fill produces only marginal improvement in the iteration count, while Table 6.5 shows that it is somewhat more expensive to implement in terms of run time as k increases. This increase in run time can be attributed to the increased sort lengths. Thus we conclude that for our particular application, which demonstrates a high degree of diagonal dominance and relatively low asymmetry, strategy 2 is to be preferred.

Comparison with Frontal Solver

We conclude this section with results for BCGS / ILUT(k) for strategy 2 and various values of k as applied to the Channel problem (both for sea level in 2D and transport in 3D) and the Square Basin problem. These results represent the best performance that could be attained for the iterative

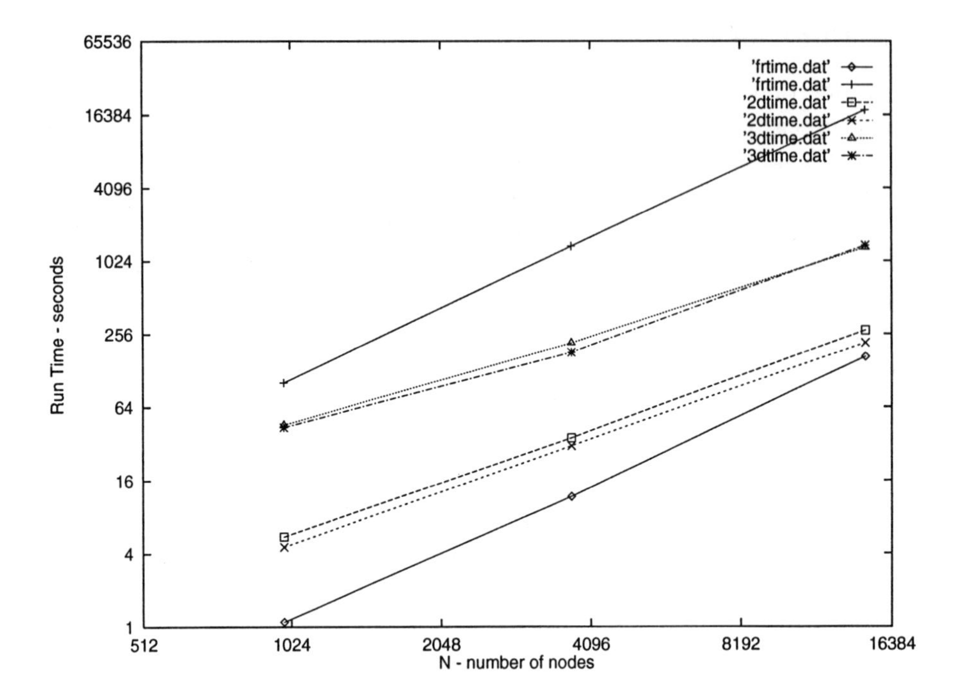

Figure 6.2: Variation of run time with fill level, k, for the Channel problem, using BCGS with ILUT(k). Results for the frontal solver are shown for comparison.

methods. They reflect the effect of several optimizations including nodal reordering, enhanced storage scheme and fill retention strategy.

Figure 6.2 shows run time results for the Channel problem, while Figure 6.3 shows similar results for the Square Basin problem. The run-time for the frontal solver is also included for comparison. As can be seen, for the sea level Channel problem, the frontal solver slightly outperforms the iterative method for $N = 14529$ and $k \leq 8$. However, the run-time evidently scales better for the iterative method and a breakeven point is expected for N somewhat greater than 15000. For the three-dimensional transport problem, the iterative methods widely outperform the frontal solver for all three problem sizes. Note that the largest problem size, $N = 14529 \times 5$ could not be run due to insufficient memory. The value shown for the run time is an estimate. In addition, the run-time scaling is substantially better. Iteration count and run time data used to construct Figure 6.2 are given in Tables 6.6, 6.7 and 6.8. (Note that 'div' indicates that the iteration diverged.)

For the Square Basin problem the breakeven point occurs for $N = 426$ and $k \geq 0$. The results of Figure 6.3 should be interpreted with caution

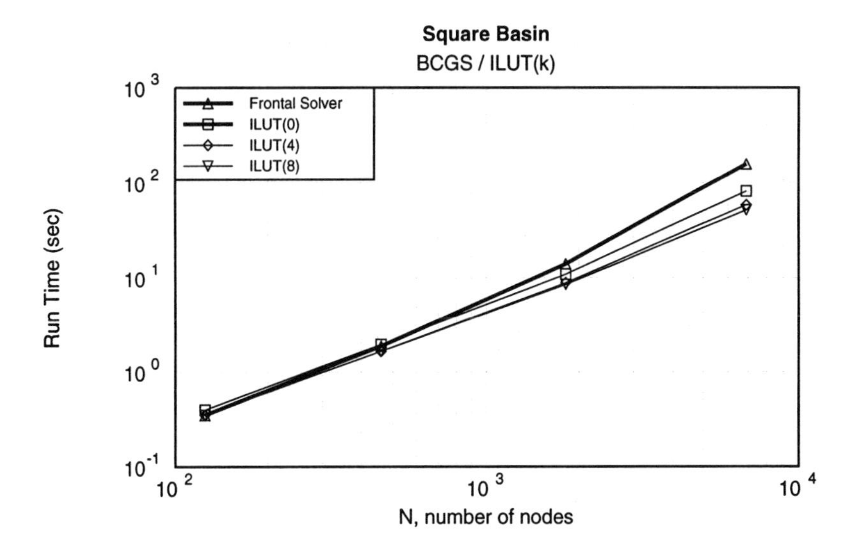

Figure 6.3: Variation of run-time with fill level, k, for the Square Basin problem, using BCGS with ILUT(k). Results for the frontal solver are shown for comparison.

Table 6.6: Iteration count for BCGS with ILUT(k) preconditioning for the Channel problem.

N	$k = 0$	$k = 4$	$k = 8$
990	46	23	14
3741	93	50	35
14529	200	115	78
990 × 5	28	12	8
3741 × 5	60	32	18
14529 × 5	div	59	45

as the Square Basin problem is certainly more favorable for the iterative methods due to the low aspect ratio of the domain. However, this problem may be a reasonable model of larger ocean basins such as the Gulf of Mexico. Tables 6.9, 6.10 and 6.11 show corresponding iteration counts and run times.

Table 6.7: Run time in CPU seconds for BCGS with ILUT(k) preconditioning for the Channel problem.

N	$k=0$	$k=4$	$k=8$
990	7.41	5.51	4.54
3741	50.10	35.89	30.93
14529	391.20	274.10	215.80
990 × 5	55.93	45.92	43.93
3741 × 5	276.77	215.89	180.84
14529 × 5	div	1314.6	1369.6

Table 6.8: Run time in CPU seconds for the frontal solver for the Channel problem.(T_{tot} is the total solution time, T_{elt} is the time to form element matrices, and T_{sol} is the time for elimination and back substitution.)

N	T_{tot}	T_{elt}	T_{sol}
990	2.92	1.79	1.13
3741	19.05	7.22	11.83
14529	195.26	28.96	166.30
990 × 5	102.0	-	-
3741 × 5	1351.5	-	-
14529 × 5	17900 est.	-	-

Code Optimizations

Implementing the ILUT(k) preconditioners for complex problems was found to be more difficult than for similar real arithmetic problems. Some code optimizations and variations that were employed should be noted. Initially, a row-wise storage scheme for A was used. That is, the nonzeros of A are stored in an $N \times w$ array where w is the maximum number of nonzeros in any row of A. This is somewhat wasteful as most rows of A contain fewer

Table 6.9: Iteration count for BCGS with ILUT(k) preconditioning for the Square Basin problem.

N	$k=0$	$k=4$	$k=8$
126	6	3	2
461	15	6	4
1796	30	14	9
6881	72	35	23

Table 6.10: Run time in CPU seconds for BCGS with ILUT(k) preconditioning for the Square Basin problem.

N	$k = 0$	$k = 4$	$k = 8$
126	0.39	0.34	0.35
461	1.95	1.62	1.61
1796	10.85	8.71	8.27
6881	79.23	56.33	50.54

Table 6.11: Run time in CPU seconds for the frontal solver for the Square Basin problem.

N	126	461	1796	6881
T	0.34	1.80	13.86	156.14

than w nonzeros. Converting to a Yale type sparse scheme [5] where these nonzeros are compressed out saves some storage, but also produces faster matrix-vector multiplies. The Yale version was roughly 25% faster. The effects of nodal reordering both on the iteration counts and the cost to form the preconditioner have already been discussed. The specifics of the N^2 scaling for a refinement ordering actually arise from the thresholding and sorting operations.

Implementing the thresholding preconditioner requires the evaluation of $|c|$ at several stages, where $c = \alpha + i\beta$ are the complex entries of a^T. Rather than using the standard definition for complex absolute value, $|c| \equiv (\alpha^2 + \beta^2)^{1/2}$, an alternative measure was used, $|c| \equiv |\alpha| + |\beta|$. This choice reduces the time to form the preconditioner by a factor of two. This may seem remarkable; however, note that the magnitude function must be applied NR times to evaluate thresholding. Given a sufficiently large bandwidth, this becomes the dominant term in computing the ILUT(k) factorization. The use of strategy 2 for fill retention further reduces the time to form the preconditioner as reflected in Table 6.5 by decreasing the sort lengths. For larger problem sizes and larger levels of fill, strategy 2 should have a greater impact.

Storage Comparisons

Thus far, the storage requirements for the preconditioned iterative methods have been ignored. The conjugate gradient type methods described here require that the sparse matrix A be stored in order to compute the matrix-vector products $Av \rightarrow w$ in the iteration. Typically one finds an average of seven nonzeros per row of A and a maximum of nine for the unstructured Channel grid in the two-dimensional sea level problem. Using a compressed row storage format requires roughly $7N$ floating point words to store the matrix, an additional $7N$ integer words to store column pointers, and N integer words to store row pointers. For the three-dimensional transport problem, the compressed storage format is used within horizontal layers and for the couplings between adjacent layers. Thus for L layers, we require $21NL$ floating point words to store the matrix. The same row and column pointers are used for each layer. The ILUT(k) preconditioner requires similar storage for the original sparsity pattern of the two-dimensional problem and an added kN floating point words to store the fill plus $(k+2)N$ integer words for various row and column pointers; all of this per horizontal layer. Thus for $k = 4, 8$ we require approximately three times the storage required for A alone or $24NL$ real and integer words of storage. This must be compared to the storage required for the frontal solver.

If the frontal solver is allocated sufficient storage to run entirely in main memory, then approximately NR words of floating point storage are needed. Table 6.1 shows that $R = 47$ for $N = 3741$, which for the 2D problem produces roughly similar storage requirements to that of the ILUT(8) iterative methods. For large N, the frontal solver would require more storage if it is to run entirely in main memory. However, the frontal solver allows the option of using much less memory if frequent disk writes can be tolerated. In that case roughly $N + R^2$ words of storage are needed, although the performance degradation due to disk writes may be severe. For the 3D problem, this estimate becomes roughly $N + N^2L^2$ words. A reasonable amount of buffer space would allow for at least 1 and perhaps 3 frontal matrices worth of storage. With this in mind, the 3D estimates become: $48NL$ words for the iterative solver and $3N^2L^2$ words for the frontal solver. Hence for $L > 16$, the frontal solver has greater storage requirements, assuming a reasonable minimum buffer space. One might be tempted to use the frontal solver within the horizontal layers as a preconditioner for the 3D problem. In this case the storage requirements would be $48NL$ words for the iterative solver with ILUT preconditioning compared to $24NL + N^{1/2}NL$ words for the iterative solver with 2D frontal preconditioning. For $N < 1000$ the frontal version might be more desirable. On current generation massively parallel machines one might expect to fit a 5,000 DOF problem on each processor. Hence, for such a machine, the ILUT solution option appears more attractive.

6.6 Conclusion

An investigation into the applicability of gradient type iterative solution methods for the harmonic decomposition version of the shallow water equations was conducted. ILUT(k) type preconditioners were developed for the resulting complex linear systems. Performance comparisons with a frontal solver as applied to a test simulation of the English Channel are presented. These results indicate that given sufficient fill, the iterative methods are almost competitive with the frontal solution method for the 2D sea level problem. A break-even point of about 15,000 degrees of freedom was found. Beyond this range it is expected that the iterative methods will offer significant savings over the frontal solver. These savings are more pronounced for basins with a lower aspect ratio than exhibited by the English Channel. Test results for a square basin indicate a much lower break-even point of about 1,000 degrees of freedom. For basins with a higher aspect ratio than the Channel it is doubtful that gradient type iterative methods could be applied competitively. For the three-dimensional transport problem, the iterative methods appear to be greatly superior to the frontal solver both in terms of run time and storage requirements for any problem with more than 1000 or so degrees of freedom and 10 - 20 vertical layers. It was also found that working with equivalent real systems of equations produces very poor performance in the iterative methods.

Acknowledgements: This work was supported by the Department of Energy and by the National Research Program of the U.S. Geological Survey. The authors would like to acknowledge the helpful suggestions and comments of R. Freund, M. Gutknecht, and D.M. Young.

References

[1] Chan, T. F., "Fourier analysis of relaxed incomplete factorization preconditioners," *SIAM JSSC*, 12, 3, 668–680, 1991.

[2] Chan, T. F. and H. Elman, "Fourier analysis of iterative methods for elliptic problems, " *SIAM Review*, 31, 1, 20-49, 1989.

[3] Datta, S. K., S. Liu and T. Ju, "The application of the conjugate gradient method to large sparse unsymmetric complex matrix equations arising from the scattering of elastic waves," Copper Mountain Conference on Iterative Methods, Copper Mountain, CO, April 1990.

[4] Duff, I. S., and G. Meurant, "The effect of ordering on preconditioned conjugate gradients," *BIT*, 29, 635–657, 1989.

[5] Eisenstat, S. C., M. Gursky, M. Schultz and A. Sherman, "Yale sparse matrix package," Research Report 114, Dept. CS, Yale Univ., 1977.

[6] Freund, R. "Conjugate gradient type methods for linear systems with complex symmetric coefficient matrices," *SIAM JSSC*, 13, 1, 425-448, 1992.

[7] Freund, R., G. H. Golub and N. Nachtigal, "Iterative solution of linear systems," *Acta Numerica* , 1, 57-100, 1992.

[8] Gustafsson, I., "A class of first order factorization methods," *BIT*, 18, 142-156, 1978.

[9] Gutknecht, M. "The unsymmetric Lanczos algorithms and their relations to Padé approximations, continued fractions and the QD algorithm," Copper Mountain Conference on Iterative Methods, Copper Mountain, CO, April 1990.

[10] Gutknecht, M. "Variants of BiCGStab for matrices with complex spectrum," IPS Research Report 91-14, ETH-Zentrum, CH-8092, Zurich, Aug. 1991.

[11] Lanczos, C. "Solution of systems of linear equations by minimized iterations", *J. Res. Nat. Bur. Std.*, 49,1, 33-53, 1952.

[12] Oppe, T. , W. Joubert and D. Kincaid, "NSPCG user's guide version 1.0," CNA Report 126, Center for Numerical Analysis, Univ. Texas, Austin, TX., April 1988.

[13] Saad, Y. "Preconditioning techniques for nonsymmetric and indefinite linear systems," *J. Comp. Appl. Math.*, 24, 89-105, 1988.

[14] Saad, Y. and M. Schultz, "GMRES: A generalized minimum residual algorithm for solving nonsymmetric linear systems," *SIAM JSSC*, 7, 3 856-869, 1986.

[15] Sauter, R. "The finite element method and ILU preconditioners," *J. Comp. Appl. Math.*, 36, 1, 91-106, 1991.

[16] Sloan, S. W. "A Fortran program for profile and wavefront reduction," *IJNME* , 28, 2651-2679, 1989.

[17] Sonneveld, P. "CGS, a fast Lanczos type solver for nonsymmetric linear systems," *SIAM JSSC* , 10, 36-52, 1989.

[18] Van der Vorst, H. "Bi-CGSTAB: A fast and smoothly converging variant of Bi-CG," *SIAM JSSC*, 10, 6, 1174-1185, 1989.

[19] Van der Vorst, H. "Iterative methods for the solution of large systems of equations on supercomputers," *Adv. Water Resources*, 13, 3, 137-146, 1990.

[20] Walters, R. A. "A model for tides and currents in the English Channel and Southern North Sea," *Adv. Water Resources*, 10, 138-148, 1987.

[21] Walters, R. A."The frontal method in hydrodynamics simulations," *Computers and Fluids*, 8, 1980, 265-272.

Chapter 7

3D Finite Element Hydrodynamic Model

M. Andreola[1], *S. Bianchi*[1], *L. Brusa*[1] *and P. Molinaro*[2]

7.1 Introduction

Water flow simulations for environmental problems often require local detailed analyses for a better understanding and accurate prediction of the fate of pollutants in water bodies. To this purpose a fully 3D model of general type is desirable since it can treat more realistic situations than other averaged models. Examples of problems which can benefit from this type of approach are those connected with the design and management of discharges into coastal waters. The complexity of fully 3D models arises from the necessity to solve the Navier-Stokes equations coupled with turbulence models and diffusive-convective equations describing the transport of pollutants. This set of non linear differential equations must be usually solved for long time intervals in large domains with complex 3D geometries. The choice of efficient numerical schemes for time and space discretization of the model equations is therefore crucial to allow realistic analyses performed with reasonable computing times. The use of implicit time stepping methods is unavoidable to describe long and relatively slow transients, such as those related to pollution problems. As far as the spatial discretization

[1]CISE Tecnologie Innovative SpA
[2]ENEL SpA - CRIS

is concerned, the finite element method is attractive because of its flexibility and ability to naturally treat complex geometries. The introduction of vector/parallel supercomputers is opening new possibilities for realistic application of fully 3D finite element models to environmental problems.

In order to explore the potentialities of this approach, CISE and ENEL are developing the MADIAN code to study thermal discharges from power plants into coastal waters. The code is based on the use of $P1 - isoP2$ elements [3] for the approximation of pressure and velocity fields, while a fully implicit operator splitting technique is applied for the solution of the non linear system of ordinary differential equations resulting from spatial discretization. The software optimization for shared memory multiprocessors is in progress; at present, the parallel features are limited to the linear equation solver which is the most time consuming task.

The outline of the chapter is as follows: First an outline of the mathematical model is given in Section 7.2. This is followed in Section 7.3 by a description of the numerical methods implemented in the code. Finally, results of some simple applications are provided to demonstrate the accuracy of the model in comparison with analytical and experimental results.

7.2 Description of the Mathematical Model

The mathematical model on which the present version of the MADIAN code is based, is formed by the following system of equations:

1. The Navier-Stokes equations for an incompressible newtonian fluid, with Reynolds averaging and the Boussinesq approximation [7];
2. The thermal energy equation;
3. The salinity equation;
4. The free surface equation.

The above-mentioned system of equations is completed using the equation of state for salt water [6] and a turbulence closure model for the computation of turbulent viscosity and diffusivity. We now provide a more formal description.

In vector notation, the Navier-Stokes equations describing the dynamics of the fluid have the form

$$\frac{\partial \boldsymbol{u}}{\partial t} + \nabla \cdot (\boldsymbol{u}\boldsymbol{u}) = -\frac{1}{\rho_0}\nabla P + \frac{1}{\rho_0}\nabla \cdot \tau + \boldsymbol{s} \tag{7.1}$$

$$\nabla \cdot \boldsymbol{u} = 0 \tag{7.2}$$

where $\boldsymbol{u} = [u_1, u_2, u_3]$ is the velocity vector, P is the pressure,$\rho = \rho(T, S)$ is the fluid density as a function of temperature and salinity, ρ_0 is the fluid density at the reference temperature T_0 and salinity S_0, and τ is the stress tensor, whose components are

$$\tau_{ij} = \varrho_o \nu_e \left(\frac{\partial u_i}{\partial x_j} + \frac{\partial u_j}{\partial x_i} \right) \tag{7.3}$$

where ν_e is the effective (molecular + turbulent) viscosity.

The source term $\boldsymbol{s}$ is defined as

$$\boldsymbol{s} = \boldsymbol{F} + \boldsymbol{G} \tag{7.4}$$

where

$$\boldsymbol{F} = \frac{\rho - \rho_0}{\rho_0} \boldsymbol{g}, \quad \boldsymbol{G} = -2\boldsymbol{\Omega} \times \boldsymbol{u}, \tag{7.5}$$

with gravity vector $\boldsymbol{g} = [0, 0, -g]$ and $\boldsymbol{\Omega}$ the earth's angular velocity vector.

The temperature field and the salinity distribution are each described by an equation of the form

$$\frac{\partial \phi^*}{\partial t} + \nabla \cdot (\boldsymbol{u}\phi^*) - \frac{1}{\alpha} \nabla \cdot \boldsymbol{q} = 0 \tag{7.6}$$

where $\phi^* = \phi - \phi_a$ is the difference between the actual value of the scalar ϕ and its ambient distribution ϕ_a. The components of the flux vector $\boldsymbol{q}$ are

$$q_i = \alpha \Gamma_e \frac{\partial \phi^*}{\partial x_i} \tag{7.7}$$

In the case of the thermal energy equation, ϕ is the temperature field T, $\alpha = \rho c_{p_0}$, where c_{p_0} is the specific heat (at constant pressure), and Γ_e is the effective thermal diffusivity. In the case of the salinity equation, ϕ is the salt concentration S, $\alpha = \varrho$ and Γ_e is the effective salinity diffusivity.

If η is the difference between the actual level of the free surface and the reference level z_0=constant, the 2-D free surface equation can be written as

$$\frac{\partial \eta}{\partial t} + u_\eta \frac{\partial \eta}{\partial x} + v_\eta \frac{\partial \eta}{\partial y} = w_\eta \tag{7.8}$$

where u_η, v_η and w_η are the velocity components at the free surface, with x, y the Cartesian coordinates in the horizontal plane.

In the present version of the MADIAN code, two turbulence models are used to evaluate local values of eddy viscosity and diffusivity. The first one is a simplified model with constant eddy coefficients; the second one is the well-known mixing-length model. In both models the vertical viscosity coefficient is corrected via empirical relations [14] to take into account the effects of buoyancy.

The velocity, temperature and salinity fields are specified as initial data at $t = t_0$. On the solid boundaries either no-slip or slip velocity conditions are imposed. In the second case the wall normal component of the velocity is equal to zero and the wall shear stress components are given by

$$\tau_{wt} = \rho_0 u^*_{wt} |\boldsymbol{u}^*_w|, \quad \tau_{wb} = \rho_0 u^*_{wb} |\boldsymbol{u}^*_w| \tag{7.9}$$

where τ_{wt} and τ_{wb} are the shear stress components along the tangent and binormal directions and $|\boldsymbol{u}^*_w|$ is the magnitude of the shear velocity. The tangent and binormal components of $\boldsymbol{u}^*_w$ are defined by

$$u^*_{wt} = \frac{u_t\sqrt{g}}{C_z}, \quad u^*_{wb} = \frac{u_b\sqrt{g}}{C_z}, \quad C_z = \frac{1}{n} h^{\frac{1}{6}} \tag{7.10}$$

where n is the Manning coefficient and h is the local depth [12]. For temperature and salinity, solid boundaries are assumed to be adiabatic and impermeable, respectively. That is, zero normal temperature and salinity gradients apply.

The free surface is described by a function of the kind $z = z_0 + \eta(x, y, t)$ where z_0 is the average level of the free surface and η is its variation with respect to z_0, obtained by solving the free surface equation (7.8). Let us suppose that, in the absence of tides, $\eta(x, y, t) << z_0$. In this case the boundary conditions can be assigned with reference to the plane described by $z = z_0$ provided that the value of the atmospheric pressure is suitably corrected. As a matter of fact, the pressure P within the fluid at the surface corresponds to the atmospheric pressure ($P = P_{atm}$). However, the fluid pressure P at the plane $z = z_0$ can be assumed, as a first approximation, to be $P = P_{atm} + \varrho_0 g \eta$. As a consequence, the normal component of the total stress is equal to the pressure value at $z = z_0$. Finally, following the above hypotheses, the shear stress due to the wind is imposed on the plane $z = z_0$ by

$$\tau_{vx} = \rho_a C_w w_x |\boldsymbol{w}|, \quad \tau_{vy} = \rho_a C_w w_y |\boldsymbol{w}| \tag{7.11}$$

where τ_{vx} and τ_{vy} are the components of the shear stress due to the wind, $|\boldsymbol{w}|$ is the magnitude of the wind velocity, ρ_a is the air density, and C_w is a drag coefficient [15].

Such assumptions lead to a computational domain invariant in time, but they restrict very strongly the validity of the model when tidal effects cannot be neglected. When, as a particular case, a "rigid lid approximation" is adopted, the boundary conditions change with respect to those stated above. In fact, the free surface equation (7.8) is replaced by $w_\eta = 0$ and this condition replaces the constraint on the total normal stress. The value of the shear stress induced by the wind is imposed as before.

Finally, regardless of the kind of model adopted for the treatment of the free surface, the net heat flow in the normal direction satisfies the mixed condition

$$q = -k(T - T_e) \tag{7.12}$$

where k is the heat exchange coefficient at the surface, T_e is the so called "equilibrium temperature" of the fluid body and T is the surface temperature of the water. The condition of zero mass flux through the free surface applies for salinity.

At the inlet open boundaries the spatial distribution of velocity is assigned. The temperature and salinity distribution of the incoming fluid are assumed to be at the environment temperature distribution T_a and salinity S_a, respectively. As for the free surface, when the "rigid lid approximation" is not adopted, the value of the level with respect to the reference plane $z = z_0$ is given.

At the inlet discharge boundaries the spatial distribution of the velocity is assigned. If effects of thermal and/or saline re-circulation are neglected between intake and discharge sections, the space distribution of temperature and salinity of the fluid entering in the computational domain through the discharge are also assigned.

The open outflow boundary conditions can be assigned as either the spatial distribution of the velocity or an assigned value of the total normal stress(zero or a given pressure distribution or the pressure distribution computed at the previous time step.) The last condition assumes that the tangent and binormal components of the velocity are zero. The relevant normal gradients for temperature T^* and salinity S^* are set to zero.

Intake boundaries are treated in exactly the same manner as for discharge: the space distribution of the flow velocity at the intake now has to be assigned. The temperature T^* and salinity S^* gradients are set equal to zero.

7.3 Description of the Numerical Method

Equations (7.1), (7.2), and (7.6) are discretized using $P1 - isoP2$ finite elements [3] which approximate pressure and velocities with linear basis functions on nested meshes, as shown in Figure 7.1. The resulting system of ordinary differential equations for the flow is

$$M\dot{U} = -(K + C)U - \frac{1}{\rho_0}BP + f \tag{7.13}$$

$$-B^T U = s \tag{7.14}$$

and for scalar transport

$$M_S\dot{\phi} = -(K_S + C_S)\phi + f_S \tag{7.15}$$

where M, M_S are the mass matrices, matrices K, K_S and C, C_S arise from the discretization of the diffusive and convective operators, respectively, and matrix B^T is obtained by discretizing the divergence operator. The components of vectors U, P, ϕ are the nodal values of velocity, pressure and scalar (temperature or salinity) fields, respectively. Vectors f, s, f_S take into account external loads and contributions from boundary conditions.

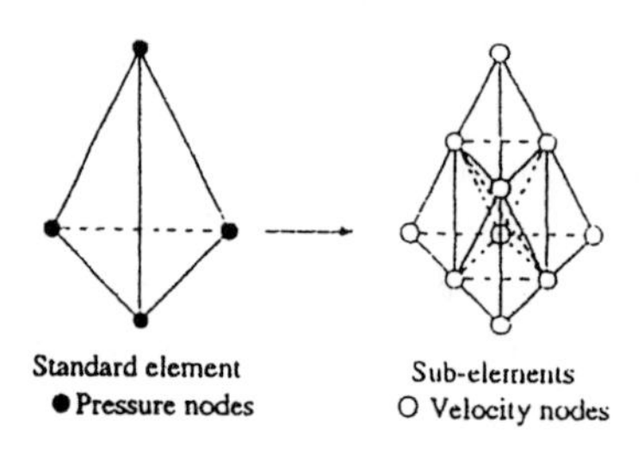

Figure 7.1: $P1 - isoP2$ finite element.

For the solution of the discretized Navier-Stokes equations in the time interval $t_n \leq t \leq t_{n+1}$, the following operator splitting scheme is applied:
1). An intermediate velocity field is computed by solving the problem

$$M\dot{U}^* = -(K^n + C^*)U^* - \frac{1}{\rho_0}BP^n + f \tag{7.16}$$

with the initial conditions

$$U^*(t_n) = U^n \tag{7.17}$$

Vectors U^n, P^n are the nodal velocities and pressures at $t = t_n$; matrix K is evaluated at $t = t_n$ while matrix C^* depends on the unknown velocity field U^*.

2). The velocity and pressure fields at $t = t_{n+1} = t_n + \Delta t$ are obtained by solving the problem

$$\boldsymbol{M}\dot{\boldsymbol{U}} = -\frac{1}{\rho_0}\boldsymbol{B}(\boldsymbol{P} - \boldsymbol{P}^n), \quad -\boldsymbol{B}^T\boldsymbol{U} = \boldsymbol{s} \tag{7.18}$$

with the initial condition

$$\boldsymbol{U}(t_n) = \boldsymbol{U}^{*n+1} \tag{7.19}$$

Equations (7.16) and (7.18) are solved by the implicit $2nd$ order Crank-Nicolson method. This leads to the following algorithm:
1). Compute $\boldsymbol{U}^{*n+1}$ by solving the nonlinear algebraic equations

$$\begin{aligned}\left[\boldsymbol{M} + \frac{\Delta t}{2}(\boldsymbol{K}^n + \boldsymbol{C}^{*n+1})\right]\boldsymbol{U}^{*n+1} \\ = \left[\boldsymbol{M} - \frac{\Delta t}{2}(\boldsymbol{K}^n + \boldsymbol{C}^n)\right]\boldsymbol{U}^n \\ -\frac{\Delta t}{\rho_0}\boldsymbol{B}\boldsymbol{P}^n + \frac{\Delta t}{2}(\boldsymbol{f}^{n+1} + \boldsymbol{f}^n)\end{aligned} \tag{7.20}$$

where matrix $\boldsymbol{C}^{*n+1}$ depends on the unknown vector $\boldsymbol{U}^{*n+1}$, while matrix $\boldsymbol{C}^n$ depends on the known initial vector $\boldsymbol{U}^n$.
2) Compute $\boldsymbol{U}^{n+1}$ and $\boldsymbol{P}^{n+1}$ by the relations

$$\boldsymbol{U}^{n+1} = \boldsymbol{U}^{*n+1} - \boldsymbol{M}^{-1}\boldsymbol{B}\boldsymbol{\psi}^{n+1} \tag{7.21}$$

$$\boldsymbol{P}^{n+1} = \boldsymbol{P}^n + \frac{2\rho_0}{\Delta t}\boldsymbol{\psi}^{n+1} \tag{7.22}$$

where $\boldsymbol{\psi}^{n+1}$ is the solution of the linear system

$$(\boldsymbol{B}^T\boldsymbol{M}^{-1}\boldsymbol{B})\boldsymbol{\psi}^{n+1} = \boldsymbol{B}^T\boldsymbol{U}^{*n+1} + \boldsymbol{s}^{n+1} \tag{7.23}$$

If $\boldsymbol{M}$ is the consistent mass matrix, $\boldsymbol{B}^T\boldsymbol{M}^{-1}\boldsymbol{B}$ is a full matrix and the solution of system (7.23) may be very expensive. In fact an iterative method, such as the conjugate gradient method (CGM), may be used and the solution of a linear system is required for each iteration. All of these difficulties are avoided if $\boldsymbol{M}$ is a lumped mass matrix. In this case matrix $\boldsymbol{B}^T\boldsymbol{M}^{-1}\boldsymbol{B}$ is a sparse matrix that can be factorized once for all and the computation of vector $\boldsymbol{\psi}^{n+1}$ is obtained by performing only one forward and one backward step for each time step. In view of this, we use a lumped mass matrix. It can be shown that the scheme (7.20)-(7.23) is stable and the truncation error is $O(\Delta t^2)$ for velocities and pressures. In many cases, however, the accuracy of the "pressure" field computed by (7.22) is unsatisfactory. This value should not be considered as the real pressure but as a scalar defined to correct $\boldsymbol{U}^{*n+1}$ in order to obtain a divergence free velocity field. Once $\boldsymbol{U}^{n+1}$ is known, the real pressure is computed using (7.13), (7.14) as

$$\boldsymbol{P}_{eff}^{n+1} = \rho_o(\boldsymbol{B}^T\boldsymbol{M}^{-1}\boldsymbol{B})^{-1}\{\boldsymbol{B}^T\boldsymbol{M}^{-1}[\boldsymbol{f}^{n+1} - (\boldsymbol{K}^n + \boldsymbol{C}^{n+1})\boldsymbol{U}^{n+1}] + \dot{\boldsymbol{s}}^{n+1}\} \tag{7.24}$$

where $\dot{\boldsymbol{s}}^{n+1}$ can be approximated by $(\boldsymbol{s}^{n+1} - \boldsymbol{s}^n)/\Delta t$. Equation (7.24) is also used to define the initial pressure value necessary to start the computations. Notice that $\boldsymbol{P}_{eff}^n$ can not replace $\boldsymbol{P}^n$ in (7.20) and (7.22) for $n \geq 1$ because the resulting scheme is unstable. The algorithm we propose is similar to the scheme described in [8] and named "Projector 2", with the following main differences:

1. Second order integration schemes are used;

2. Convection is treated in a fully implicit way, whereas the explicit approximation of the convective terms proposed in [8] requires the use of stabilizing terms added to the viscosity;

3. A lumped mass matrix is used for (7.13), since, in our experience the mixed approach suggested in [8] degrades the accuracy;

4. The pressure field is computed by the discretized equilibrium equations (7.13), (7.14), once the velocity field is known.

The main computational task of the proposed scheme is the solution of (7.20). For this problem the following iterative scheme is used:

$$\begin{aligned}(\boldsymbol{M} + \frac{\Delta t}{2}\boldsymbol{K}^n)\boldsymbol{X}_{i+1} &= -\frac{\Delta t}{2}\boldsymbol{C}^{*n+1}\boldsymbol{U}_i^{*n+1} + \boldsymbol{R}^n \\ \boldsymbol{U}_{i+1}^{*n+1} &= \alpha\boldsymbol{X}_{i+1} + (1-\alpha)\boldsymbol{U}_i^{*n+1}\end{aligned} \tag{7.25}$$

where i is the iteration index, $\boldsymbol{R}^n$ is the RHS of (7.20) and α is a relaxation parameter. The linear systems (7.25) may be solved by applying a frontal technique [10] or by means of a conjugate gradient method with a diagonal preconditioning.

It can be proved that the scheme converges for any value of Δt if $0 < \alpha < 1$, $\boldsymbol{M}$ and $\boldsymbol{K}$ are symmetric positive definite matrices and $(\boldsymbol{C} + \boldsymbol{C}^T)$ is a positive semidefinite matrix. However the convergence may be slow for large values of Δt.

The scalar transport equation (7.15) is solved by means of the Crank-Nicolson scheme

$$[\boldsymbol{M}_S + \frac{\Delta t}{2}(\boldsymbol{K}_S^n + \boldsymbol{C}_S^{n+1})]\boldsymbol{\phi}^{n+1} = [\boldsymbol{M}_S - \frac{\Delta t}{2}(\boldsymbol{K}_S^n + \boldsymbol{C}_S^n)]\boldsymbol{\phi}^n + \frac{\Delta t}{2}(\boldsymbol{f}_S^{n+1} + \boldsymbol{f}_S^n) \tag{7.26}$$

where matrix $\boldsymbol{K}_S$ is evaluated at $t = t_n$, while matrices $\boldsymbol{C}_S^{n+1}$ and $\boldsymbol{C}_S^n$ depend on the velocity fields $\boldsymbol{U}^{n+1}$ and $\boldsymbol{U}^n$, respectively. The nonsymmetric linear system (7.26) may again be solved by applying a frontal algorithm [9] or by means of the iterative scheme (7.25) where $\boldsymbol{C}_i^{*n+1}$ is replaced by $\boldsymbol{C}_S^{n+1}$.

7.4 Examples

Two examples were computed to validate the 3D code. The first case study was Jeffery-Hamel flow. Consider the flow of an incompressible fluid between two intersecting infinite vertical plates as represented in Figure 7.2.

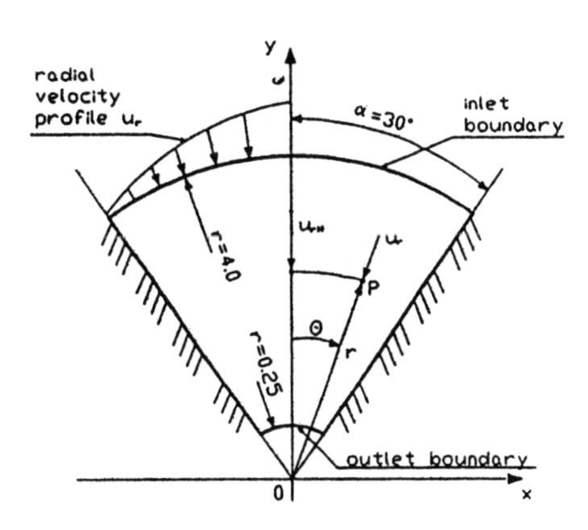

Figure 7.2: Geometry of the Hamel problem.

Introducing a cylindrical system of coordinates and imposing the boundary conditions

$$u_r(r,\theta) = 0 \text{ at } \theta = \pm\alpha, \; u_r(r,\theta) = u_{rM} \; or \; \frac{\partial u_r(r,\theta)}{\partial \theta} = 0 \; at \; \theta = 0 \tag{7.27}$$

the asymptotic analytic solution of the Navier-Stokes equations for large Re is [2,4,5]

$$u_r(r,\theta) = \frac{\nu Re\lambda}{\alpha r}, \quad p(r,\theta) = \frac{\rho\nu^2 Re}{\alpha r^2}\left(2\lambda - \frac{Re}{2\alpha} - 1\right) \tag{7.28}$$

with

$$\lambda = 3tanh^2\left[\sqrt{-\frac{\alpha Re}{2}}\left(1 - \frac{\theta}{\alpha}\right) tanh^{-1}\sqrt{\frac{2}{3}}\right] - 2; \quad Re = \frac{\alpha r u_{rM}}{\nu} \tag{7.29}$$

where Re represents the Reynolds number and u_{rM} is the centerline velocity (< 0).

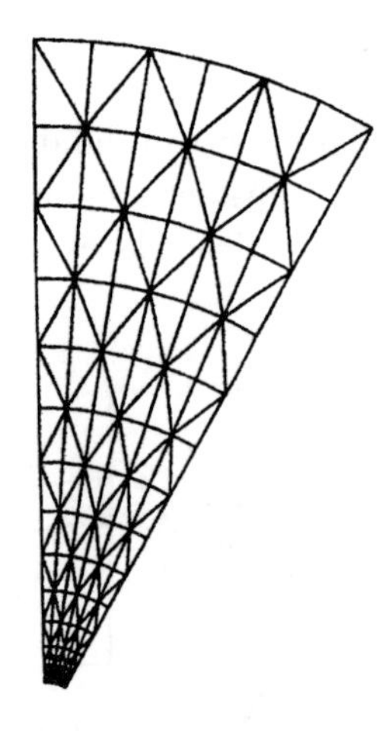

Figure 7.3: Computational domain

In order to solve the described problem with the MADIAN code, a 3D computational domain $\{r(0.25, 4); \theta(0°, 30°); z(0, 0.1)\}$ is discretized. The resulting mesh contains $2016 P1 - isoP2$ elements which correspond to 3651 velocity nodes and 714 pressure nodes (see Figure 7.3).

Assuming $\rho = 1000 kg/m^3$, $v = 2 \times 10^{-3} m^2/s$ and $Re = 1088$, the solution of the problem is computed starting from the solution of the associated Stokes problem and imposing the following boundary conditions: no slip at the wall; exact velocity profile deduced from (7.28) at the inlet ($r = 4$) and outlet ($r = 0.25$) boundaries; symmetry condition on the plane at $\theta = 0$. The main characteristics of the transient computation, run on a *CRAY-YMP*, are: $\Delta t = 5 \times 10^{-4}$ for $0 \leq t \leq 0.005$ and $\Delta t = 1.5 \times 10^{-3}$ for $0.005 \leq t \leq 0.1055$; iteration tolerance $\varepsilon = 10^{-4}$; the calculation required a mean number of iterations for Δt of 8.6; the solution of the linear systems (7.25) by means of the preconditioned conjugate gradient method required approximately 2.5 seconds per timestep while the mean CPU time per timestep was 8 seconds. The computed results are compared with the analytical solution (7.28) in Figure 7.4. The differences are approximately 4% for the velocity field and 2% for the pressure field.

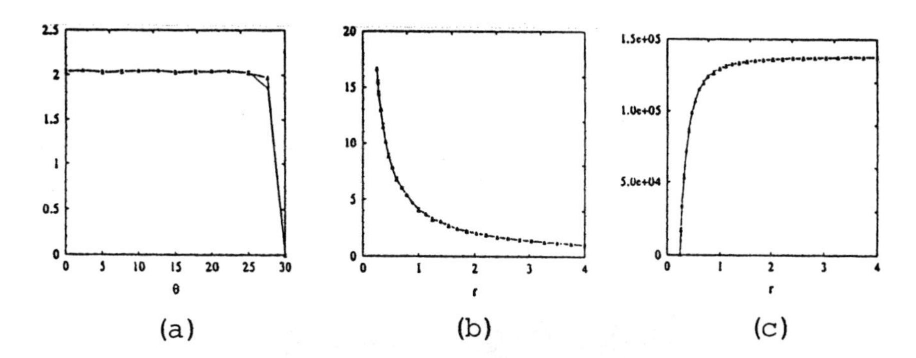

Figure 7.4: Comparison between the computed results(x symbol) and the analytical solution (solid line): (a) azimuthal velocity profile at $r = 2.0357$; (b) radial velocity profile at $\theta = 0°$; (c) radial pressure profile at $\theta = 0°$.

It should be noted that only a small fraction of the CPU time is spent on the solution of the linear systems. This is due both to the optimization of the solver and to the moderate size of the problem and of the related matrices. Better performance should be obtained after the completion of the code optimization.

The second test case corresponds to a 2D surface buoyant jet represented in Figure 7.5. The computational mesh used to solve the problem with the MADIAN code is formed by 2151 $P1 - isoP2$ elements which correspond to 4438 velocity nodes and 797 pressure nodes (see Figure 7.6). Assuming $\rho_0 = 1000 kg/m^3$ and $\nu_0 \approx 10^{-6} m^2/s$ at $T_0 = 20°C$, a transient simulation has been performed starting from rest and imposing the following boundary conditions: $u = 0.11 m/s$ at the inlet discharge; zero total normal stress at the outlet; rigid lid approximation for the free surface; slip condition at the wall; symmetry condition on the remainder of the vertical plane. The Prandtl model with a constant mixing length l_0 in neutral conditions has been used for turbulence closure. The main characteristics of the transient computation, run on an Alliant FX-80 with 3 concurrent CPUs, are $\Delta t = 0.01s$ for $0 \le t \le 24s$ with iteration tolerance again $\varepsilon = 10^{-4}$; The mean number of iterations per timestep is 2.7. The solution of linear systems (7.25) and (7.26) by means of the preconditioned conjugate gradient method

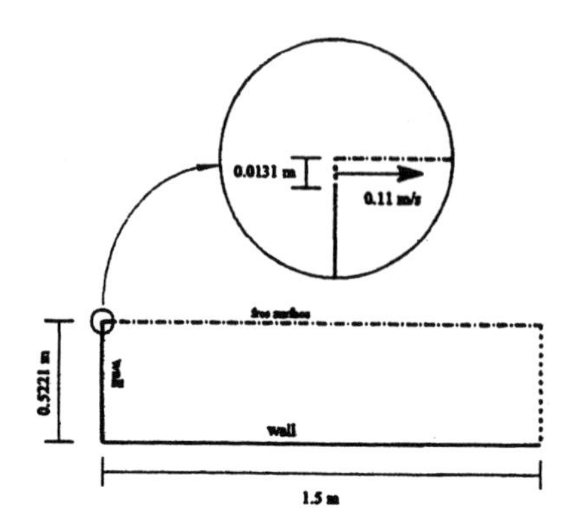

Figure 7.5: Geometry of the problem

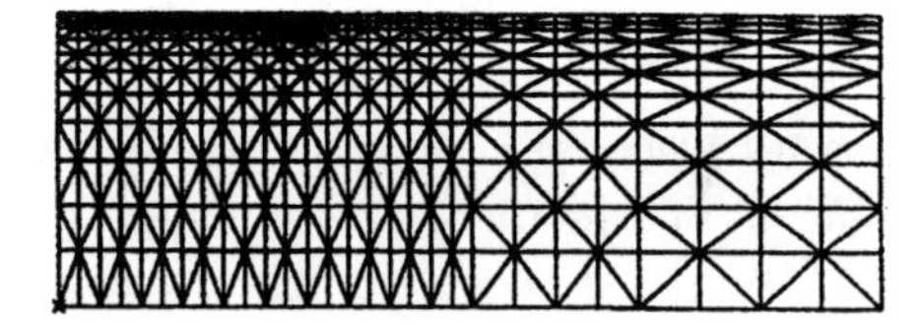

Figure 7.6: Computational mesh

required approximately 33 seconds per timestep while the mean CPU time per timestep was 118 seconds. The computed results at $x = 0.608m$ are compared with experimental data [1, 13] in Figure 7.7. The differences are probably due to the turbulence model and in particular to the choice of a constant l_0 instead of increasing l_0 with the jet width.

7.5 Conclusions

The finite element method provides a viable approach for the study of fully 3D flows. Efficient numerical schemes, capable of exploiting the capabilities of vector and parallel supercomputers, are essential to practical computations. Our future work will deal with optimization of the software for distributed memory computers, use of more accurate and complex turbulence models, and the application of the code to a wide class of environmental problems.

A contribution to this last item is now in progress within the MED-COAST project, partly founded by Directorate General XII/E of the Com-

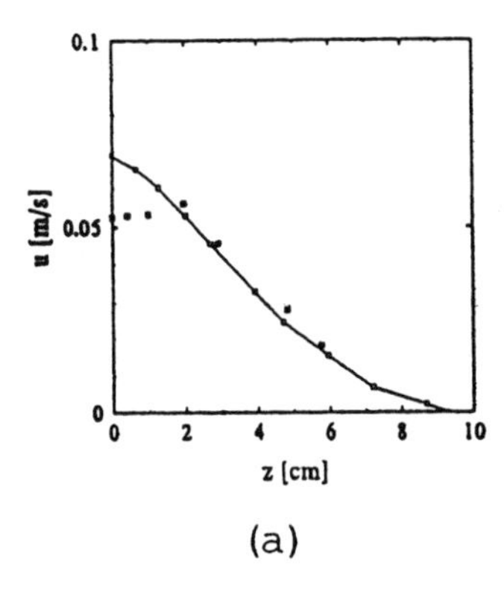

(a)

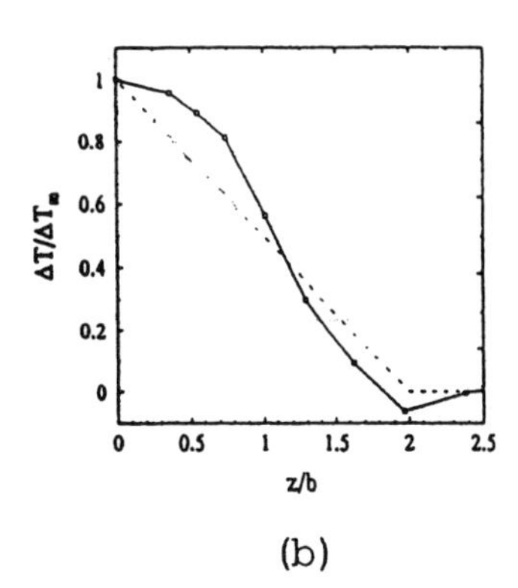

(b)

Figure 7.7: Comparison between the computed results (solid lines) at $x = 0.608m$ and experimental data: (a) velocity vertical profile; (b) normalized temperature vertical profile.

mission of the European Communities through the MAST 2 programme. Here the MADIAN code is to be applied to the simulation of complex hydrodynamic situations occuring in deep Mediterranean coastal waters.

The present work is currently being applied in studies of the deep coastal area of the western Mediterranean Sea from an oceanographic point of view. Preliminary results have been obtained on the CRAY-YMP using initial and boundary data from experimental measurements. These computations indicate a longshore current in geostrophic equilibrium under the combined influence of coriolis force with vertical and horizontal temperature and salinity gradients. These results are in good agreement with data and results from computations with other methods. Further application studies are in progress and will be reported later.

Acknowledgments: The authors would like to thank C. Amerio and E. Bon for their substantial work in the MADIAN code development.

References

[1] Baddour, R.E and Farghaly, H.A., "Surface thermal plume in channel", *J. Hydr. Eng.*, 115,7, 869-886, 1989.

[2] Batchelor, G.K., *An Introduction to Fluid Dynamics*, Cambridge University Press, 1977.

[3] Bercovier, M. and Pironneau, O., "Error estimate for finite element method solution of the Stokes problem in the primitive variables", *Num. Math.*, 33, 211-224, 1979.

[4] Engelman, M.S., Sani, R.L. and Gresho, P.M., "The implementation of normal and/or tangential boundary conditions in finite element codes for incompressible fluid flows", *Int. J. Num. Methods Fluids*, 2, 225-238, 1982.

[5] Gartling, D.K., Nickell, R.E. and Tanner, R.I., "A finite element convergence study for accelerating flow problems", *Int. J. Num. Methods Fluids*, 11, 1155-1174, 1977.

[6] Gill, A.E., *Atmosphere-Ocean Dynamics*, Academic Press, London, 1982.

[7] Gray, D.D. and Giorgini, A., "The validity of the Boussinesq approximation for liquids and gases", *Int. J. Heat Mass Transfer*, 19, 545-551, 1976.

[8] Gresho, P.M., "On the theory of semi-implicit projection methods for viscous incompressible flow and its implementation via a finite element method that also introduces a nearly consistent mass matrix. Part 1: Theory; Part 2: Implementation", *Int. J. Num. Meth. in Fluids*, 11, 587-620 (Part 1), 621-659 (Part 2), 1990.

[9] Hood, P., "Frontal solution program for unsymmetric matrices", *IJNME*, 10, 379-399, 1976.

[10] Irons, B.M., "A frontal solution program for finite element analysis", *IJNME*, 2,5-32, 1970.

[11] Janin, J.M., Lepeintre, F. and Péchon, P., "TELEMAC-3D: A finite element code to solve 3D free surface flow problems", in *Computer Modelling of Seas and Coastal Regions*, Patridge P.W. (ed.), Computational Mechanics Publications, Southampton, 1992.

[12] Orlob, G.T., *Mathematical Modelling of Water Quality: Stream, Lakes and Reservoirs*, John Wiley & Sons, 1983.

[13] Rajaratnam, N. and Humphries, J.A., "Turbulent non-buoyant surface jets", *J. Hydr. Res.*, 22,2, 103-115 1984.

[14] Rodi, W., *Turbulence Models and their Applications in Hydraulics - A State of the Art Review*, IAHR Publication, Delft, The Netherlands, 1980.

[15] Wu, J., "Wind-stress coefficients over sea surface near neutral conditions - a revisit", *J. Phys. Oceanogr.*, 10, 727-740, 1980.

Chapter 8

Po River Delta Flow

V. Pennati[1] *and S. Corti*[1]

8.1 Introduction

There are several very important environmental issues associated with flow and transport at the mouth of a river. These include movement of suspended sediments, the discharge distribution among the river branches, the simulation of the intrusion of sea water due to tidal waves, the dispersion of residual heat of power plants and modeling the transport of pollutants. Fundamental to analysis of the above phenomena is simulation of the flow. This is very often based on the two-dimensional shallow water equations.

Numerous investigators have contributed to the development of two-dimensional models for the numerical approximation of the shallow water equations both in conservation and non-conservation forms [27]. The most popular methods used for the discretization of the problem are based on finite differences (e.g. see [1, 7, 9–11, 13, 14, 18, 21, 25, 28, 31, 32]). Recently alternative approaches such as finite volume or spectral element methods have been proposed [4, 8, 15, 20, 24, 33]. There currently exist a considerable number of finite element studies of this subject (e.g. see [2, 3, 12, 17, 19, 23, 26, 29]). Finite element schemes are particularly appropriate because it is possible to represent very general domains, such as the basin of a river or lakes.

In this chapter we present some numerical results related to the flow in the Po river delta obtained by means of a shallow water finite element model and compare these results with field measurements.

[1]ENEL-Cris, Via Ornato 90/14, 20162 Milano, Italy

8.2 Problem Formulation

The numerical results of this study have been obtained by solving the conservation form of the shallow water equations written in terms of the unit width discharges

$$\frac{\partial p}{\partial t}+\frac{\partial}{\partial x}\left(\frac{p^2}{h}+\frac{1}{2}gh^2\right)+\frac{\partial}{\partial y}\left(\frac{pq}{h}\right)+S_x-gh$$
$$-\frac{\partial T_{xx}}{\partial x}-\frac{\partial T_{yx}}{\partial y}-gh\frac{\partial z}{\partial x}=0 \qquad (8.1)$$

$$\frac{\partial q}{\partial t}+\frac{\partial}{\partial x}\left(\frac{pq}{h}\right)+\frac{\partial}{\partial y}\left(\frac{q^2}{h}+\frac{1}{2}gh^2\right)+S_y$$
$$-\frac{\partial T_{xy}}{\partial x}-\frac{\partial T_{yy}}{\partial y}-gh\frac{\partial z}{\partial y}=0 \qquad (8.2)$$

$$\frac{\partial h}{\partial t}+\frac{\partial p}{\partial x}+\frac{\partial q}{\partial y}=0 \qquad (8.3)$$

where density ρ has been assumed constant, g is the acceleration due to gravity, h is the water depth, z is the water level from a reference plane, p and q are the unit width discharges in the x and y directions respectively. S_x and S_y represent the bottom friction terms in the x and y directions respectively, and T_{xx}, T_{xy}, T_{yx} and T_{yy} are the components of the stress tensor of the internal forces due to turbulence.

The bottom friction is estimated using the equations

$$S_x = \frac{gp}{C^2h^2}\left(p^2+q^2\right)^{\frac{1}{2}} \qquad (8.4)$$

$$S_y = \frac{gq}{C^2h^2}\left(p^2+q^2\right)^{\frac{1}{2}} \qquad (8.5)$$

where C is the Chezy coefficient.

The internal stress components are defined by

$$T_{xx} = 2h\nu\frac{\partial}{\partial x}\left(\frac{p}{h}\right) \qquad (8.6)$$

$$T_{yy} = 2h\nu\frac{\partial}{\partial x}\left(\frac{q}{h}\right) \qquad (8.7)$$

$$T_{yx}=T_{xy} = h\nu\left[\frac{\partial}{\partial y}\left(\frac{p}{h}\right)+\frac{\partial}{\partial y}\left(\frac{q}{h}\right)\right] \qquad (8.8)$$

where ν is the turbulent viscosity. In the present model ν can be defined in several ways: e.g. (1) in terms of the rate of change of the momentum in the direction of the flow, or in the transverse direction; or (2) proportional to the local average velocity times a characteristic subdomain length; or (3) finally, by means of a large eddy model. From a mathematical point of view equations (8.1) - (8.3) form an incomplete parabolic system and the number and type of boundary conditions to be imposed depend on the nature of the flow and on the physical features of the boundaries.

In the limiting case $\nu = 0$, we assign at the inflow boundary two conditions for subcritical flow, i.e. the discharges p and q or p_n and q_t (the discharge components normal and tangential to the boundary, respectively) or q_t and ξ. In the case of supercritical inflow, three conditions have to be imposed: p and q (or p_n and q_t) and the elevation ξ.

At the outflow boundary only one condition has to be imposed in the case of subcritical flow. This can be the elevation ξ, or the discharge p_n, or a product of p_n and h or a radiation condition. For supercritical flow no outflow condition is required. On a closed boundary, such as a coast or a channel wall, we impose a slip condition p_n=0 or that the discharges p and q (or p_n and q_t) are null.

If we consider also the diffusion terms, a further condition is to be prescribed on the boundary; e.g., the normal and tangential components of the viscous stresses are equal to zero where p_n and q_t are unknown. Whenever a component of the discharge is fixed the viscous stresses are not computed because the corresponding algebraic equation is eliminated from the final system. This simplifies the calculation, but momentum conservation is not guaranteed near the boundary. The assignment of discharges not parallel to the coordinate axis is made by means of a suitable rotation matrix.

The physical domain can be partitioned using triangular or quadrilateral elements and in each isoparametric element the dependent variables are approximated by polynomial shape functions in the usual way. In particular, the elevation is approximated by piecewise-linear polynomials and the velocity by piecewise-quadratics to prevent the occurrence of spurious waves [30]. A semidiscrete Galerkin finite element formulation of the problem then leads to a system of ordinary differential equations in time which is integrated by means of the θ method with $0.5 \leq \theta \leq 1$. Numerical integration on the reference element is carried out by means of the Gauss or Hammer methods for quadrilateral or triangular elements respectively.

The system of non-linear algebraic equations generated at each timestep is solved iteratively using a Newton-Raphson technique with Frontal elimination for the nonsymmetric Jacobian matrix system. A modified Newton scheme in which the Jacobian matrix is constant through the timestep or is evaluated periodically is also available. Numerical experience has shown that the method converges if the time steps are sufficiently small and the flow changes slowly in time. The above scheme is implemented in the code MONOS [6,16].

8.3 The Po River Delta

In the study presented here we simulate the flow in the last section of the Po river, i.e. the last 20 km of its course. The average width of the river in this section is about 400 m and the mouth is comprised of five branches. The depth varies between 20 m and a few centimeters and in some zones the river bed presents steep variations. After the first branch the river has a sharp elbow where the river width diminishes and the greatest depth (bathymetry) is attained which implies that velocities and discharges may assume large values.

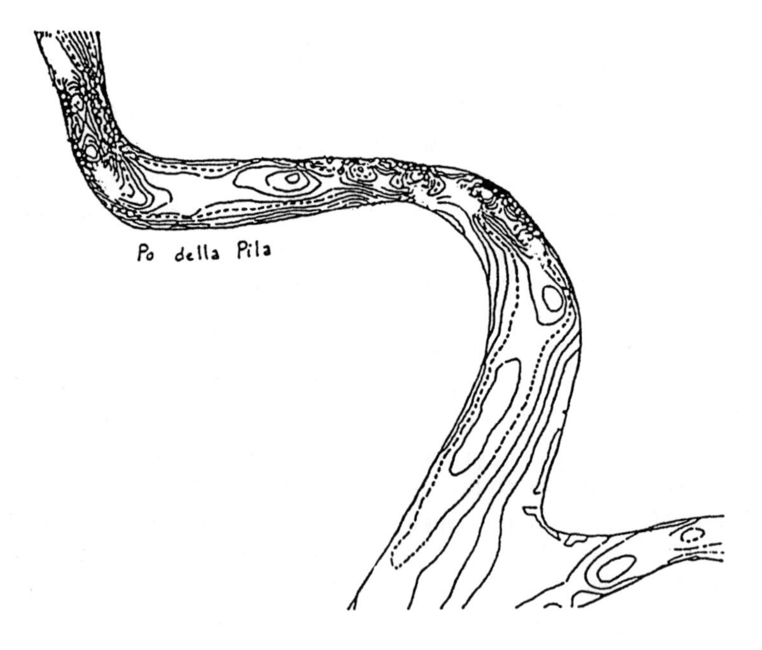

Figure 8.1: Detail of the bathymetry.

The discharges can vary between 300 m^3/s when the flow rate is lowest and several thousand m^3/s when the flow rate is a maximum. In the situation corresponding to low discharge, the combined action of the river flow and of the tide can lead to flow reversal at some of the delta mouths, so that instead of outflow there is inflow. For this reason the salt wedge can extend over a long section. For example, when the discharge is 400 m^3/s the salt wedge can extend as far as 17 km upstream.

We studied the transient behavior for a 72 hour period from midday 12-17-1974 to midday 12-20-1974. The discharge was approximately 500 m^3/s. A large amount of recorded field data, related to velocities, discharges, elevations, temperatures and salt concentration, was also available for this time. By means of this data it was possible to validate the numerical simulation scheme and program.

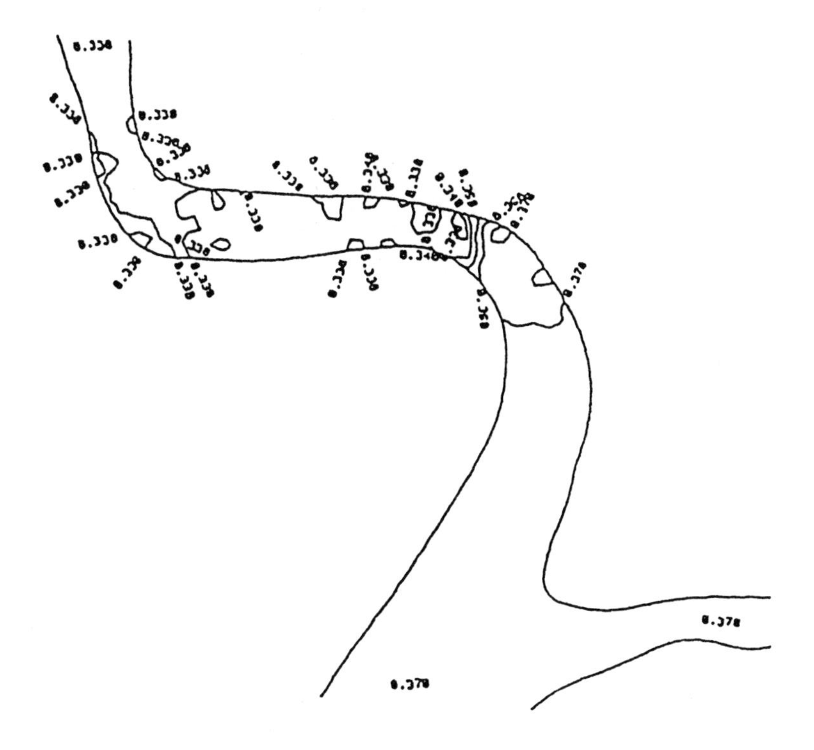

Figure 8.2: Elevation contours near the elbow of the Po river delta at 12 noon 12-18-1974.

8.4 Numerical Study

For the spatial discretization we used a mesh of 10602 triangular elements and 23563 nodes. The depth was automatically assigned at every node us-

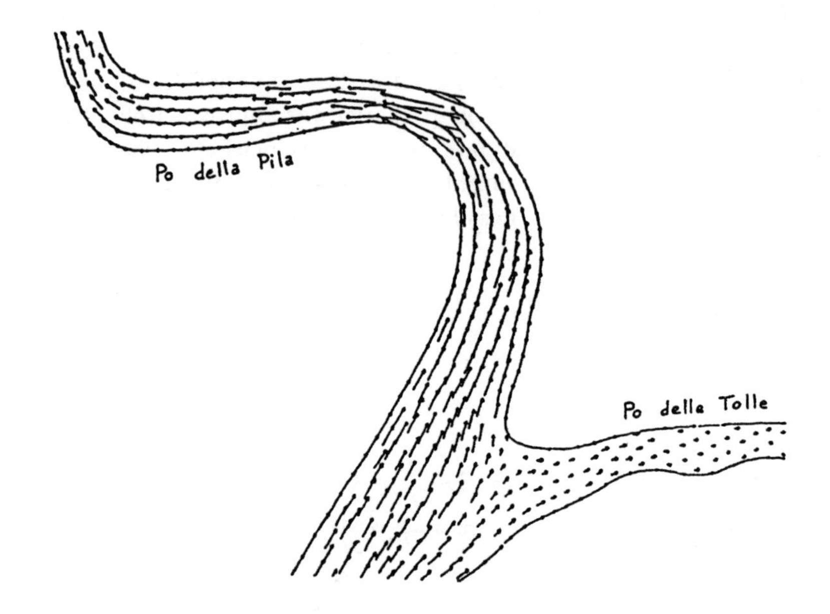

Figure 8.3: Velocity field near the elbow of the Po river delta at 12 noon 12-18-1974.

ing the Amitie code [22], using a least square approximation of a surface sampled at a finite scattered number of points by means of second order triangular and quadrilateral finite elements. The input for Amitie was the depth of the river-bed measured along 240 traversal sections, in 1972. In Figure 8.1 some details of the bathymetry are shown. Boundary conditions were chosen to correspond to the evolution of the flow as governed by the tidal variation with measured elevations approximated by means of truncated Fourier expansions at the open boundaries. On the inflow boundary the tangential discharge q_t was set to be zero. On the wall we assigned a full slip condition; i.e. the normal discharge p_n is zero. The initial condition was obtained by simulating a transient over 18000 s, developing from absolute rest [5]. Time integration parameter θ was 0.7 with the time step 900 s. The Chezy coefficient was 50 $m^{\frac{1}{2}}/s$ and viscosity coefficient ν=10 m^2/s; the latter value was chosen with the stability conditions required by the discretization of convective terms in mind.

In Figure 8.2 the elevation contours calculated by MONOS are reported. These results correspond to the physical measurements at 8 p.m. on 12-18-1974 for the region in Figure 8.1. As can be easily seen, they show a steep variation of the elevation, which suggests the presence of high values for the velocity. For comparison, in Figure 8.3 the velocity field at the same instant for the zone of Figure 8.2 is given and agrees with expectations.

8.5 Conclusions

In this chapter we have presented some results for the Po river delta flow obtained using the finite element code MONOS. The numerical results are in good agreement with measured elevations. Furthermore, the velocity field appears qualitatively correct. However we emphasize the intrinsic difficulty for the shallow water equations in treating a strongly irregular bed. Improvements would be expected if the bottom friction was better modeled by changing suitably the Chezy coefficient. Work is in progress for modeling the salt and heat transport.

Acknowledgments: We wish to thank our colleagues ENEL-Cris-Uii and ENEL-Cris-Uat for their collaboration.

References

[1] Abbot, M. B., *Computational hydraulics: Elements of the theory of free-surface flows*, Pitman, London, 1979.

[2] Agoshkov, V. I., D. Ambrosi, V. Pennati, A. Quarteroni and F. Saleri, "Mathematical and numerical modeling of shallow water flow", *Computational Mechanics*, 11, 280-299, 1993.

[3] Agoshkov, V. I., D. Ambrosi, E. Ovtchinnikov, V. Pennati and F. Saleri, "Finite element, finite volume and finite differences approximation to shallow water equations", in *Finite Elements in Fluids*, K. Morgan et al. (eds.), Pineridge Press, 1993.

[4] Alcrudo, F., and P. Garcia Navarro, "Computing two-dimensional flood propagation with a high resolution extension of Mc Cormack's method", in *Modeling of flood propagation over initially dry areas*, P. Molinaro et al. (eds.), New York, 1994.

[5] Ambrosi, D., S. Corti, V. Pennati and F. Saleri, " A two- dimensional numerical simulation of the Po river delta flow", in *Modeling of flood propagation over initially dry areas*, P. Molinaro et al. (eds.), New York, 1994.

[6] Barberi, A., "Simulazione di Onde a Fronte Ripido Su Fondo Asciutto", ENEL-Cris, Internal Report 3492, 1987.

[7] Bellos, C. V., J. V. Soulis and J.G. Sakkas, "Computation of two-dimensional dam-break induced flows", *Adv. Water Resources*, 14, 1, 31-41, 1991.

[8] Benkhaldoun, F., and L. Monthe, "An adaptive nine points finite volume Roe scheme for two-dimensional Saint-Venant equations", in *Modeling of flood propagation over initially dry areas*, P. Molinaro et al. (eds.), New York, 1994.

[9] Benque, J. P., J. A. Cunge, J. Feuillet, A. Hauguel and F. M. Holly, "New methods for tidal current computation", *J. ASCE*, 108,396-417, 1982.

[10] Casulli, V., "Semi-implicit finite difference method for the two- dimensional shallow water equations", *J. of Comp. Physics*, 86, 56-73, 1990.

[11] Cetina, R., and R. Rajar, "Two-dimensional modeling of flow in the river Sava", in *Modeling of flood propagation over initially dry areas*, P. Molinaro et al. (eds.), New York, 1994.

[12] Chen, C. L., and K. K. Lee, "Great lakes river estuary hydrodynamic finite element model", *J. of Hydraulic Engineering*, 117, 11, 1531-1550, 1991.

[13] Cunge, J. A., F. M. Holly and A. Verwey, *Practical aspects of computational river hydraulics*, Pitman, London, 1980.

[14] De Goede, E. D., "A time splitting method for the three-dimensional shallow water equations", *Int. J. Num. Meth. Fluids*, 13, 519-534, 1991.

[15] Di Gianmarco, P., and E. Todini, "A control volume finite element method for the solution of two-dimensional overland flow problems", in *Modeling of flood propagation over initially dry areas*, P. Molinaro et al. (eds.), New York, 1994.

[16] Di Monaco, A., and P. Molinaro, "A finite element two-dimensional model of free surface flows: verification against experimental data for the problem of the emptying of a reservoir due to dam-breaking", in *Adv. Water Resources*, D. Ouazar et al. (eds.), Morocco, 1988.

[17] Galland, J. C., N. Goutal and J. M. Hervouet, "Telemac: a new numerical model for solving shallow water equations", *Adv. Water Resources*, 14, 3, 138-148, 1991.

[18] Hauser, J., H. G. Paap, D. Eppel and A. Mueller, "Solution of shallow water equations for complex flow domains via boundary fitted coordinates", *Int. J. Num. Meth. Fluids*, 5, 727-744, 1985.

[19] Katopodes, N. D., "Two-dimensional surges and shocks in open channels", *J. Hydraulic Engineering*, 110, 6, 794-812, 1984.

[20] Ma, H., "A spectral element basin model for the shallow water equations", *J. Comp. Physics*, 109, 133-149, 1993.

[21] Montefusco, L., and A. Valiani, "A comparison between computed and measured bed evolution in a river bend", in *Modeling of flood propagation over initially dry areas*, P. Molinaro et al. (eds.), New York, 1994.

[22] Pennati, V., *Amitie: Approssimazione Ai Minimi Quadrati Per Mezzo di Elementi Finiti Isoparametrici del Secondo Ordine di una Funzione Definita su di una Regione Del Piano*, PhD Thesis, Università di Milano, Milano, 1978.

[23] Pironneau, O., *Méthodes des éléments finis pour les fluides*, Masson, 1988.

[24] Priestly, A., "A quasi Riemannian method for the solution of one-dimensional shallow water flow", *J. Comp. Physics*, 106, 139-146, 1993.

[25] Szymkiewicz, R., "Oscillation-free solution of shallow water equations for non staggered grid", *J. Hydraulic Engineering*, 119, 10, 1118-1137, 1993.

[26] Tabuenca, P., and J. Cardona, "Numerical model for the study of hydrodynamics on bays and estuaries", *Applied Mathematical Modeling*, 16, 78-85, 1992.

[27] Tan, W., *Shallow water hydrodynamics*, Elsevier, Amsterdam, 1992.

[28] Vreugdenhil, C. B., *Numerical Methods for Shallow Water Flow*, Von Karman Institute for Fluid Dynamics, Lecture series 1990-03, 1990.

[29] Walters, R. A., "Numerically induced oscillations in finite element approximations to the shallow water equations", *Int. J. Num. Meth. Fluids*, 3, 591-604, 1983.

[30] Westerink, J. J., R. A. Luettich Jr., J. K. Wu and R. L. Kolar, "The influence of normal flow boundary conditions on spurious modes in finite element solutions to the shallow water equations", *Int. J. Num. Fluids*, 18, 1021-1060, 1994.

[31] Wilders, P., T. L. Van Stijn, G. S. Stelling and G. A. Fokkema, "A fully implicit splitting method for accurate tidal computations", *IJNME*, 26, 2707- 2721, 1988.

[32] Wubs, F. W., *Numerical Solution of the Shallow Water Equations, PhD Thesis*, Amsterdam, 1987.

[33] Zhao, D. H., H. W. Shen, G. Q. Tabios III, J. S. Lai and W. Y. Tan, "Finite volume two-dimensional unsteady flow model for river basins", *J. Hydraulic Engineering*, 120, 7, 863-883, 1994.

Chapter 9

Shallow-Water and Transport Modeling of the Sydney Deepwater Outfalls

B.A. O'Connor[1] *and B. Cathers*[2]

9.1 Introduction

The two most commonly employed methods in computational hydraulics today are the finite difference technique (with its point-wise approximations to spatial derivatives) and the finite element technique (with its piecewise approximations to spatial derivatives). Progress in finite element methodology post-dates that in finite differences by about two decades. In 1973 Grotkop formulated a finite element model based upon the shallow water equations to investigate tidal motion in the North Sea. Since then many models have been devised mostly based upon the shallow water equations or the wave equation. However, it is only in relatively recent times that the finite element method has been applied extensively to practical environmental problems involving free surface flows.

In this chapter several aspects of practical interest are examined using a commercially available finite element program in an environmental investi-

[1]Department of Civil Engineering, University of Liverpool, Brownlow Street, PO Box 147, Liverpool, L69 3BX, UK.

[2]Department of Civil and Mining Engineering, University of Wollongong, Northfields Avenue, Wollongong, NSW 2522, Australia.

gation. We then discuss some theoretical work concerned with the spurious numerical effects associated with finite element mesh selection.

9.2 The Finite Element Model

The particular finite element code employed is known as RMA-10 and employs a three dimensional model which is capable of simulating stratified flows [7]. There are varying grades of "three-dimensionality" in shallow water models so that model capability should be carefully specified. More specifically, RMA-10 solves two horizontal momentum equations, a vertically integrated continuity equation, the transport diffusion equation for salinity and the equation of state which links density to salinity and temperature. Pressure in the vertical direction is hydrostatic. The dependent variables are the two horizontal velocity components, the depth and the salinity. The vertical velocity is recovered from the 3D version of the continuity equation.

An attractive feature of the model is the capability to reduce the dimensionality from three dimensions to two or even one. Hence in those areas where the flows and the density structure are homogeneous, the dimension can be appropriately reduced. This facility offers the potential for significant savings in run time.

In the present investigation, the model was used in 2D depth-averaged mode: The dependent variables are the two velocity components and the depth. The governing equations are the shallow water equations which include the lateral shear stress terms in the two horizontal momentum equations. As experience with the models progresses and as more data becomes available, it is planned to incorporate model operation in the 3D mode at a later stage of the study.

9.3 The Sydney Deepwater Outfalls

As part of an environmental monitoring program for Sydney's three deepwater outfalls (at North Head, Bondi and Malabar), a numerical modeling investigation was undertaken (see Figure 9.1). A suite of models was employed to simulate the transport and diffusion of the effluent. Near-field models predicted the concentrations and movement of the plumes of effluent from a depth of about 60m to the water surface or to a level of neutral buoyancy at the trapping level. The far-field models were then used to transport the plumes by the typically shore-parallel East Australia Current.

The far-field model utilized the finite element code RMA-10 in 2D depth-averaged mode.

Data on the near shore ocean was provided by an instrumented buoy (referred to as the Ocean Reference Station or ORS) located in a depth of 65m, 3km off the coast. The ORS was tethered in a location which was central to the region of interest near the Bondi outfall (see Figure 9.1a) and due to its cost of construction and operation, was the only one built. This buoy was equipped with two anemometers (wind speed and direction), a thermistor string (water temperatures between depths of 15m and 50m) and a current meter (ocean currents at depths of 15m and 50m). The aforementioned parameters were and are logged continuously.

One of the finite element grids for simulating the far field movement of the effluent is also indicated in Figure 9.1a. The modeling domain has three open boundaries and one land boundary on the western side. A desirable property of a finite element mesh is that it be isotropic and homogeneous with respect to the Courant number $C = \frac{\sqrt{gh}\Delta t}{\Delta s}$ where h is the depth, Δt is the time increment and Δs is a measure of the size of the 2D finite element. That is, if the ratio $\frac{\sqrt{h}}{\Delta s}$ is uniform throughout the mesh, the Courant number will be spatially uniform and hence also the numerical properties of the solution (such as phase speed, wave amplification and group velocity). This point is further discussed in Section 9.5.

A number of field verification exercises were carried out in order to validate the output from the various near and far field models. These exercises involved boats deployed in the modeling region collecting data on the plume location and tracer concentrations. This data was supplemented with information on the currents which was remotely sensed.

Praagman [8] has determined the required boundary conditions for a 2D depth-averaged model based on the number of characteristics entering the modeling domain. He has shown that a valid set of boundary conditions may be specified as follows: Along closed boundaries, the normal velocity is zero. Along open boundaries with an outflow, only the water level is given. Along open boundaries with an inflow, two variables should be specified, namely the water level and the tangential velocity. As is usually the case in practice, a Froude number which is much less than unity signifies negligible convective accelerations and in these circumstances, the inflow open boundary condition simplifies to specification of the water surface only.

In the present study, velocities were defined along the 'closed' or land boundary and water surface elevation was defined around the 'open' boundaries. Along the western land boundary, the velocity normal to the boundary was set to zero and the tangential velocity was left free (full slip). The non-slip condition along this boundary was also tested but found to excessively retard the flow. In other words, the coastal boundary layer is much smaller than the typical (cross-shelf) element size and could not be resolved by the finite element mesh.

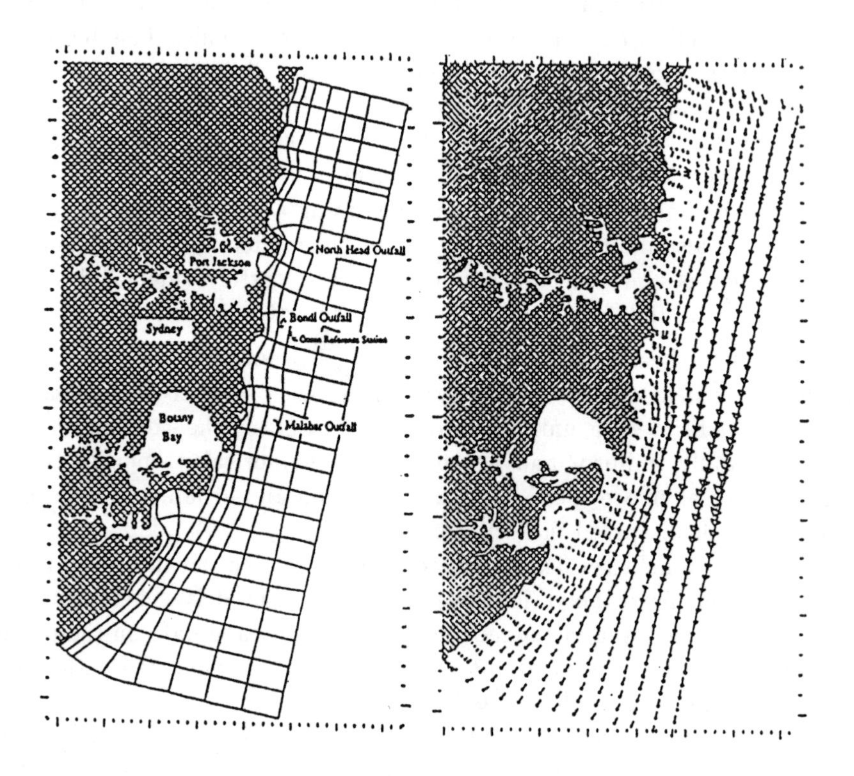

Figure 9.1: (a) Location of Sydney's deepwater outfalls and finite element mesh. (b) Computed velocity field.

For the routine, daily operation of the models, only data from the ORS was available. The finite element model was driven by prescribed surface elevations around the open edges of the mesh. It was necessary to extrapolate the point data from the ORS to the limits of the modeling domain through the use of the linearized, depth-averaged, shallow water momentum equations. This method is somewhat "heavy handed" but was considered satisfactory when the model predictions were compared with the measured field results.

At each time step, the surface elevation of every open boundary node was calculated as the sum of two small vertical offsets from contributions of the cross- and long-shelf water surface gradients applied over their distance from the ORS. Figure 9.1b shows the velocity field predicted by the model on the mesh depicted in Figure 9.1a.

9.4 Model Dimensionality

In order to provide a reasonably accurate simulation of the flow, a model must have the appropriate number of spatial dimensions. Traditionally, most engineering applications have adopted one of the following representations: A 1D cross-sectionally averaged formulation such as would be used in a river, a 2D depth-averaged approach for an estuary, or a 2D laterally averaged approach for a stratified river flow.

With the increasingly available computer memory and speed as well as the demands to apply modern technology to environmental problems, there are signs of a 'dimensional creep' in which increasingly complex models are being developed and employed. For example, the depth-averaged approach which has been in widespread use for some two decades, is not appropriate for upstream reaches where the salt and freshwaters in a tidal waterway may be stratified rather than well-mixed. A layered model should be used. Nor is the 2D depth-averaged approach suitable in deeper (especially wind driven waters), where the fluid velocities are significantly different in either magnitude or direction through the water column. Here, three-dimensional models are increasingly being applied.

In the deepwater outfalls project, approximately one year of data was available from the ORS. Recall that in a depth of about 65m, the ORS was equipped with the two S4 current meters 15m below the water surface and 15m above the sea bed. Also, 11 thermistors were vertically spaced through the middle half of the water column between the two current meters. Data from the two current meters and the thermistor string was processed in order to gauge the appropriateness of a 2D model formulation. Figure 9.2 shows temperatures from the thermistor string which indicate that the condition of the water column changes from stratified to unstratified [1].

Consideration was given to the difference between the following parameters measured at the ORS : Two current speeds (ΔU), two current directions ($\Delta \theta_U$), and the water temperatures near the location of the two current meters (ΔT).

If the current is uniform with depth and co-flowing through the water

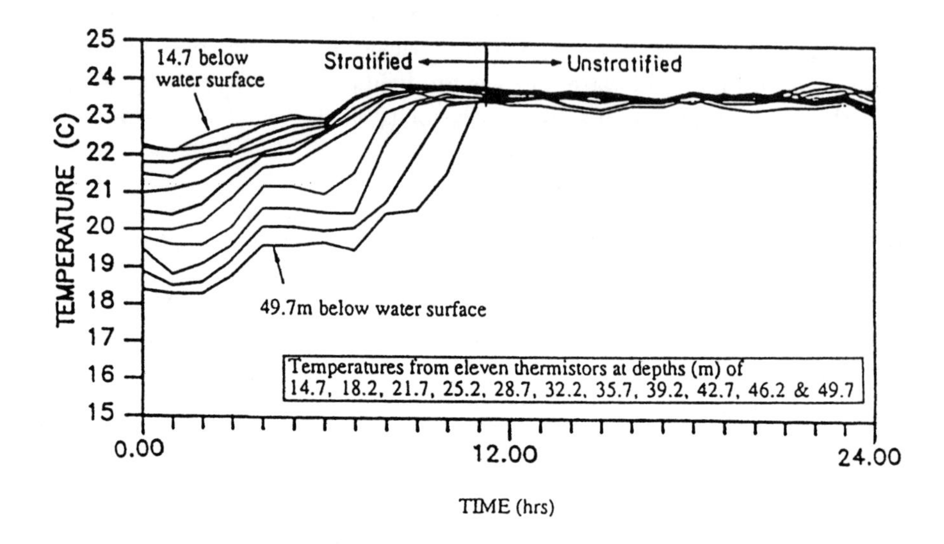

Figure 9.2: Change from stratified to homogeneous conditions through the water column.

column, a depth-averaged formulation is justified. These conditions were deemed to be satisfied if the following criteria hold: $\Delta U < 0.1$m/s, $\Delta\theta_U < 45^o$, and $\Delta T \leq 1^oC$.

When the ORS database was searched, it was found that the above conditions were simultaneously met only 7% of the time. This result was then used as part of the justification for initiating a 3D modeling approach to the nearshore oceanic flows near Sydney. The 3D modeling work is still preliminary and ongoing. The capability of the RMA-10 model to be also operated in 3D mode is a very useful feature. In general terms, the actual finite elements are changed from 2D triangular or quadrilateral elements to layers of triangular or quadrilateral prisms.

The RMA-10 model can also handle a mixture of 2D and 3D elements within the modeling domain. Hence, in future simulations, the nearshore ocean can be modeled using 3D elements. However, in the more shallow regions of bays and river entrances where the depths are less than about 30m, there are no thermoclines present and a 2D depth-averaged approach may well be adequate.

9.5 Mesh Considerations

In environmental modeling, the meshes in 2D numerical models are usually of two main types: regular grids of rectangles or topologically regular grids in curvilinear coordinates (typically employed in finite difference models), or irregular grids of triangles or quadrilaterals (typically in finite element and finite volume models). In most investigations, there is a region of particular interest in which finer spatial resolution is required. This can be accomplished by locally refining the mesh. In finite elements, this is easily accomplished, particularly if the elements are triangles. It is desirable to maintain element shapes which are regular and have aspect ratios which are (very) approximately unity in all directions. Such elements are more robust whereas slender elements tend to introduce inaccuracies and may promote numerical instability.

There are essentially two types of local meshing: passive and dynamic. In passive local meshing, the results from a simulation with a coarse grid are subsequently used to drive the refined mesh by providing boundary conditions for the refined mesh. Hence, a model run is required for both the coarse mesh and the fine mesh. The domain of the refined mesh overlaps a portion of the coarse mesh. The flow of information is one-way, being from the coarse mesh to the fine mesh. This approach is adequate provided: the processes being resolved in the refined mesh do not significantly affect the flows in the coarse mesh, and the boundary conditions provided by the coarse mesh are sufficiently accurate and well removed from the local area of interest within the fine mesh region.

In dynamic local meshing, the refined mesh is dynamically linked to the coarse mesh and the flow of information is two-way between both meshes. For this approach, only one model run is needed. Dynamic local meshing fits naturally with finite element methodology and it is tempting to think that it is always the best method and that there are no numerical repercussions. This, however, is not the case. A series of simple, mainly 1D numerical investigations based upon the Crank-Nicolson finite element scheme has been carried out to improve our understanding and to quantify some of the side-effects of dynamic local meshing [2–6]. The following discussion also applies to finite difference models containing different mesh sizes, but is more relevant to finite element models. It was mentioned in Section 9.3 that it is desirable that the ratio of $\frac{\sqrt{h}}{\Delta x}$ be approximately uniform in a FE mesh. It is clear that when a mesh is dynamically locally refined, there may be a sudden change in the value of this ratio and hence of the Courant number. This being the case, there will also be changes in the numerical parameters (phase speed, wave amplitude and group velocity) from the outside region to the inside, refined region.

Numerical Reflection and Transmission

The solution can be expressed as the summation of Fourier components. When a Fourier mode encounters a change in the mesh, there are a number of consequences. Consider the situation of an incident "wave" impinging on the interface between two 1D regions, each characterized by a uniform but different nodal spacing and which have a common node at the junction or interface. Physically, both regions are indistinguishable (i.e. same flow depth and same fluid). Apart from the incident wave, the particular system of reflected and transmitted waves which is established depends upon the form of the dispersion relation of the numerical scheme.

In the dispersion relation, the angular frequency is a function of the wavenumber of the Fourier components. Generally, the dispersion relations can be categorized as either 'hill-shaped' (Figure 9.3a) in which there are typically two wavenumbers corresponding to each angular frequency and there is a maximum angular frequency which marks the peak of the dispersion relation), or monotonic increasing (Figure 9.3c) in which there is just one wavenumber corresponding to each angular frequency.

Depending on the shape of the dispersion relation for the numerical scheme, the resulting wave system will consist of two waves apart from the incident wave. When the dispersion relation is 'hill-shaped', there will be two transmitted waves each with the same angular frequency. When the dispersion relation is monotonic increasing, a reflected wave and a transmitted wave will be established.

Figures 9.3b and 9.3d illustrate the various wave systems. If the behavior of the numerical model was ideal in that it perfectly matched the behavior of waves in the continuum, the incident wave would be transmitted across the interface. That is, there would be no reflected waves and only one transmitted wave. Hence local meshing or abruptly changing the nodal spacing can give rise to inaccurate numerical solutions from the model. Solution degradation is worsened the larger the change in mesh spacing from one side of the junction to the other and the closer the wavelengths of the incident Fourier modes are to either of the two mesh spacings. Good solutions which are relatively unaffected by the nesting or changes in mesh size, should have predominant Fourier modes with wavelengths which are an order of magnitude longer (say) than the nodal spacing on either side of the junction.

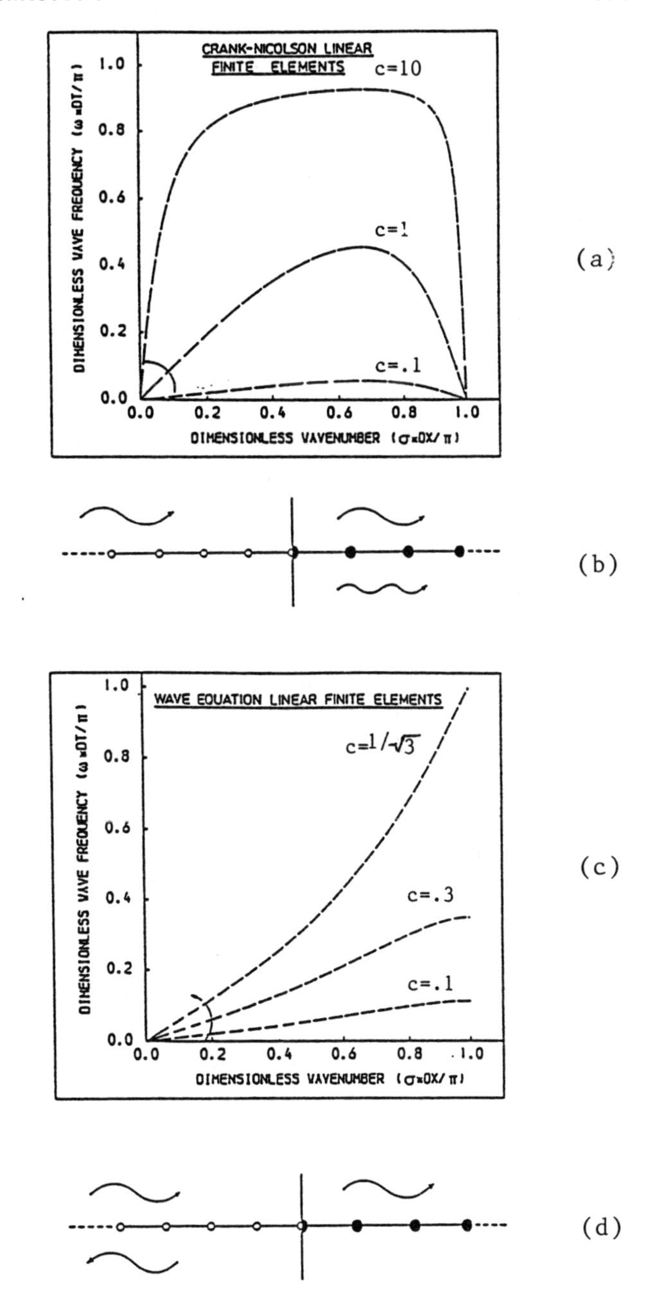

Figure 9.3: Dispersion relations and their corresponding wave systems due to a change in mesh size.

One particularly undesirable situation can arise for those numerical schemes with 'hill-shaped' dispersion relations. If the angular frequency of the incident wave is too high for the downstream region to resolve (since the angular frequency of the incident wave exceeds the maximum fequency for the dispersion relation in the downstream region), total internal wave reflection will occur. In this case, all the incident wave energy will be reflected due solely to the 'numerics' rather than the physics and hence such a result is entirely spurious.

Numerical Refraction

In 2D when a fine mesh is nested within a coarser mesh there is an additional wave process present and that is wave refraction [6]. Once again consider the case where the flow depths throughout the two regions are constant and equal. Due to the different mesh spacing in the coarse and fine meshes, the Courant numbers will be different. Since the speed of the waves depends upon the Courant number, the waves will travel with different phase speeds in the two regions. The result is that a wave impacting the nested region will undergo spurious numerical refraction.

The extent of the refraction depends upon several factors. These include the particular numerical scheme which will also depend upon the form of the shallow water equations, the approach angle with respect to the interface, the wavelength of the incident wave with respect to the mesh spacings in both regions, and the relative change in the mesh spacing.

Once again, the worst case possible is that, depending upon the factors just listed, the incident wave may undergo total internal reflection. This pathological case may occur when the angle of the approach of the incident waves is increased beyond that which results in the refracted wave moving parallel to the interface.

When wave refraction occurs, calculations for an energy budget across the interface show that kinetic plus potential energy is conserved. The wave energy is propagated in the direction of the wave ray which is defined by the two components of the numerical group velocity. This direction is distinct from that of the wave orthogonal which is in the direction of wave propagation and is orthogonal to the wave crest (see Figure 9.4). In the continuum, these two directions are the same; i.e. the wave orthogonal and the wave ray are in the same direction. That they are different in the discrete system, means that there is a spurious numerical diffraction which is present in both the fine and the coarse meshes.

Another consequence of the mesh nesting alluded to above is that when

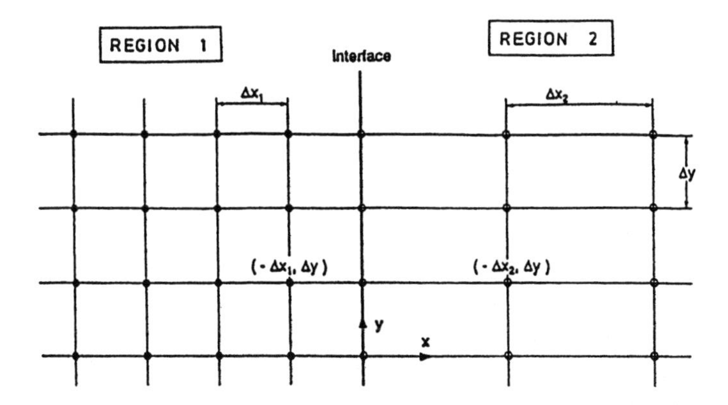

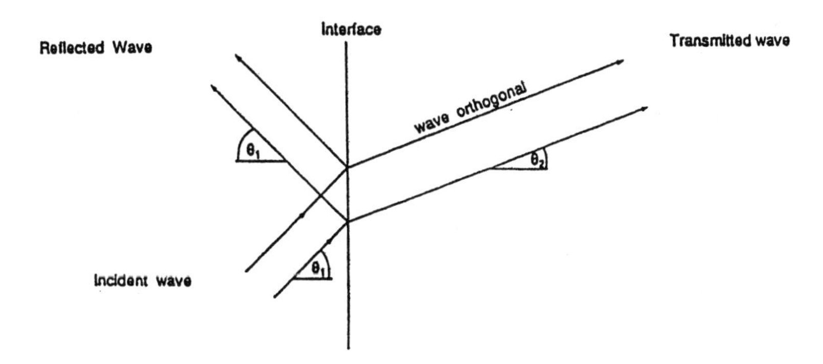

Figure 9.4: Unwanted numerical refraction and reflection of the incident wave in 2D due to a change in mesh size.

the Fourier wave passes through the intertace, the direction of *energy* propagation is altered. This process is akin to wave refraction in which the direction of wave propagation changes across the interface.

9.6 Composite Approaches

The governing equations for hydrodynamic numerical models applied to estuaries and nearshore oceans are usually based upon the continuity and momentum equations throughout the entire domain. Although models applied separately to river or flood flows are often based on the 1D versions of the continuity and momentum equations, this is not always appropriate.

Some flow situations may require the inclusion of hydraulic structures such as weirs, bridge piers, culverts, flap gates, energy dissipators or constrictions. Where these structures occur, the usual approach is to retain the continuity equation as one of the governing equations but to replace the momentum equation by a rating curve or an energy equation which is customized to the particular structure or flow situation. Such relations often contain a blend of empiricism and theory.

Rating curves are required to model weirs (submerged or not), culverts (whose hydraulic behavior can be grossly characterized as varying between a weir, pipe or orifice flow regime) and flap gates (one-way orifice flow). The energy equation or energy loss relation is used for flow through bridge piers, energy dissipators, flow constrictions and river confluences (particularly where there is a gross change in cross-sectional flow area or sharp changes in flow direction are involved).

Commercial software packages based upon finite difference formulations and which are able to cope with the flows mentioned above, already exist and there appear to be fewer composite finite element implementations that handle hydraulic structures in this manner. Of course, when 1D finite element models are used to simulate the hydrodynamics and water quality of a branched waterway, the connectivity of the waterway is automatically handled in a finite element formulation. All confluences are modeled based upon the continuity and momentum equations. This implies appropriate cross-sectional flow areas must be assigned at the mesh points, but no special internal boundary conditions (based on energy levels) are needed. In 2D, the same comments apply except that there is obviously no difficulty encountered with the flow cross-sections since only the flow depths are needed.

9.7 Conclusions

The advantages of the finite element formulation are well known. These include their capability to fit flow boundaries, to refine the mesh locally in areas of interest, and to handle branched channels routinely. Finite element codes are characterised by a high degree of modularity as contrasted wth finite difference codes. This means that 2D and 3D modules (for modeling regions of 2D and 3D flow) may be contained within the one model with common portions of the code being used by both modules. One aspect of the finite element approach which requires additional work is the use of different governing equations in different subregions (e.g. locally incorporating energy relations in place of the momentum equation).

Acknowledgements: The investigations described herein were supported by the NSW EPA and the Sydney Water Board and were carried out by AWACS Pty Ltd and Computational Mechanics International P/L. Their permission to include the project work is gratefully acknowledged. The project work described herein was mainly undertaken by W. Peirson, I. King, G. Tong, C. King and the second author.

References

[1] Cathers, B. and W.L. Peirson,"Sydney Deepwater Outfalls Environmental Monitoring Program, Commissioning Phase, Numerical Modeling", AWACS Report 91/01, 1992.

[2] Cathers, B., *Numerical Aspects of a Hydrodynamic Finite Element Model*, PhD Thesis, University of Manchester, 1986.

[3] Cathers, B., S. Bates and B.A. O'Connor, "Internal wave reflections and transmissions arising from a non-uniform mesh. Part I: An analysis for the Crank-Nicolson linear finite element scheme", *Int. J. Num. Meth. Fluids*, 9, 783-810, 1989.

[4] Cathers, B., S. Bates, R. Penoyre and B.A. O'Connor, "Internal wave reflections and transmissions arising from a non-uniform mesh. Part II: A generalized analysis for the Crank-Nicolson linear finite element scheme", *Int. J. Num. Meth. Fluids*, 9, 811-832, 1989.

[5] Cathers, B., S. Bates and B.A. O'Connor, "Internal wave reflections and transmissions arising from a non-uniform mesh. Part III: The occurrence of evanescent waves in the Crank-Nicolson linear finite element scheme", *Int. J. Num. Meth. Fluids*, 9, 833-853, 1989.

[6] Cathers, B. and S. Bates, "Spurious Numerical Refraction", submitted to *Int. J. Num. Meth. Fluids*, 1994.

[7] King, I.P., "A Finite Element Model for Three-Dimensional Density Stratified Flow, Post-Commissioning Phase", AWACS Interim Report 93/01/04, 1993.

[8] Praagman, N., *A Finite Element Solution of the Shallow Water Equations*, PhD Thesis, Delft University of Technology, 1979.

Chapter 10

Circulation and Salinity Intrusion in Galveston Bay, Texas

R.C. Berger[1]

10.1 Introduction

Galveston Bay is the largest and most productive estuary on the Texas Gulf coast (Figure 10.1). It is wide and shallow, predominately less than 6 ft in depth, and is incised by a 40-ft- deep, 400-ft-wide navigation channel from the Gulf of Mexico to the Port of Houston. The mean diurnal tide range is 1.4 ft, though wind strongly influences the actual tide. Freshwater inflow from its several tributaries averages about 13,000 cfs. The largest is the Trinity River, though future demands of the City of Houston will result in ever-increasing freshwater discharge to the bay via Buffalo Bayou. Mixing conditions range from partly to well-mixed, with winds significantly affecting both circulation and mixing processes. Navigation traffic demands have led to a proposed enlargement of the navigation channel by 10 ft in depth and 200 ft in width in two stages. Salinity and circulation changes are evaluated using a three-dimensional (3D) numerical model of the bay system. The model results are then used to drive an ecosystem model to predict the impact upon oyster production.

[1]USAE Waterways Experiment Station, Vicksburg, MS.

This chapter reviews the hydrodynamic and salinity modeling portion of this study. The initial section contains general background on navigation channels in an estuarine environment. The next section describes the RMA10-WES 3D numerical code used to produce the hydrodynamic results. The actual application is then described including the discretization, and the results.

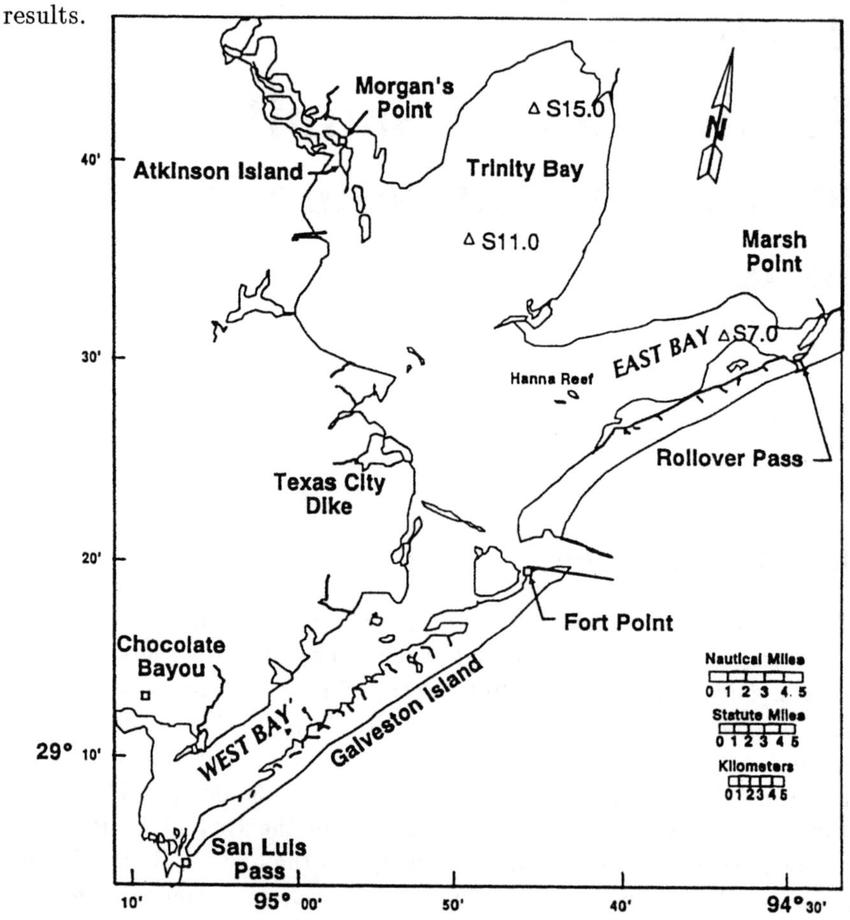

Figure 10.1: Location of Galveston Bay and three of the field data collection locations.

10.2 Estuarine Processes

Estuarine circulation is the result of tides, winds, upland discharge, density gradients and their interactions. The currents are principally driven by the rise and fall of significant diurnal and semi-diurnal tidal periods. Near new and full moon these are in-phase and produce large tide ranges, termed spring tides. At the first and third quarters the tide range is small, and these are termed neap tides. Currents that are directed upland are referred to as "flood" currents, and seaward currents are "ebb" currents. The lateral distribution of currents often contains areas that are either flood or ebb dominant, simply as a result of bathymetric variations or in combination with density gradients. Density variations tend to produce flood-dominant currents near bed, and ebb-dominant near the surface. Figure 10.2 shows a highly-stratified condition of an arrested salinity wedge. Here the near-bed currents are flood at all times and the near-surface are ebb at all times. There is a thin interface in which the saltwater and freshwater mix.

Most U.S. estuaries are partly-mixed to well-mixed with some tendency for the currents to flood near bottom and ebb at the surface but for most of the time they are near the same direction. Estuaries are more stratified during high freshwater inflow, low tide range and low wind magnitude. Conversely, they become more well-mixed for lower freshwater inflow, larger tide range and greater wind speed. These vertical circulation patterns can result in lateral variations as well, with more flood-dominant channels in deeper areas and ebb-dominance in the shallows. The presence of a deep trough from the Gulf in the form of a navigation channel then provides a path for saltwater to intrude upstream. Consequently, the navigation channel becomes a flood-dominant channel and the shallows more ebb-dominated. Figure 10.3 shows the isohalines from field records for the Gaudalupe estuary in Texas in the Fall of 1986. This bay contains no deep-draft navigation channel and the salinity concentration drops directly with distance upstream. The concentration is laterally uniform. Contrast this with the Galveston Bay isohalines for Fall 1986, also from field data (Figure10.4). The channel is a high salinity circulation pathway and the isohalines are parallel to the channel. Concentrations drop off most rapidly perpendicular to the channel. Therefore, any deepening might be expected to increase bay salinity, necessitating a model study to quantify the expected impact.

10.3 The Hydrodynamic Code

The geometric complexity of this estuary, with its navigation channel, multiple inlets, and many proposed disposal islands, requires a numerical model that relies upon an unstructured computational mesh. The code chosen is the Galerkin finite element model RMA10-WES, which is a Waterways

Experiment Station (WES) adaptation of the RMA-10 code [4]. This code computes time-varying open channel flow and salinity/temperature transport in 1D, 2D, and 3D. It invokes the hydrostatic pressure assumption. Vertical turbulence is supplied using a Mellor-Yamada Level 2 [5] $k - l$ approach modified for stratification by the method in [1]. The salinity/density relationship is based upon [7].

The full three-dimensional equations are reduced to a set of two momentum equations, an integrated continuity equation, a convection-diffusion equation and an equation of state. The simplification is a result of the hydrostatic pressure approximation. This leads to the system

$$\rho \frac{Du}{Dt} - \nabla \cdot \sigma_x + \frac{\partial P}{\partial x} - \Gamma_x = 0 \tag{10.1}$$

$$\rho \frac{Dv}{Dt} - \nabla \cdot \sigma_y + \frac{\partial P}{\partial y} - \Gamma_y = 0 \tag{10.2}$$

$$\frac{\partial h}{\partial t} + u_\zeta \frac{\partial \zeta}{\partial x} - u_a \frac{\partial a}{\partial x} + v_\zeta \frac{\partial \zeta}{\partial y} - v_a \frac{\partial a}{\partial y} + \int_a^\zeta \left(\frac{\partial u}{\partial x} + \frac{\partial v}{\partial y} \right) dz = 0 \tag{10.3}$$

$$\frac{Ds}{Dt} - \frac{\partial}{\partial x}\left(D_x \frac{\partial s}{\partial x} \right) - \frac{\partial}{\partial y}\left(D_y \frac{\partial s}{\partial y} \right) - \frac{\partial}{\partial z}\left(D_z \frac{\partial s}{\partial z} \right) = 0 \tag{10.4}$$

$$\rho = F(s) \tag{10.5}$$

Elevation-related terms are defined in Figure 10.5.

In the above equations, P = pressure, ρ = density, $u, v, w = x, y, z$ velocity components, h = depth, a = bed elevation, ζ = water surface elevation and s = salinity. The x, y velocity components at the water surface are denoted u_ζ, v_ζ and the x, y velocity components at the bed u_a, v_a. The stress components are given by

$$\sigma_x = \begin{bmatrix} E_{xx} & \dfrac{\partial u}{\partial x} \\ E_{xy} & \dfrac{\partial u}{\partial y} \\ E_{xz} & \dfrac{\partial u}{\partial z} \end{bmatrix} \qquad \sigma_y = \begin{bmatrix} E_{yx} & \dfrac{\partial v}{\partial x} \\ E_{yy} & \dfrac{\partial v}{\partial y} \\ E_{yz} & \dfrac{\partial v}{\partial z} \end{bmatrix} \tag{10.6}$$

and

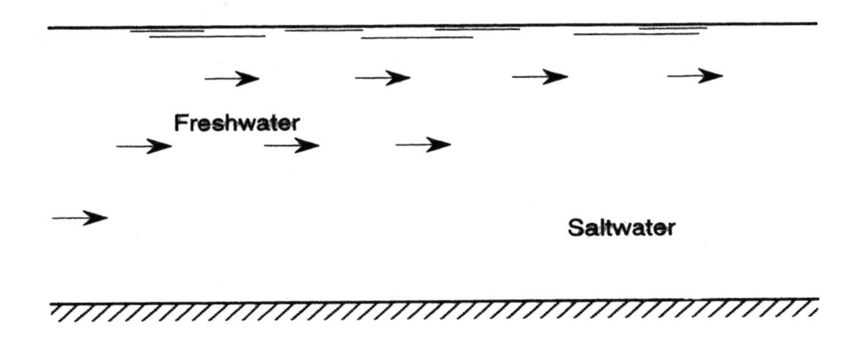

Figure 10.2: Description of arrested salinity wedge.

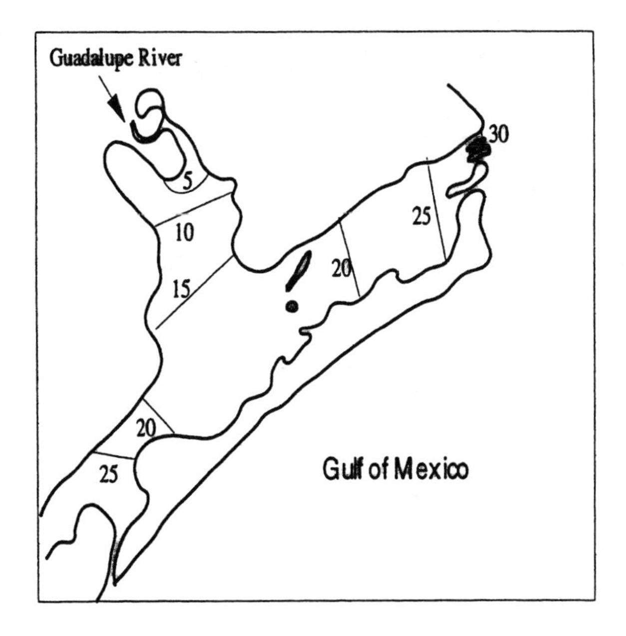

Figure 10.3: Gaudalupe estuary Fall isohalines (modified from [6]).

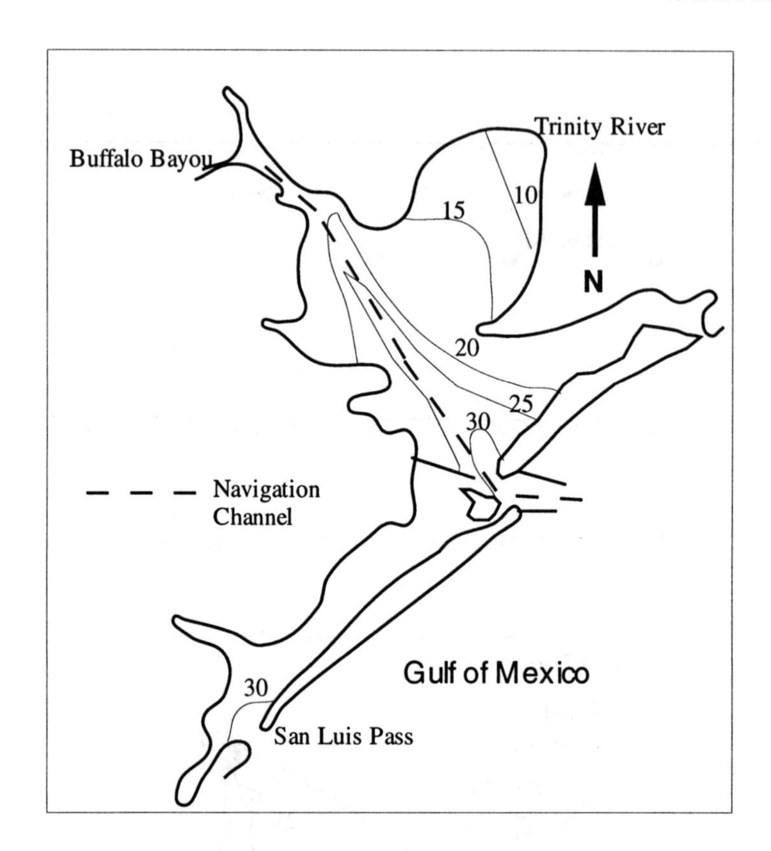

Figure 10.4: Galveston Bay Fall isohalines (modified from [6].)

Figure 10.5: Definitions for elevation terms.

$$\Gamma_x = \rho\Omega v - \frac{\rho g u_a (u_a^2 + v_a^2)^{\frac{1}{2}}}{C^2} + \psi W^2 \cos(\Theta) \tag{10.7}$$

$$\Gamma_y = -\rho\Omega u - \frac{\rho g v_a (u_a^2 + v_a^2)^{\frac{1}{2}}}{C^2} + \psi W^2 \sin(\Theta) \tag{10.8}$$

with $\Omega = 2\omega \sin(\phi)$, where ω is the rate of angular rotation of the earth, and ϕ is the local latitude; W is the wind speed, Θ is the wind direction counterclockwise from Easterly, ψ = a coefficient from [11] and C comes from the Chezy or Manning friction formulation.

The continuity equation

$$\frac{\partial u}{\partial x} + \frac{\partial v}{\partial y} + \frac{\partial w}{\partial z} = 0 \tag{10.9}$$

is included as a second part of each solution step to calculate the vertical velocity: Equation (10.9) is converted to an appropriate boundary value problem through differentiation with respect to z. After rearrangement it takes the form

$$\frac{\partial^2 w}{\partial z^2} = -\frac{\partial}{\partial z}\left(\frac{\partial u}{\partial x} + \frac{\partial v}{\partial y}\right) \tag{10.10}$$

subject to boundary conditions

$$w_\zeta = u_\zeta \frac{\partial \zeta}{\partial x} + v_\zeta \frac{\partial \zeta}{\partial y} + \frac{\partial h}{\partial t} \quad \text{at the water surface} \tag{10.11}$$

and

$$w_a = u_a \frac{\partial a}{\partial x} + v_a \frac{\partial a}{\partial y} \quad \text{at the bed} \tag{10.12}$$

Note that in these equations the values of u and v are known at all locations from the previous part of the solution step. Values of w from this solution step are used in the next iteration for u, v, h and s.

The geometric system varies with time; i.e., the water depth h varies during the simulation. In order to develop an Eulerian form for the solution it is desirable to transform this system to one that can be described with a constant geometric structure. Early development of the model [2] used a σ-transformation in which the bed and the water surface are transformed to constants. In a later analysis of this method [3], it was pointed out that at locations where a sharp break in bottom profile occurs the transformation is not unique and momentum in the component directions may not be correctly preserved. An alternative transformation that preserves the bottom profile as defined, but transforms the water surface to a constant elevation is now used (z^{∇} transformation).

This transformation is defined by

$$x^{\nabla} = x, \quad y^{\nabla} = y, \quad z^{\nabla} = a + (z - a)\frac{b - a}{h} \tag{10.13}$$

where b is the fixed vertical location to which the water surface will be transformed. Equations (10.1)-(10.11) are transformed accordingly.

Another advantage of this transformation is that it produces z^{∇} = constant lines which are close to horizontal; i.e., z = constant lines. This reduces fictitious density driven currents near bed profile breaks [8]. Since stratification-related phenomena are usually nearly horizontal, it is important that the transformation leave constant surfaces that are nearly horizonal. If one considers the pressure gradient (due to the density gradient) in this transformation we have

$$\frac{\partial P}{\partial x} = \frac{\partial P}{\partial x^{\nabla}} + \frac{\partial P}{\partial z^{\nabla}}\frac{\partial z^{\nabla}}{\partial x} \tag{10.14}$$

In a strongly stratified stagnant system this pressure gradient should be zero. However, note that in (10.14) with the transformed system we are dependent upon two terms (each of which could be large) to cancel each other. This could cause artificial currents due to truncation and roundoff error. A transformation in which $\frac{\partial z^{\nabla}}{\partial x} \approx 0$ i.e., $z^{\nabla} \approx z$, will reduce this problem. Figure 10.6 shows an example for a case similar to the Galveston project in which a 40 ft-deep channel passes through an 8 ft-deep bay. Here b is chosen to be an elevation of 0 and ζ is 2 ft. Near the break in the bed profile $\frac{\partial z^{\nabla}}{\partial x}$ is fairly small, or z^{∇} surfaces are nearly horizontal. Contrast this with the σ transformation in Figure 10.7. The σ = constant surfaces are far from horizontal along the channel side slopes, and the truncation and roundoff errors tend to drive fictitious currents which cause the denser saltwater to leave the channel. The z^{∇} transformation results in

$$\frac{\partial z^{\nabla}}{\partial x} = O\left(\zeta - b\right) \tag{10.15}$$

whereas the σ transformation implies

$$\frac{\partial \sigma}{\partial x} = O\left(h\right) \tag{10.16}$$

which is much larger.

The Galerkin finite element approximation of (10.1) -(10.4) and (10.10) utilizes a quadratic element basis for u, v, w and s and a linear element basis for h and P. The non-linearity is addressed by Newton Raphson iteration at each time step. Generally the iteration process is split into calculation of (10.1) - (10.3), then (10.10), followed by (10.4). This sequence is repeated until convergence is reached.

10.4 Application and Verification

The model application includes mesh generation and boundary specification, model verification, and testing. The grids used in this study are shown in Figures 10.8 and 10.9. The first is the grid for existing conditions and the second for the enlarged channel. In addition to the bathymetric change of an enlarged channel the plan grid also includes 17 proposed disposal and mitigation islands. The computational meshes have about 12000 nodes and 5100 elements which provide a horizontal node spacing of 70 ft (in the channel) to 600 ft (in tributary bays) and vertical resolution of 3 to 6 ft. Subsequent discussion covers model verification, results, and numerical problems involved in this study.

The verification process is intended to show that the model is appropriate for the task undertaken. In this case one would like to see the model reproduce salinity and circulation fields for previous channel dimensions. Unfortunately, there is insufficient field data available for this comparison. However, the model can be shown to reliably reproduce results for present bathymetry over a range of hydrodynamic conditions. The model must be able to show that it not only behaves like the natural system, but that it does so as a result of the physics represented in the numerical code and not as a result of the adjustment process alone.

In this study the model is adjusted to a short set of data (1-2 weeks) and then run with no further adjustment in comparison to 6 months of tide, velocity and salinity data collected throughout the bay and channel. A sensitivity analysis concluded that uncertainty in the Gulf salinity boundary and the freshwater inflow results in a minimum probable uncertainty of about 1.5 ppt. This is used to indicate that no additional adjustment is needed. The adequacy of the model for driving the oyster model is based upon the uncertainty of other input parameters used in that ecosystem model, which appear to be less restrictive than that of the boundary uncertainty. Additionally, the model residual circulation, flow and salinity patterns have been

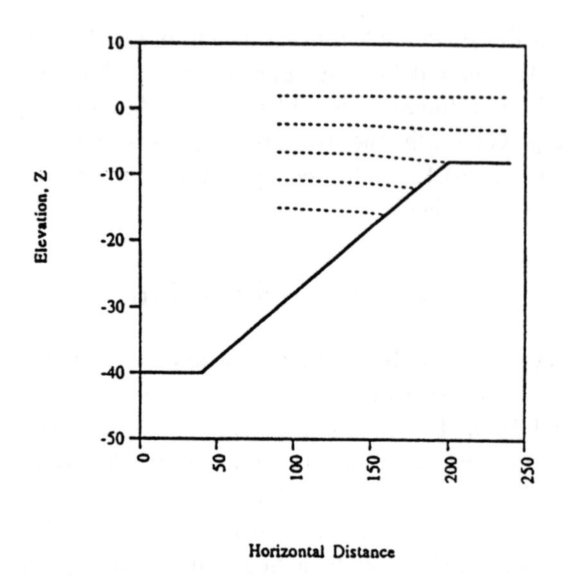

Figure 10.6: Lines of constant z^{∇} near a significant grade change.

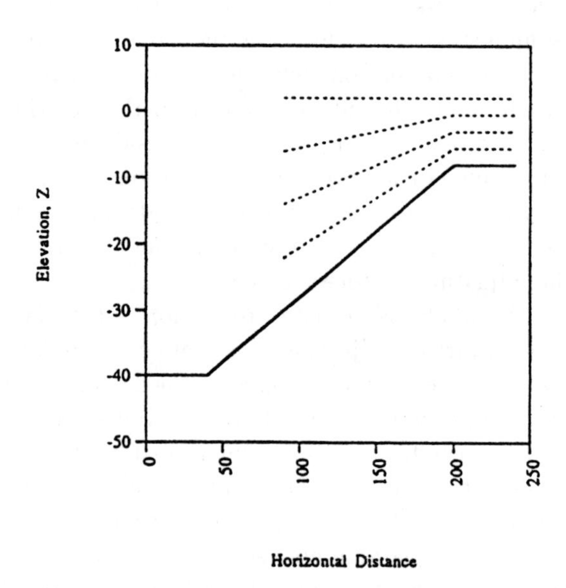

Figure 10.7: Lines of constant σ near a significant grade change.

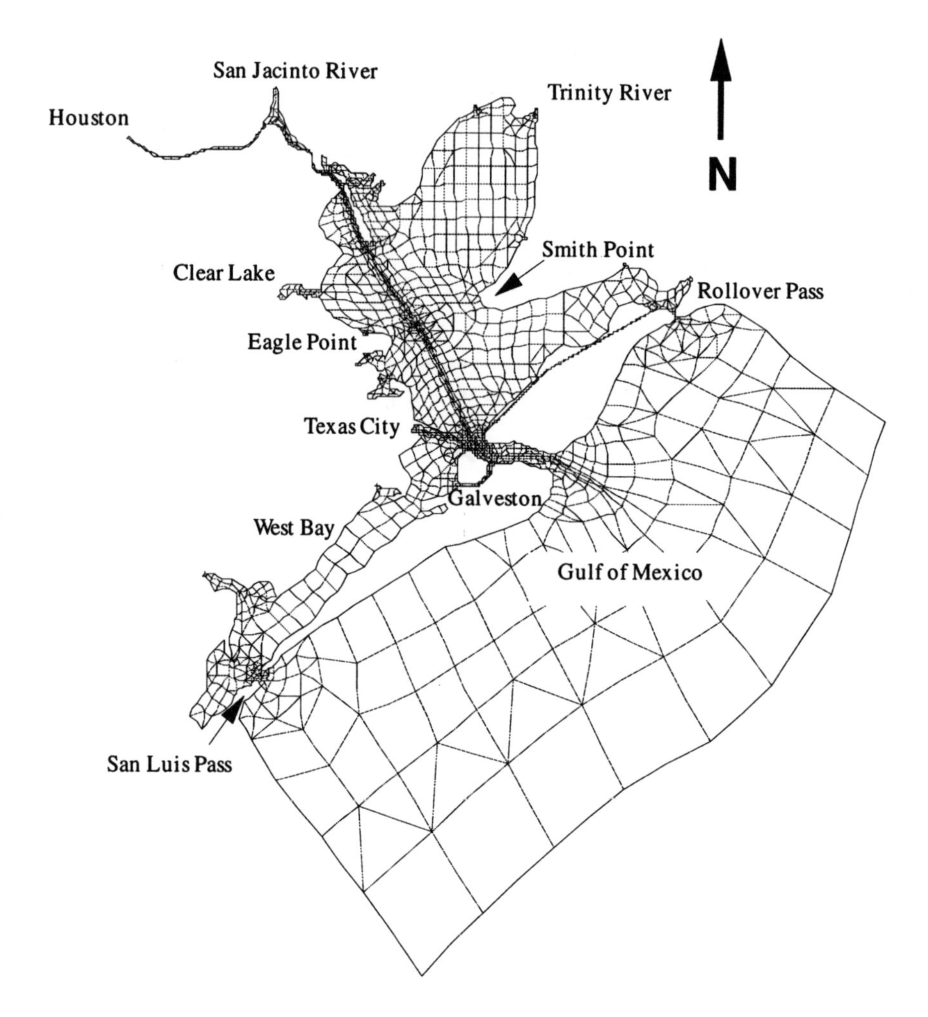

Figure 10.8: Computational mesh for existing geometry.

compared to the characteristic behavior of the natural bay described in the literature and through discussions with pilots and field personnel.

The actual field data collected for this effort included some 21 stations for tides, current velocity, and salinity. Figure 10.1 shows the location of three of these stations. The results at these stations for the model and field observations are shown in Figure 10.10. The data record begins June 19, 1990 just after a major flood in the Trinity Basin. The model results are shown as a solid line and the moored field gage as dashed. The discrete symbols represent hand-held field measurements made over several depths.

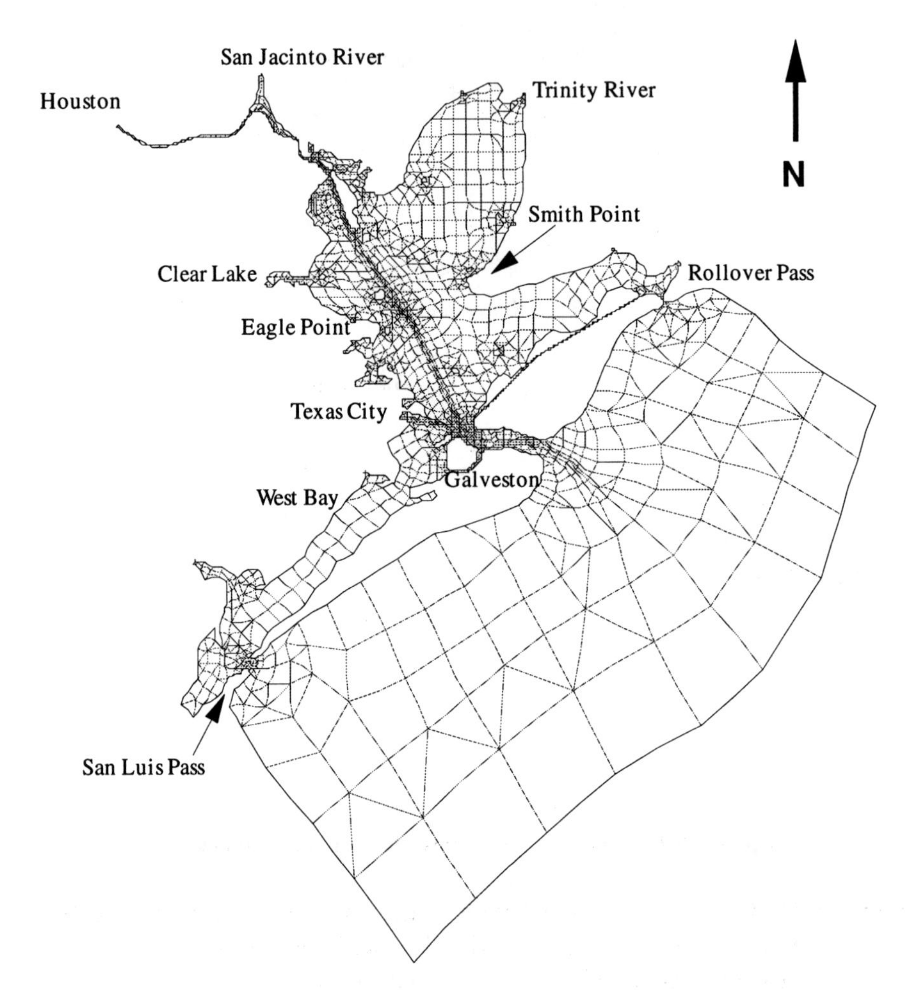

Figure 10.9: Computational mesh for enlarged channel geometry.

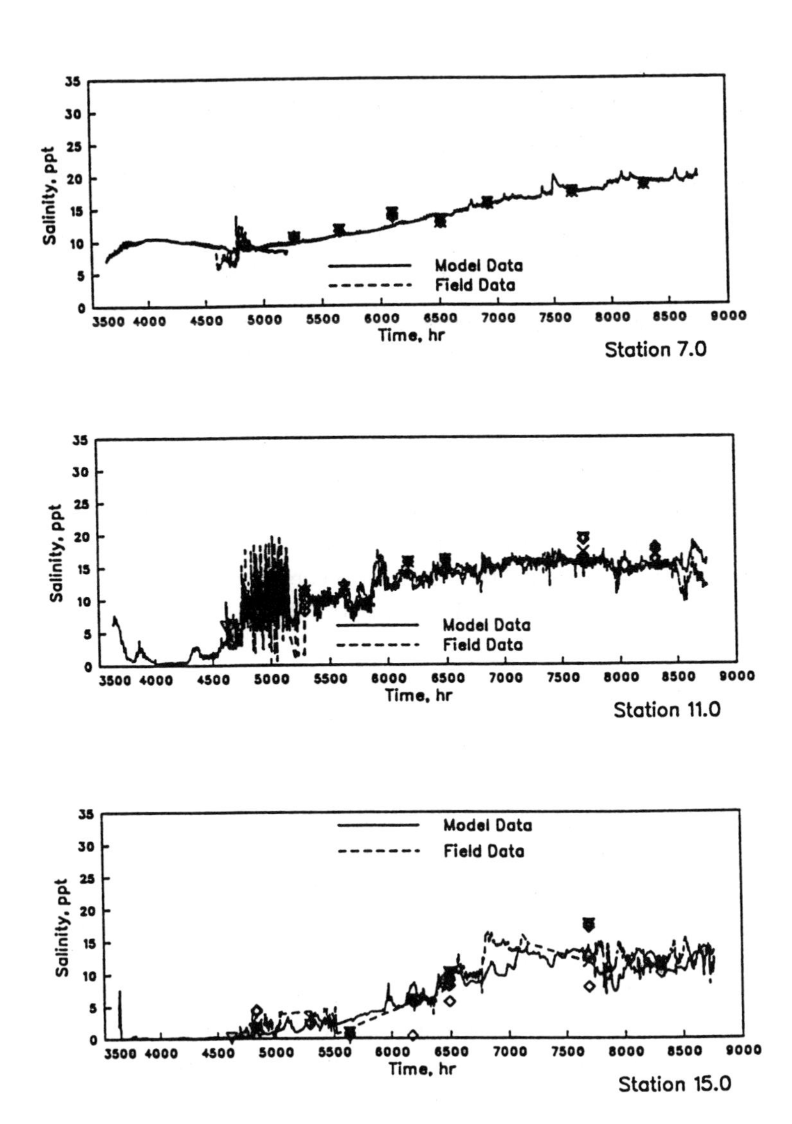

Figure 10.10: Comparison of field observations and model results. Discrete symbols represent field observations over depth during meter service.

Table 10.1: Long term salinity verification statistics.

Station	Mean, ppt. (Model-Prototype)	MAE, ppt	Correlation	d
7.0*	−0.3	0.8	0.97	0.98
11.0	−0.3	1.3	0.85	0.91
15.0	−0.8	1.8	0.89	0.94
* Statistics at this station based upon hand-held field measurements.				

The figure shows that the model can follow a large salinity rebound quite accurately. Some statistics that quantify this comparison are shown in Table 10.1. MAE is the mean of the absolute value of the error between model and field readings. The mean and MAE are indicators of a shift in salinity from prototype to model. The correlation coefficient is an indicator of how well the model follows trends. The statistic [9, 10]

$$d = 1 - \left[\sum_i (M_i - P_i)^2 / \sum_j \left(|M_j^\nabla| - |P_j^\nabla| \right)^2 \right], \quad 0 \le d \le 1 \qquad (10.17)$$

has been found to be a good indicator of the quality of the comparison for both a shift and trends. In (10.17) M_i is the model reading i, P_i is the prototype reading i at the corresponding time, M_j^∇ is the model reading j minus the average prototype value, and P_j^∇ is the prototype reading j minus the average prototype value.

Throughout the bay the difference between model and field values is generally less than the probable uncertainty produced by the boundary conditions alone. This indicates that no further adjustments would be worthwhile. Of course, this does not necessarily mean that the transport model is sufficiently accurate to drive the oyster model. If the hydrodynamic model "error" compared to prototype is of the same order as the uncertainty in the parameters used to "tune" the oyster model, then one can argue that the hydrodynamic model results are sufficient. The model parameters are not clearly delineated in the literature. However, it appears that the oyster model requirements are less restrictive than the sensitivity conditions.

10.5 Results

The model has been used to evaluate existing bay conditions, as well as several geometries based on proposed channel enlargement plans. Also included in the series of tests are the variations in the distribution of freshwater inflow within the bay system. These distributions are referred to as "hydrologies". It is anticipated that as the City of Houston grows, water from within the Bay drainage system may have to be diverted or brought in from outside the drainage basin. These conditions have been tested for existing and enlarged channel configurations. The simulations are usually for the period of January through September. This covers the critical period for oysters plus spin-up for the model.

As an example of the model salinity results, Figure 10.11 is typical. This is a comparison of base (existing geometry) and Phase I (45 ft deep by 530 ft wide channel). These results are the near bed month average isohalines for July. The strong channel salinity intrusion is apparent in the bottom depth isohaline figure. The comparison from existing to Phase I conditions shows that the most notable increase due to deepening occurs in the upper bay west of the channel.

10.6 Conclusions

This chapter discusses the application of a 3D finite element model for hydrodynamics and salinity to the Galveston Bay estuary. The model is used to evaluate the impact of a proposed enlargement of the navigation channel to Houston. The code is a Galerkin-based model using quadratic element basis for velocity and salinity with linear basis for depth. Pressure is assumed to be hydrostatic. The model uses a z^{∇}-transformation over depth. This transforms the water surface to a constant but the bed elevation is unchanged; this is shown to be superior to a σ-transformation in which both surface and bed are constants. A model validation study is presented in which model results are compared with collected field data. The uncertainty in model freshwater and Gulf salinity boundaries provides the criteria for judging the suitability of the model verification. Model results are shown for existing and enlarged channel geometry. The model indicates that the deepening should increase the upper bay salinities west of the channel more substantially than in Trinity and East Bays.

Acknowledgements: This study was sponsored by the U.S. Army Engineer District, Galveston, Texas. Permission was granted by the Chief

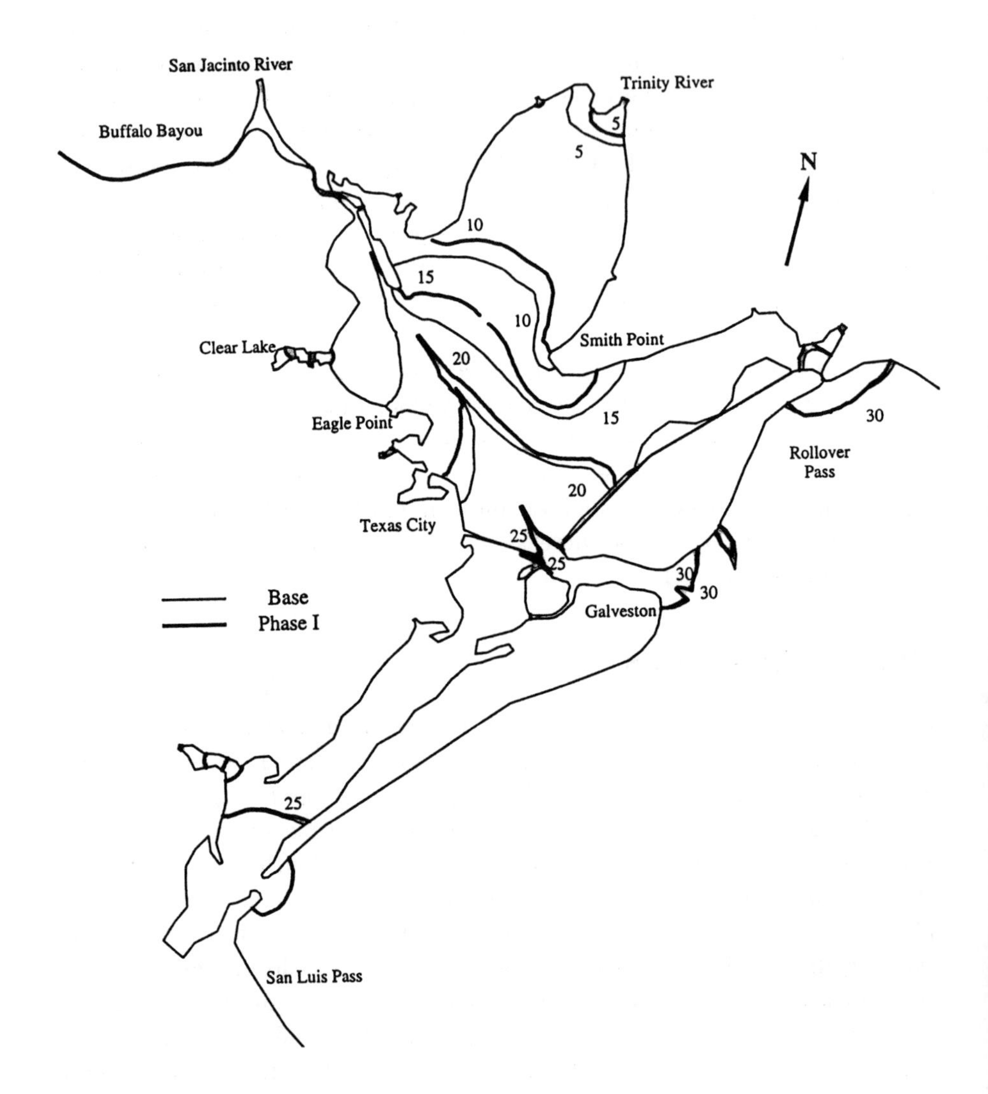

Figure 10.11: Model results showing monthly average isohalines for July under low flow existing hydrologic distribution.

of Engineers to publish this information.

References

[1] Henderson-Sellers, B., "A simple formula for vertical eddy diffusion coefficients under conditions of nonneutral stability," *J. of Geophysical Research*, 87,C8, 5860-5864,1984.

[2] King, I. P., "A Finite Element Model for Three Dimensional Flow," Resource Management Associates, Lafayette, CA., for USAE Waterways Experiment Station, Vicksburg, MS, 1982.

[3] King, I. P., "Strategies for finite element modeling of three dimensional hydrodynamic systems," *Adv. Water Resources*, 8, 69-76, 1985.

[4] King, I. P., *RMA-10, A Finite Element Model for Three Dimensional Density Stratified Flow*, Dept. of Civil and Environmental Engineering, Univ. of California, Davis, 1993.

[5] Mellor, G.L. and T. Yamada, "Development of a turbulence closure model for geophysical fluid problems," *Reviews of Geophysics and Space Physics*, 20,4, 851-875, 1982.

[6] Orlando Jr., S. P., L. P. Rozas, G. H. Ward, and C. J. Klein, *Analysis of Salinity Structure and Stability for Texas Estuaries*, Strategic Assessment Branch, NOS/NOAA, Rockville, MD, 25-33, 1991.

[7] Pritchard, D. W., *A Summary Concerning the Newly Adopted Practical Salinity Scale, 1978 and the International Equation of State of Seawater, 1980*, Marine Sciences Research Center, State University of New York, Stony Brook, New York, 1980.

[8] Stelling, G. S. and J. A. van Kester, "Horizontal gradients in sigma transformed bathymetries with steep bottom slopes," *Hydraulic Engineering '93, Proceedings of the 1993 Conference, ASCE*, 2123- 2134, 1993.

[9] Willmott, C. J., "Some comments on the evaluation of model performance," *Bulletin American Meteorological Society*, 63,11, 1309- 1313, 1982.

[10] Willmott, C. J., S. G. Ackleson, R. E. Davis, J. J. Feddema, K. M. Klink, D. R. Legates, J. O'Donnell and C. M. Rowe, "Statistics for the

evaluation and comparison of models," *J. Geophysical Research*, 90,C5, 8995-9005, 1985.

[11] Wu, J., "Wind-stress coefficients over sea surface near neutral conditions – a revisit," *J. Physical Oceanography*, 10,5, 727-740, 1980.

Chapter 11

Sentinels and Parameter Identification

T. Männikkö[1]

11.1 Introduction

In this chapter we consider parameter identification in systems which contain some unknown perturbations. We introduce a mathematical model which describes, for example, pollution in a sea gulf. The data concerning the sources of pollution and the initial concentration of the pollution is only partially known. The object is to determine the contribution of each source in spite of all the uncertainties. We present the method of sentinels, introduced by J.L. Lions [8–11], which is used to identify the unknown amplitudes appearing in the model.

The idea of the method of sentinels is to construct a functional which is sensitive to the sources of pollution but insensitive to the uncertainties. The method can be easily applied to various different kinds of environmental problems, such as monitoring the pollution in rivers [1, 2, 4], or in lakes [5, 13, 14]. A good presentation on the applications can be found in [6], and the algorithmic point of view is well presented in [3].

In this chapter we restrict ourselves to the case where the unknown perturbations can be approximated by a linear combination of some known

[1] Laboratory of Scientific Computing, University of Jyväskylä, Finland.

functions. As a consequence, the calculation of the sentinel consists essentially in solving a linear system of equations. We also present the finite element discretization of the method, and give one numerical example.

11.2 The Problem Statement

We consider the following model problem: Let Ω be an open bounded domain in $\mathbb{R}^2$ with continuous boundary Γ. Let $]0,T[$ be the time interval during which the system is studied, and let us denote $Q = \Omega\times]0,T[$ and $\Sigma = \Gamma\times]0,T[$. Moreover, let Γ_0 be a nonempty subset of Γ and let us denote $\Gamma_1 = \Gamma \setminus \Gamma_0$, $\Sigma_0 = \Gamma_0\times]0,T[$, and $\Sigma_1 = \Gamma_1\times]0,T[$. The state y, a function of space variable $x \in \Omega$ and time $t \in]0,T[$, solves the parabolic system

$$\frac{\partial y}{\partial t} + \mathcal{A}y = f(x,t) + \sum_{i=1}^{N} \lambda_i \delta(x - a_i) s_i(t) \ \text{ in } Q \tag{11.1}$$

with

$$y = g(x,t) \text{ on } \Sigma_0, \ \frac{\partial y}{\partial n} = 0 \ \text{ on } \Sigma_1 \tag{11.2}$$

and initial data

$$y(\cdot,0) = y^0(x) + \sum_{j=1}^{M} \tau_j \hat{y}_j^0(x) \ \text{ in } \Omega. \tag{11.3}$$

Here $\mathcal{A}$ denotes an elliptic second order differential operator with regular coefficients—in this case we consider $\mathcal{A} = -\Delta + \sigma I$, where $\Delta = \partial^2/\partial x_1^2 + \partial^2/\partial x_2^2$ is the Laplace operator, $\sigma \geq 0$ is a constant, and I is the identity operator. Furthermore, δ is the Dirac distribution and $a_1, \ldots, a_N$ are some points of Ω. The functions $f \in L^2(Q)$, $s_i \in L^2(]0,T[)$, $g \in L^2(\Sigma_0)$, and $y^0, \hat{y}_j^0 \in L^2(\Omega)$ are all assumed to be known, whereas the real parameters λ_i and τ_j are unknown. We also assume that the distributions $\{\delta(\cdot - a_i)s_i \mid 1 \leq i \leq N\}$, as well as the functions $\{\hat{y}_j^0 \mid 1 \leq j \leq M\}$, are linearly independent.

It can be shown that, for given λ_i and τ_j, the system (11.1) admits a unique solution in $L^2(Q)$. Since the right-hand side of the equation contains Dirac distributions, this solution has to be defined in a weak sense by using the method of transposition. For details we refer to [7].

This problem has the following physical interpretation: The domain Ω is a sea gulf and Γ_0 represents the part of the boundary which is open to the sea. The state y describes the concentration of the pollution in the gulf. There are N (point) sources of pollution, $a_1, \ldots, a_N$, and their "shape functions" $s_i \geq 0$ are known, but the amplitudes of pollution, λ_i, are unknown. On

Γ_1 (the shoreline) there is no flux into or out from the gulf, and on Γ_0 the concentration is assumed to be known. The initial condition is only partially known; y^0 represents the "most probable" concentration and $\tau_j \hat{y}_j^0$ are some small unknown perturbations. The same (or very similar) model has been studied in [12] and [15].

We assume that the state y can be observed in some "observatory" $\omega \subset \Omega$ during the time interval $]0, T[$. The object is then to identify the amplitudes λ_i from the observation of y, without trying to identify the coefficients τ_j.

Since the state y is affine with respect to the unknowns, it can be expressed as

$$y = \sum_{i=1}^{N} \lambda_i \psi_i + \sum_{j=1}^{M} \tau_j \varphi_j + y_c, \tag{11.4}$$

where $\psi_i = \psi_i(x,t)$, $\varphi_j = \varphi_j(x,t)$, and $y_c = y_c(x,t)$ solve the systems

$$\frac{\partial \psi_i}{\partial t} + \mathcal{A}\psi_i = \delta(x - a_i) s_i(t) \text{ in } Q \tag{11.5}$$

with

$$\psi_i = 0 \text{ on } \Sigma_0, \quad \frac{\partial \psi_i}{\partial n} = 0 \text{ on } \Sigma_1, \quad \psi_i(\cdot, 0) = 0 \text{ in } \Omega, \tag{11.6}$$

$$\frac{\partial \varphi_j}{\partial t} + \mathcal{A}\varphi_j = 0 \text{ in } Q \tag{11.7}$$

with

$$\varphi_j = 0 \text{ on } \Sigma_0, \quad \frac{\partial \varphi_j}{\partial n} = 0 \text{ on } \Sigma_1, \quad \varphi_j(\cdot, 0) = \hat{y}_j^0(x) \text{ in } \Omega, \tag{11.8}$$

and

$$\frac{\partial y_c}{\partial t} + \mathcal{A}y_c = f(x,t) \text{ in } Q \tag{11.9}$$

with

$$y_c = g(x,t) \text{ on } \Sigma_0, \quad \frac{\partial y_c}{\partial n} = 0 \text{ on } \Sigma_1, \quad y_c(\cdot, 0) = y^0(x) \text{ in } \Omega. \tag{11.10}$$

respectively. The expression (11.4) means in fact that y belongs to some finite dimensional (affine) subspace of $L^2(Q)$.

11.3 The Method of Sentinels

We shall now present a method which can be used to identify the unknown amplitudes. This method utilizes sentinels, which were introduced in [8]–[10], and studied more thoroughly in [11].

Let $y(\lambda,\tau) = y(\lambda,\tau;x,t)$ denote the solution of system (11.1) with parameters $\lambda = (\lambda_1,\ldots,\lambda_N)$ and $\tau = (\tau_1,\ldots,\tau_M)$. We define the *sentinel*

$$\mathcal{S}(\lambda,\tau) = \int_{\omega\times]0,T[} (h_0 + w) y(\lambda,\tau)\, dx\, dt, \tag{11.11}$$

where $h_0 = h_0(x,t)$ is a given function from $L^2(\omega\times]0,T[)$ such that

$$h_0 \geq 0, \quad \int_{\omega\times]0,T[} h_0\, dx\, dt = 1, \tag{11.12}$$

and $w = w(x,t)$, $w \in L^2(\omega\times]0,T[)$, is to be determined. Eventually, the sentinel $\mathcal{S}$ will be evaluated by using the observation of the state y, instead of the "true" solution $y(\lambda,\tau)$.

We want the sentinel $\mathcal{S}$ to be selectively sensitive to the variations of the amplitudes λ_i and insensitive to the variations of the coefficients τ_j. More precisely, we want to find a function w such that

$$\begin{aligned} \frac{\partial \mathcal{S}(0,0)}{\partial \lambda_i} &= c_i, \qquad 1 \leq i \leq N, \\ \frac{\partial \mathcal{S}(0,0)}{\partial \tau_j} &= 0, \qquad 1 \leq j \leq M, \end{aligned} \tag{11.13}$$

where $c_i \geq 0$ are some predefined constants; in what follows, we have $c_1 = 1$ and $c_i = 0$ if $i \neq 1$. Furthermore, we want w to be as small as possible, i.e.,

$$\|w\|_{L^2(\omega\times]0,T[)} \quad \text{is a minimum.} \tag{11.14}$$

Using the expression (11.4) for y, we notice that the conditions (11.13) are equivalent to

$$\begin{aligned} \int_{\omega\times]0,T[} (h_0 + w)\psi_i\, dx\, dt &= c_i, \qquad 1 \leq i \leq N, \\ \int_{\omega\times]0,T[} (h_0 + w)\varphi_j\, dx\, dt &= 0, \qquad 1 \leq j \leq M. \end{aligned} \tag{11.15}$$

The object is now to find w satisfying (11.14) and (11.15). Equivalently, this can be formulated as follows:

Problem 11.1 *Minimize the cost function*

$$J(w) = \frac{1}{2} \int_{\omega\times]0,T[} w^2\, dx\, dt \tag{11.16}$$

over $L^2(\omega\times]0,T[)$ with respect to the constraints (11.15). □

Since the cost function is quadratic and the constraints are linear equality constraints, it is evident that there exists a solution to Problem 11.1, provided we can show that there is some feasible point. First we prove the following result:

Proposition 11.1 *The functions* $\{\psi_i|_{\omega\times]0,T[}, \varphi_j|_{\omega\times]0,T[} \mid 1 \leq i \leq N,\ 1 \leq j \leq M\}$ *are linearly independent.*

Proof: Define $\theta = \sum_{i=1}^N \alpha_i\psi_i + \sum_{j=1}^M \beta_j\varphi_j$. The result is proved if we show that $\theta|_{\omega\times]0,T[} = 0$ implies $\alpha_i = \beta_j = 0$ for all i and j. Let us denote $\tilde{\Omega} = \Omega\setminus\{a_1,\ldots,a_N\}, \tilde{\omega} = \omega\setminus\{a_1,\ldots,a_N\}$ and $\tilde{Q} = \tilde{\Omega}\times]0,T[$. Furthermore, let $\tilde{\theta}$ denote the restriction of θ to $\tilde{Q}$. Then, due to (11.5)–(11.7), we have

$$\frac{\partial\tilde{\theta}}{\partial t} + \mathcal{A}\tilde{\theta} = 0 \text{ in } \tilde{Q} \tag{11.17}$$

with

$$\tilde{\theta} = 0 \text{ on } \Sigma_0, \quad \frac{\partial\tilde{\theta}}{\partial n} = 0 \text{ on } \Sigma_1, \quad \tilde{\theta}(\cdot,0) = \sum_{j=1}^M \beta_j\hat{y}_j^0(x) \text{ in } \tilde{\Omega}. \tag{11.18}$$

If $\theta|_{\omega\times]0,T[} = 0$, then $\tilde{\theta}|_{\tilde{\omega}\times]0,T[} = 0$, and the classical uniqueness theorem of Mizohata [16] implies that $\tilde{\theta} \equiv 0$ in $\tilde{Q}$. But, since the set $\{a_1,\ldots,a_N\}$ has a zero measure, we have in fact $\theta = 0$ a.e. in Q. Therefore, again due to (11.5)–(11.7), we have

$$\begin{aligned} \frac{\partial\theta}{\partial t} + \mathcal{A}\theta &= \sum_{i=1}^N \alpha_i\delta(x-a_i)s_i(t) = 0 && \text{in } Q, \\ \theta(\cdot,0) &= \sum_{j=1}^M \beta_j\hat{y}_j^0(x) = 0 && \text{in } \Omega. \end{aligned} \tag{11.19}$$

Distributions $\{\delta(\cdot-a_i)s_i \mid 1 \leq i \leq N\}$, as well as functions $\{\hat{y}_j^0 \mid 1 \leq j \leq M\}$, were assumed to be linearly independent, and hence (11.19) implies that $\alpha_i = 0$ for all i and $\beta_j = 0$ for all j. □

Now we can show the following:

Proposition 11.2 *There exists some* $w \in L^2(\omega\times]0,T[)$ *satisfying (11.15).*

Proof: Let us denote the scalar product in $L^2(\omega\times]0,T[)$ by $(\cdot,\cdot)$. We first show that there is some $v \in L^2(\omega\times]0,T[)$ such that $(v,\psi_1) \neq 0$ and $(v,\psi_i) = (v,\varphi_j) = 0$ for $2 \leq i \leq N$ and $1 \leq j \leq M$.

Let E denote the subspace generated by the functions $\psi_i|_{\omega\times]0,T[}$, $2 \leq i \leq N$, and $\varphi_j|_{\omega\times]0,T[}$, $1 \leq j \leq M$. Let us define $v = (\psi_1 - P\psi_1)|_{\omega\times]0,T[}$, where P is an orthogonal projection into E. Then v is orthogonal to E, i.e., $(v,\psi_i) = (v,\varphi_j) = 0$ for all $i \neq 1$ and for all j. Moreover, we have

$$(v,\psi_1) = (\psi_1 - P\psi_1, \psi_1) = (\psi_1 - P\psi_1, \psi_1 - P\psi_1) = \|v\|^2 \neq 0,$$

since, due to Proposition 11.1, ψ_1 does not belong to E.

Now we can define $w = v/(v,\psi_1) - h_0$, and we see that this w satisfies the constraints (11.15). □

Next we shall indicate how to find the optimal w. The Lagrange (or Kuhn-Tucker) conditions imply that there exist Lagrange multipliers $\alpha = (\alpha_1,\ldots,\alpha_N)$ and $\beta = (\beta_1,\ldots,\beta_M)$ such that the optimal w has the form

$$w = \sum_{i=1}^{N} \alpha_i \psi_i|_{\omega\times]0,T[} + \sum_{j=1}^{M} \beta_j \varphi_j|_{\omega\times]0,T[}, \tag{11.20}$$

providing that this w is a regular point with respect to the constraints; i.e., the gradient vectors of the constraints are linearly independent. But this is a direct consequence of Proposition 11.1. We note also, that the expression (11.20) means in fact that w belongs to a finite dimensional subspace of $L^2(\omega\times]0,T[)$, and the functions $\{\psi_i|_{\omega\times]0,T[}, \varphi_j|_{\omega\times]0,T[} \mid 1 \leq i \leq N,\ 1 \leq j \leq M\}$ form the basis of that subspace.

Substituting the expression (11.20) into (11.15), we notice that the multipliers α and β are obtained by solving the matrix equation

$$\begin{pmatrix} A_{\psi\psi} & A_{\psi\varphi} \\ A_{\varphi\psi} & A_{\varphi\varphi} \end{pmatrix} \begin{pmatrix} \alpha \\ \beta \end{pmatrix} = \begin{pmatrix} b_\psi \\ b_\varphi \end{pmatrix} + \begin{pmatrix} c \\ 0 \end{pmatrix}, \tag{11.21}$$

where, e.g.,

$$\begin{aligned} A_{\psi\varphi} &\in \mathbf{R}^{N\times M}, & (A_{\psi\varphi})_{ij} &= \int_{\omega\times]0,T[} \psi_i \varphi_j \, dx\, dt, \\ b_\psi &\in \mathbf{R}^{N}, & (b_\psi)_i &= -\int_{\omega\times]0,T[} h_0 \psi_i \, dx\, dt, \end{aligned} \tag{11.22}$$

and $c = (c_1,\ldots,c_N)$.

Of course, in order to use this method, we have to be sure that the matrix in (11.21) is really nonsingular. Using Proposition 11.1, it is easy to show that the rows (or columns) of the matrix in (11.21) are linearly independent, which implies the nonsingularity.

When the multipliers α and β are solved from (11.21), we may use (11.20) in the definition of the sentinel, and we have

$$\mathcal{S}(\lambda,\tau) = \sum_{i=1}^{N} \lambda_i c_i + \int_{\omega\times]0,T[} (h_0 + w) y_c \, dx \, dt = \sum_{i=1}^{N} \lambda_i c_i + \mathcal{S}(0,0). \quad (11.23)$$

Recalling how the constants c_i were chosen, we notice that

$$\lambda_1 = \mathcal{S}(\lambda,\tau) - \mathcal{S}(0,0), \quad (11.24)$$

where the first term in the right-hand side is "observed"—i.e., we observe $y|_{\omega\times]0,T[}$ and then evaluate the sentinel with this observed state—and the latter term is "calculated"—i.e., we solve y_c (which equals $y(0,0) = y(0,0;x,t)$) from (11.9) and evaluate the sentinel with this calculated state.

All the other amplitudes λ_i are approximated similarly; thus, we need to construct N sentinels in order to identify all the unknown sources.

11.4 Discretization of the Method

We shall now briefly describe the numerical implementation of our method. Let us begin with the discretization of (11.5)–(11.9). The discretization is performed using a finite element discretization in space and finite differencing with respect to time. We consider here only the equation (11.9); all the other equations are handled similarly.

Let $\mathcal{T}_h$ denote a regular triangulation of Ω (with triangles T), where h is a grid parameter corresponding to the length of the triangle edges, and let x_i, $1 \le i \le N_x$, be the nodal points of $\mathcal{T}_h$. We assume that the nodes are numbered in such a way that the first P nodes are the nodes lying on Γ_0. Furthermore, let ϑ_i, $1 \le i \le N_x$, denote continuous piecewise linear basis functions satisfying $\vartheta_i(x_j) = \delta_{ij}$ (Kronecker's delta symbol).

The time is discretized such that we have

$$0 = t^0 < t^1 < \cdots < t^{K_t} = T, \quad \Delta t = t^k - t^{k-1}. \quad (11.25)$$

The function y_c is then approximated by $\sum_{i=1}^{N_x} y_i^k \vartheta_i(x)$, where y_i^k denotes the value of y_c at node x_i and time t^k. Moreover, let us denote $\bar{Y}^k = (y_1^k, \ldots, y_P^k)$ and $Y^k = (y_{P+1}^k, \ldots, y_{N_x}^k)$. The functions f, g and y^0 are approximated similarly by vectors $F^k = (f_1^k, \ldots, f_{N_x}^k)$, $G^k = (g_1^k, \ldots, g_P^k)$ and $Y_0 = (y_1^0, \ldots, y_{N_x}^0)$, which contain the nodal values of those functions.

Due to the initial and boundary conditions in (11.9), we have simply

$$\begin{aligned} (\bar{Y}^0; Y^0) &= Y_0, \\ \bar{Y}^k &= G^k, \qquad 1 \le k \le K_t. \end{aligned} \tag{11.26}$$

Thus, it remains to calculate Y^k, $1 \le k \le K_t$.

Let us multiply equation (11.9) by ϑ_i, where $P+1 \le i \le N_x$, and integrate over Ω. Then, using the Green's formula (integration by parts), we obtain

$$\int_\Omega \frac{\partial y_c}{\partial t} \vartheta_i \, dx + \int_\Omega \nabla y_c \cdot \nabla \vartheta_i \, dx + \sigma \int_\Omega y_c \vartheta_i \, dx = \int_\Omega f \vartheta_i \, dx. \tag{11.27}$$

The time derivative $\partial y_c / \partial t$ is approximated with a fully implicit scheme, and the integrations over Ω are performed by using the quadrature rule

$$\int_T h \, dx \approx I_T(h) = \frac{\operatorname{meas}(T)}{3} (h(z_1) + h(z_2) + h(z_3)), \tag{11.28}$$

where z_i, $1 \le i \le 3$, are the midpoints of the edges of the triangle T. Thus, the "discrete form" of (11.27) is

$$\begin{aligned} &\sum_{j=1}^{N_x} \frac{y_j^k - y_j^{k-1}}{\Delta t} \sum_{T \in \mathcal{T}_h} I_T(\vartheta_j \vartheta_i) + \sum_{j=1}^{N_x} y_j^k \sum_{T \in \mathcal{T}_h} I_T(\nabla \vartheta_j \cdot \nabla \vartheta_i) \\ &\quad + \sigma \sum_{j=1}^{N_x} y_j^k \sum_{T \in \mathcal{T}_h} I_T(\vartheta_j \vartheta_i) = \sum_{j=1}^{N_x} f_j^k \sum_{T \in \mathcal{T}_h} I_T(\vartheta_j \vartheta_i), \quad 1 \le k \le K_t. \end{aligned} \tag{11.29}$$

Let us define the mass matrix M and the stiffness matrix A by

$$\begin{aligned} M &= (m_{ij}) \in \mathbf{R}^{(N_x - P) \times N_x}, \quad m_{ij} = \sum_{T \in \mathcal{T}_h} I_T(\vartheta_i \vartheta_j), \\ A &= (a_{ij}) \in \mathbf{R}^{(N_x - P) \times N_x}, \quad a_{ij} = \sum_{T \in \mathcal{T}_h} I_T(\nabla \vartheta_i \cdot \nabla \vartheta_j), \end{aligned} \tag{11.30}$$

and let us denote $\hat{M} = M / \Delta t$. Then, if $i = P+1, \ldots, N_x$ in (11.29), we obtain the matrix equation

$$\hat{M} \left(\begin{pmatrix} \bar{Y}^k \\ Y^k \end{pmatrix} - \begin{pmatrix} \bar{Y}^{k-1} \\ Y^{k-1} \end{pmatrix} \right) + A \begin{pmatrix} \bar{Y}^k \\ Y^k \end{pmatrix} + \sigma M \begin{pmatrix} \bar{Y}^k \\ Y^k \end{pmatrix} = M F^k, \quad 1 \le k \le K_t. \tag{11.31}$$

Let us decompose M and A by $M = (M_1; M_2)$ and $A = (A_1; A_2)$, where the matrix M_1 (resp. A_1) contains the first P columns of M (resp. A). Then, using (11.26), equation (11.31) can be written as

$$\begin{aligned}(\hat{M}_2 + A_2 + \sigma M_2)Y^k &= MF^k + \hat{M}_2 Y^{k-1} + \hat{M}_1 G^{k-1} \\ &\quad -(\hat{M}_1 + A_1 + \sigma M_1)G^k, \quad 1 \le k \le K_t,\end{aligned} \tag{11.32}$$

where $G^0 = \bar{Y}^0$. Now, starting from the known $Y_0 = (\bar{Y}^0; Y^0)$, we go through all time levels, and solve for Y^k, $1 \le k \le K_t$, from (11.32).

Since equation (11.32) is linear, it can be solved with the usual successive over relaxation (SOR) method. If we define matrices $B^k = (b^k_{ij})$ and vectors $H^k = (h^k_{P+1}, \ldots, h^k_{N_x})$ by

$$\begin{aligned} B^k &= \hat{M}_2 + A_2 + \sigma M_2, \\ H^k &= MF^k + \hat{M}_2 Y^{k-1} + \hat{M}_1 G^{k-1} - (\hat{M}_1 + A_1 + \sigma M_1)G^k, \end{aligned} \tag{11.33}$$

then (11.32) can be written as

$$B^k Y^k = H^k, \quad 1 \le k \le K_t, \tag{11.34}$$

and the SOR-method can be expressed in the following way (for the sake of notational simplicity, Y^k is denoted by Y):

Algorithm 11.1 *Calculation of the state Y.*

(0) *Assign the initial guess $Y^{(0)} := Y^{k-1}$. Set $n := 0$.*

(1) *For every $i = P+1, \ldots, N_x$ calculate the components $y_i^{(n+1)}$ as follows:*

   (1.1) *Set $y_i^{(n+1/2)} := (h_i^k - \sum_{j<i} b^k_{ij} y_j^{(n+1)} - \sum_{j>i} b^k_{ij} y_j^{(n)})/b^k_{ii}$.*

   (1.2) *Assign $y_i^{(n+1)} := y_i^{(n)} + \gamma(y_i^{(n+1/2)} - y_i^{(n)})$, where γ is an appropriate relaxation parameter.*

(2) *If $\|Y^{(n+1)} - Y^{(n)}\| < \varepsilon \|Y^{(n+1)}\|$, where ε is some small positive number, then assign $Y := Y^{(n+1)}$ and stop; otherwise set $n := n+1$ and continue from step (1).*

The norms in step (2) are maximum norms (i.e., $\|Y\| = \max_{P+1 \le i \le N_x} |y_i|$).

The next step is to construct the matrix equation (11.21). This requires integrations over $\omega \times]0, T[$. As an example we give the discrete form of the matrix element $(A_{\psi\varphi})_{ij}$.

Let $\Psi^k = (\psi_1^k, \ldots, \psi_{N_x}^k)$, $1 \le k \le K_t$, denote the vectors containing the nodal values of ψ_i, and similarly, let $\Phi^k = (\varphi_1^k, \ldots, \varphi_{N_x}^k)$, $1 \le k \le K_t$, denote the vectors containing the nodal values of φ_j. Moreover, let $\tilde{M}$ denote the mass matrix associated with the "observatory", i.e.,

$$\tilde{M} = (\tilde{m}_{ij}) \in \mathbb{R}^{N_x \times N_x}, \quad \tilde{m}_{ij} = \begin{cases} \sum_{T \in \mathcal{T}_h} I_T(\vartheta_i \vartheta_j), & \text{if } x_i, x_j \in \omega \ , \\ 0, & \text{otherwise} \ . \end{cases} \tag{11.35}$$

Then we use the approximation

$$\int_{\omega \times]0,T[} \psi_i \varphi_j \, dx \, dt \approx \Delta t \sum_{k=1}^{K_t} (\Psi^k)^T \tilde{M} \Phi^k. \tag{11.36}$$

The resulting symmetric positive definite system is solved using the Cholesky method.

11.5 Numerical Example

In our numerical example we consider the following problem: Let us take $\Omega =]0,1[\times]0,1[, \Gamma = \partial\Omega$, $\Gamma_0 =]0,1[\times\{1\}$, and $\Gamma_1 = \Gamma \setminus \Gamma_0$. Moreover, $T = 1$, and thus, $Q = \Omega\times]0,1[$, $\Sigma = \Gamma\times]0,1[$, $\Sigma_0 = \Gamma_0\times]0,1[$, and $\Sigma_1 = \Gamma_1\times]0,1[$. Furthermore, we have two sources of pollution, namely $a_1 = \{(1/8, 1/8)\}$ and $a_2 = \{(7/8, 1/8)\}$, and the observatory is $\omega = [3/8, 5/8]\times]3/8, 5/8[$.

We assume that the state y is governed by the system

$$\frac{\partial y}{\partial t} - \Delta y + y = \sum_{i=1}^{2} \lambda_i \delta(x - a_i) s_i(t) \text{ in } Q \tag{11.37}$$

with

$$y = 0 \text{ on } \Sigma_0, \ \frac{\partial y}{\partial n} = 0 \text{ on } \Sigma_1, \ y(\cdot, 0) = y^0(x) \text{ in } \Omega \tag{11.38}$$

where the initial state is

$$y^0(x) = \frac{1}{10}(-6x_1^4 + 14x_1^3 - 9x_1^2 + 2)(2x_2^3 - 3x_2^2 + 1) \tag{11.39}$$

and the "shape functions" are

$$s_1(t) = \begin{cases} 4t, & 0 \le t \le 0.5 \ , \\ -4t + 4, & 0.5 < t \le 1 \ , \end{cases} \quad s_2(t) = \begin{cases} 1.5, & 0 \le t \le 0.5 \ , \\ 0.5, & 0.5 < t \le 1 \ . \end{cases} \tag{11.40}$$

For test purposes we construct the observation in the observatory such that $\lambda_1 = 2$, $\lambda_2 = 1$ will be the exact solution; i.e., we solve the system

Table 11.1: Results of the method of sentinels.

M	λ_1	λ_2
4	1.93353	1.12743
8	1.96484	1.07041
12	2.00386	0.99289
16	2.00247	0.99420
20	1.99955	1.00091
24	2.00024	1.00000

(11.37) with these values and use $y|_{\omega\times]0,1[}$ as an observation of y. We shall then calculate the amplitudes assuming that the initial data (function y^0) is not known.

Let $\hat{y}_j^0$, $j = 1, 2, \ldots$, denote some basis of $L^2(\Omega)$. In this example, $\hat{y}_j^0$ are the eigenfunctions of operator $-\Delta$ with the boundary conditions of system (11.37); i.e., they have the form

$$\hat{y}_j^0(x) = \cos(j_1\pi x_1)\cos\left((j_2 + \frac{1}{2})\pi x_2\right) \tag{11.41}$$

with some integers j_1 and j_2. We then approximate the function y^0 by a truncated Fourier series $\sum_{j=1}^{M} \tau_j \hat{y}_j^0$, where τ_j are unknown, and thus, we replace the initial condition of system (11.37) by

$$y(\cdot, 0) = \sum_{j=1}^{M} \tau_j \hat{y}_j^0(x) \quad \text{in } \Omega. \tag{11.42}$$

Our aim is to try to find out how many terms in the series are needed in order to get good approximations for the unknown amplitudes.

In the numerical implementation we have $9 \times 9 = 81$ nodes in the triangulation of Ω, and the time step is $\Delta t = 1/16$. The method of sentinels has been found to be very insensitive to the choice of function h_0, and thus, we use the simplest possible choice, namely, $h_0 \equiv 1/(T \operatorname{meas}(\omega)) = 16$.

The results are given in Table 11.1. The computed values of the amplitudes are given as functions of M, the number of terms in the series (11.42). As we see, there is good convergence towards the exact solution $\lambda_1 = 2$, $\lambda_2 = 1$.

More numerical test results concerning this same example can be found in [12], where there also is a comparison with the classical method of least squares, and an account on how the noise in the observation can be eliminated.

Acknowledgements: This chapter is based on work done at Compiègne University of Technology, France, in the research group of Professor Jean-Pierre Kernévez.

References

[1] Aïnseba, B.E., Kernévez, J.P., and Luce, R., "Application des sentinelles à l'identification des pollutions dans une rivière", *RAIRO Modél. Math. Anal. Numér.*, 28, 297–312, 1994.

[2] Aïnseba, B.E., Kernévez, J.P., and Luce, R., "Identification de paramètres dans les problèmes non linéaires à données incomplètes", *RAIRO Modél. Math. Anal. Numér.*, 28, 313–328, 1994.

[3] Bodart, O., *Application de la méthode des sentinelles à l'identification de sources de pollution dans un système distribué. Contrôles insensibilisants, PhD Thesis*, Compiègne University of Technology, Division of Applied Mathematics, 1994.

[4] Bodart, O., and Kernévez, J.P., "Sentinels in rivers", in *Jornadas Hispano-Francesas sobre Control de Sistemas Distribuidos*, Proceedings, Málaga, Spain, October 25–26, 1990, Grupo de Análisis Matemático Aplicado de la Universidad de Málaga, 69–76, 1990.

[5] Bodart, O., Kernévez, J.P., and Männikkö, T., "Sentinels for distributed environmental systems", in *Modelling, Identification and Control*, (ed. Hamza, M.H.), Proceedings of the 11th IASTED International Conference, Innsbruck, Austria, February 10–12, 1992, Acta Press, Zurich, 219–222, 1992.

[6] Kernévez, J.P., "Méthodes numériques de calcul de sentinelles. Application à l'identification de sources de pollution", *Rech. Math. Appl.*, Masson, Paris (to appear).

[7] Lions, J.L., *Optimal Control of Systems Governed by Partial Differential Equations*, Springer-Verlag, Berlin–Heidelberg, 1971.

[8] Lions, J.L., "Sur les sentinelles des systèmes distribués. Le cas des conditions initiales incomplètes", *C. R. Acad. Sci. Paris, Sér. I Math.*, 307,819–823, 1988.

[9] Lions, J.L., "Sur les sentinelles des systèmes distribuès. Conditions frontières, termes sources, coefficients incomplètement connus", *C. R. Acad. Sci. Paris, Sér. I Math.*, 307, 865–870, 1988.

[10] Lions, J.L., "Furtivité et sentinelles pour les systèmes distribués à données incomplètes", *C. R. Acad. Sci. Paris Sér. I Math.*, 311, 691–695, 1990.

[11] Lions, J.L., "Sentinelles pour les systèmes distribués à données incomplètes", *Rech. Math. Appl. 21*, Masson, Paris, 1992.

[12] Männikkö, T., "Method of sentinels for eliminating uncertainties and noise in the observation", *Adv. Math. Sci. Appl.*, 2, 485–502, 1993.

[13] Männikkö, T., and Bodart, O., "Parameter identification with the method of sentinels", in *Proceedings of the Workshop on Optimization and Optimal Control*, (ed. Neittaanmäki, P.), Jyväskylä, Finland, September 28–29, 1992, Report 58, University of Jyväskylä, Department of Mathematics, 97–107, 1993.

[14] Männikkö, T., Bodart, O., and Kernévez, J.P., "Numerical methods to compute sentinels for parabolic systems with an application to source terms identification", in *System Modelling and Optimization* (eds. Henry, J., and Yvon, J.P.), Proceedings of the 16th IFIP Conference, Compiègne, France, July 5–9, 1993, *Lecture Notes in Control and Inform. Sci. 197*, Springer-Verlag, Berlin–Heidelberg, 670-679, 1994.

[15] Männikkö, T., Kernévez, J.P., and Lions, J.L., "Sentinels for monitoring the environment", in *Intelligent Process Control and Scheduling, Modelling and Control of Water Resources Systems and Global Changes*, (eds. Kerckhoffs, E.J.H., Koppelaar, H., Van Der Beken, A., and Vansteenkiste, G.C.), Proceedings of the 1991 European Simulation Symposium, Ghent, Belgium, November 6–8, 1991, SCSI, Ghent, 229–234, 1991.

[16] Mizohata, S., "Unicité du prolongement des solutions pour quelques opérateurs différentiels paraboliques", *Mem. Coll. Sc. Univ. Kyoto, Ser. A Math.*, 31 1958, 219–239.

Chapter 12

Mathematical and Numerical Modeling of Pollution of Lakes

P. Neittaanmäki[1], *V. Rivkind*[2] *and L. Rukhovets*[3]

12.1 Introduction

This article deals with mathematical and numerical modeling of the pollution of lakes and Finnish Gulf. The material is divided into three parts. The first is devoted to models of geophysical hydrodynamics in lakes, gulfs and similar bodies of water. The second concerns the modeling of pollution and biological processes in lakes. In the third part we consider some problems in hollows (which occur very often in lakes and rivers). We apply a discrete Galerkin method which corresponds to a finite element method with a specific form of numerical integration. Results of numerical simulations are presented.

[1]University of Jyväskylä, Department of Mathematics, P.O. Box 35, FIN–40351 Jyväskylä, Finland
[2]St. Petersburg State University, Bibliotechnaja squ 2, Stary Petershof 198904, St. Petersburg, Russia
[3]St. Petersburg Economical and Mathematical Institute of Russian Academy of Science, Ul. Chaikovskogo 1, St. Petersburg 191187, Russia

12.2 Modeling of Water Exchange

The investigation of water exchange is usually reduced to one significant problem of basin dynamics (lakes, gulfs, reservoirs, border-seas), from the problem of total circulation of water masses. Let us consider this task for a basin with islands. Solving the problem in terms of the stream function in the case of a multi-bounded domain we must satisfy along with common boundary conditions some additional constraints which ensure the uniqueness of the evolution of basin surface level.

The total circulation of water masses can be described by the system of equations

$$\frac{\partial U}{\partial t} - \ell V + g\frac{\partial \xi}{\partial x} = -\frac{k(x,y)}{H}U + \frac{\tau_x}{H\rho_w} - f_x \tag{12.1}$$

$$\frac{\partial V}{\partial t} + \ell U + g\frac{\partial \xi}{\partial y} = -\frac{k(x,y)}{H}V + \frac{\tau_y}{H\rho_w} - f_y \tag{12.2}$$

$$\frac{\partial (UH)}{\partial x} + \frac{\partial (VH)}{\partial y} = 0 \tag{12.3}$$

Here $\boldsymbol{V} = (U, V)$ is the vector for the average velocity with respect to the depth, ξ is the deviation of the basin surface from the equilibrium state; H is the depth of the basin (we always assume that $H(x,y) \geq H_0 > 0$); ρ_w is the average density of water; f_x, f_y are terms describing the density stratification influence upon the liquid motion; $k(x,y)$ is the dimensional resistance coefficient near the bottom; (τ_x, τ_y) is the vector of wind tangential stress; ℓ is the Coriolis parameter and g is the acceleration due to gravity.

The above system of equations is used in numerous three-dimensional mathematical models of geophysical hydrodynamics in order to obtain the barotropic component of the velocity field [8–10, 14, 15]. For example, it appears as a result of an averaging procedure applied with respect to the depth parameter to equations of large-scale ocean dynamics. The system (12.1)–(12.3) is considered in a plane domain Ω coinciding with the basin surface. Inside the basin there exist a number of islands. We denote by S_0 the outer boundary of Ω, and $S_1, S_2, \ldots, S_m$ the inner boundaries corresponding to the islands which are denoted respectively by $\omega_1, \omega_2, \ldots, \omega_m$. In case of stable runoff-wind streams, we set the time derivatives to zero in (12.1)–(12.3). The boundary conditions for the system (12.1)–(12.3) are assumed to be of the form:

$$\boldsymbol{V}_n|_{S_0} = v_0(x,y), \quad \boldsymbol{V}_n|_{S_k} = 0, \quad k = 1, 2, \ldots, m. \tag{12.4}$$

Here $\boldsymbol{V}_n$ is the projection of the velocity vector in the direction of the outward normal to the domain boundary, $v_0(x, y)$ is a given function on S_0 (and equal to zero at a solid boundary). From mass conservation, $v_0(x, y)$ must satisfy the following condition

$$\int_{S_0} v_0 H \, ds = 0. \tag{12.5}$$

The stream function is introduced in the usual way by

$$UH = \frac{\partial \Psi}{\partial y}, \quad VH = -\frac{\partial \Psi}{\partial x}. \tag{12.6}$$

which implies

$$\begin{aligned} & - \frac{\partial}{\partial x}\left(\frac{k(x,y)}{H^2}\frac{\partial \Psi}{\partial x}\right) - \frac{\partial}{\partial y}\left(\frac{k(x,y)}{H^2}\frac{\partial \Psi}{\partial y}\right) + \frac{\partial}{\partial x}\left(\frac{\ell}{H}\frac{\partial \Psi}{\partial y}\right) - \frac{\partial}{\partial y}\left(\frac{\ell}{H}\frac{\partial \Psi}{\partial x}\right) \\ & = \frac{\partial}{\partial x}\left(\frac{\tau_y}{H\rho_w} - f_y\right) - \frac{\partial}{\partial y}\left(\frac{\tau_x}{H\rho_w} - f_x\right) \end{aligned} \tag{12.7}$$

The corresponding boundary conditions for stream function can be derived from (12.4) and (12.6) and we have

$$\left.\frac{\partial \Psi}{\partial s}\right|_{S_0} = H v_0, \qquad \left.\frac{\partial \Psi}{\partial s}\right|_{S_1 \cup S_2 \cup \ldots \cup S_m} = 0 \tag{12.8}$$

where $\partial / \partial s$ denotes the tangential derivative on the boundary curve S_k. In order to guarantee the unique solvability of (12.7)–(12.8), we set the following boundary conditions on S_0

$$\Psi|_{S_0} = \int_0^s H v_0 \, ds \equiv \varphi_0(x, y) \tag{12.9}$$

where a direction of traversal for boundary S_0 and a starting point $s = 0$ are assumed specified with $\Psi = 0$ at the starting point. From the second condition in (12.8), the boundary conditions on S_k are

$$\Psi|_{S_k} = \gamma_k, \quad k = 1, 2, \ldots, m, \tag{12.10}$$

where $\gamma_1, \gamma_2, \ldots, \gamma_k$ are unknown constants.

Let us now consider the question of determining the unknown constants γ_i, $i = 1, \ldots, k$. First, recall that the function $\xi(x, y)$, which is the deviation of the basin surface from equilibrium state, can be obtained by integrating

$d\xi = \frac{\partial \xi}{\partial x} dx + \frac{\partial \xi}{\partial y} dy$. Taking into account the multi-bounded character of domain Ω, we also have the constraints

$$\int_{S_k} d\xi = 0, \quad k = 1, 2, \ldots, m. \tag{12.11}$$

These constraints make it possible to determine values of constants $\gamma_1, \gamma_2, \ldots, \gamma_m$. From (12.1)–(12.2), (12.5) and (12.11) it follows that [14,15]

$$\int_{S_k} \left[\frac{k(x,y)}{H^2} \frac{\partial \Psi}{\partial n} + \left(\frac{\partial \tau_s}{H \rho_w} - f_s \right) \right] ds = 0, \quad k = 1, 2, \ldots, m, \tag{12.12}$$

where $\tau_s = \tau_x \cos(n, y) - \tau_y \cos(n, x)$, $f_s = f_x \cos(n, y) - f_y \cos(n, x)$, and $\partial \Psi / \partial n$ is the derivative in the direction of the outward normal.

Finally, we pose the boundary value problem (12.7), (12.9), (12.10), (12.12) as a modified Dirichlet problem [16] to determine $\Psi(x, y)$.

The method of fictitious domains (FDM, or the method of embedded domains) is an effective approach for constructing an approximation (e.g. see [7,13,19,22]) Physically, the fictitious domain method can be interpreted as if the islands are considered "flooded" to a small depth.

The FEM for solving the modified Dirichlet's problem is constructed on the basis of piecewise linear functions (see [11,17,18]). The grid domain Q^h is constructed for the single connected domain $Q \equiv \Omega \cup \overline{\omega}_1 \cup \overline{\omega}_2 \cup \ldots \cup \overline{\omega}_m$ with boundary S_0. This grid generation is performed in such a way that $Q^h \subset Q$ and the usual regularity conditions for the grid are satisfied [11,17,18]. Let S_0^h denote the boundary of Q^h. We triangulate domains ω_k^h (approximating ω_k, $k = 1, 2, \ldots, m$) such that the usual regularity conditions are satisfied. Moreover, it is natural to demand the appropriate correspondence between triangulations of ω_k^h and triangulation of domain $Q^h \setminus (\cup_{k=1}^m \omega_k^h)$.

Next, let us introduce the notations

$$\Omega^h \equiv Q^h \setminus \left(\bigcup_{k=1}^m \omega_k^h \right), \qquad D^h \equiv Q^h \setminus \left(\bigcup_{k=1}^m \omega_k \right). \tag{12.13}$$

We denote by B_h the finite dimensional subspace of $W_2^1(Q^h)$ containing continuous piecewise-linear functions in Q^h. We next denote by $\dot{B}_h$ the subspace of B_h containing functions being equal to zero on S_0^h. We assume functions from $\dot{B}_h$ are extended by zero on the strip $Q \setminus Q^h$ so that one can regard $\dot{B}_h$ to be a subspace of $\dot{W}_2^1(Q)$. We shall denote by $\tilde{\varphi}$ a function $\varphi \in B_h$ to emphasize the fact that this function belongs to subspaces B_h and $\dot{B}_h$.

Let us denote by $\tilde{v}$ the continuous piecewise-linear function defined on S_0^h which coincides with the values of function

$$V_{S_0^h} = \int_0^s H v_m \, ds \tag{12.14}$$

at the points of S_0 which correspond to the nodes. (This correspondence is established with the help of normals to S_0).

The approximate solution $\widetilde{\Psi} \in B_h$ for the problem (12.7), (12.9), (12.10), (12.12) satisfies the integral identity

$$\begin{aligned} &\int_{\Omega^h} \frac{k(x,y)}{H^2} \nabla\widetilde{\Psi} \cdot \nabla\tilde{\varphi} \, dx \, dy + \frac{1}{h^\kappa} \sum_{k=1}^{m} \int_{\Omega_k^h} \nabla\widetilde{\Psi} \cdot \nabla\tilde{\varphi} \, dx \, dy \\ &+ \int_{D^h} \frac{\ell}{H} \left(\frac{\partial\widetilde{\Psi}}{\partial x} \frac{\partial\tilde{\varphi}}{\partial y} - \frac{\partial\widetilde{\Psi}}{\partial y} \frac{\partial\tilde{\varphi}}{\partial x} \right) dx \, dy \\ &+ \int_{D_h} \left[\left(\frac{\tau_y}{H\rho_w} - f_y \right) - \left(\frac{\tau_x}{H\rho_w} - f_x \right) \right] dx \, dy = 0, \end{aligned} \tag{12.15}$$

for all $\tilde{\varphi} \in \dot{B}_h$.

Under the assumption $\Psi \in W_2^2(\Omega)$ and for $k = 1$ in (12.15) the piecewise linear approximation satisfies

$$\left\{ \int_{\Omega^h} \frac{k(x,y)}{H^2} |\nabla(\Psi - \widetilde{\Psi})|^2 \, d\Omega \right\}^{1/2} \leq ch. \tag{12.16}$$

Under the stronger assumption that $\Psi \in W_2^3(\Omega)$ with $k = 3/2$ the estimate improves to

$$\left\{ \int_{\Omega^h} \frac{k(x,y)}{H^2} |\nabla(\Psi - \widetilde{\Psi})|^2 \, d\Omega \right\}^{1/2} \leq ch^{3/2}. \tag{12.17}$$

The estimate (12.17) is proved in the usual way (see [11, 12, 19, 23–25]).

In Figure 12.1, results of calculations using the fictitious domain method are shown by the continuous lines (FDM) and the standard Finite Element Method with dashed lines [13]. We see that results of both methods agree closely.

The problem of water circulation in a basin possessing islands, can be solved not only in terms of the stream function but also in terms of a function determining the level value of the basin surface. In order to obtain the corresponding boundary-value problem we solve the equation system (12.1)–(12.3) with respect to UH and VH [15]. Then after substituting

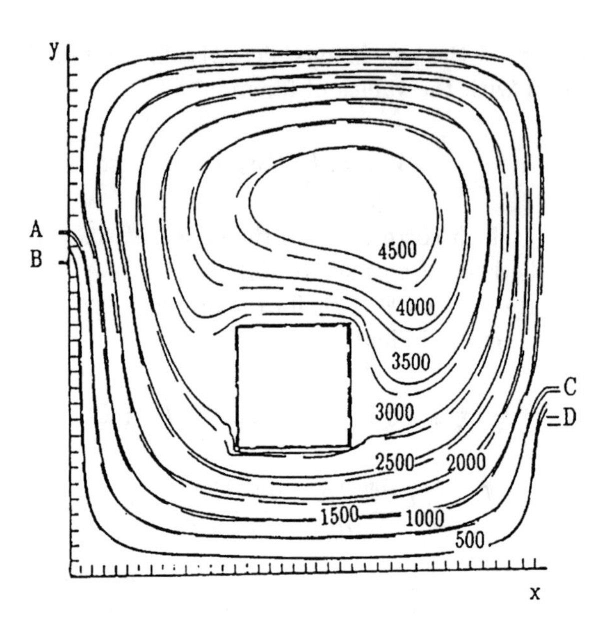

Figure 12.1: Comparison of fictitious domain solution(solid line) with standard FEM (dashed).

the expressions for (UH) and (VH) from equations (12.1), (12.2) into the continuity equation (12.3),

$$\begin{aligned}
&-\frac{\partial}{\partial x}\left(\frac{gkH^2}{k^2+\ell^2H^2}\frac{\partial\xi}{\partial x}\right)-\frac{\partial}{\partial y}\left(\frac{gkH^2}{k^2+\ell^2H^2}\frac{\partial\xi}{\partial y}\right)\\
&-\frac{\partial}{\partial x}\left(\frac{g\ell H^3}{k^2+\ell^2H^2}\frac{\partial\xi}{\partial y}\right)+\frac{\partial}{\partial y}\left(\frac{g\ell H^3}{k^2+\ell^2H^2}\frac{\partial\xi}{\partial x}\right)\\
&=-\frac{\partial}{\partial x}(kg_x+\ell Hg_y)-\frac{\partial}{\partial y}(-\ell Hg_x+kg_y)
\end{aligned}\tag{12.18}$$

where

$$g_x \equiv \frac{H\tau_x}{\rho_w}-H^2f_x, \qquad g_y \equiv \frac{H\tau_y}{\rho_w}-H^2f_y. \tag{12.19}$$

Because it is natural to regard the solution of the problem (12.1)–(12.4) to be continuous up to the boundary, expressions for UH and VH can be used to construct the boundary conditions for equation (12.18). From (12.4) and (12.18), the boundary condition can be expressed in the form

$$\left[\frac{gkH^2}{k^2+\ell^2H^2}\frac{\partial\xi}{\partial x}+\frac{g\ell H^3}{k^2+\ell^2H^2}\frac{\partial\xi}{\partial y}-\frac{(k_xg_x+\ell Hg_y)}{k^2+\ell^2H^2}\right]\cos(n,x)$$

$$\left[\frac{gkH^2}{k^2+\ell^2H^2}\frac{\partial\xi}{\partial y}-\frac{g\ell H^3}{k^2+\ell^2H^2}\frac{\partial\xi}{\partial x}\right.$$
$$\left.-\frac{(-\ell Hg_x+kg_y)}{k^2+\ell^2H^2}\right]\cos(n,y)\bigg|_{S_0\cup S_1\cup S_2\cup\ldots\cup S_m}=-HV_n \qquad (12.20)$$

where $V_n|_{S_0}=v_0,\quad V_n|_{S_k}=0,\quad k=1,2,\ldots,m.$

This completes the description of the boundary-value problem for $\xi(x,y)$. The variational form of the problem (12.18), (12.20) follows as: find $\xi\in W_2^1(\Omega)$ which satisfies the integral identity

$$\int_\Omega\frac{gkH^2}{k^2+\ell^2H^2}\nabla\xi\cdot\nabla\varphi\,dx\,dy+\int_\Omega\frac{g\ell H^3}{k^2+\ell^2H^2}\left(\frac{\partial\xi}{\partial y}\frac{\partial\varphi}{\partial x}-\frac{\partial\xi}{\partial x}\frac{\partial\varphi}{\partial y}\right)dx\,dy$$
$$=\int_\Omega\frac{1}{k^2+\ell^2H^2}\left[(kg_x+\ell Hg_y)\frac{\partial\varphi}{\partial x}+(-\ell Hg_x+kg_y)\frac{\partial\varphi}{\partial y}\right]dx\,dy$$
$$-\int_{S_0}Hv_0\varphi\,ds, \qquad (12.21)$$

for all $\varphi\in W_2^1(\Omega)$.

The above formulation of the boundary value problem (12.18), (12.20) reveals that boundary condition (12.20) enters as a natural boundary condition. In order to have a unique solution, ξ must also satisfy the condition

$$\int_\Omega\xi\,dx\,dy=0. \qquad (12.22)$$

The approximate solution to this problem is constructed on a grid in a simply–connected domain as in the previous case. In contrast to the problem for the stream function, the finite element scheme can now be constructed on a rectangular grid due to the natural boundary condition (12.20).

As before, let us denote by Q^h the grid of triangles with $Q^h\subset Q$. We denote by B_h the subspace of $W_2^1(Q^h)$ containing continuous, piecewise–linear functions on Q^h. The function $\tilde{\xi}(x,y)\in B_h$ is the approximate solution of the problem (12.18), (12.20) if it satisfies the integral identity

$$L_\Omega(\tilde{\xi},\tilde{\varphi})+H^2\int_{Q\backslash\Omega}\nabla\tilde{\xi}\cdot\nabla\tilde{\varphi}\,dx\,dy$$
$$=\int_\Omega\frac{1}{k^2+\ell^2H^2}\left[(kg_x+\ell Hg_y)\frac{\partial\tilde{\varphi}}{\partial x}+(-\ell Hg_x+kg_y)\frac{\partial\tilde{\varphi}}{\partial y}\right]dx\,dy$$
$$-\int_{S_0}Hv_0\tilde{\varphi}\,ds \qquad (12.23)$$

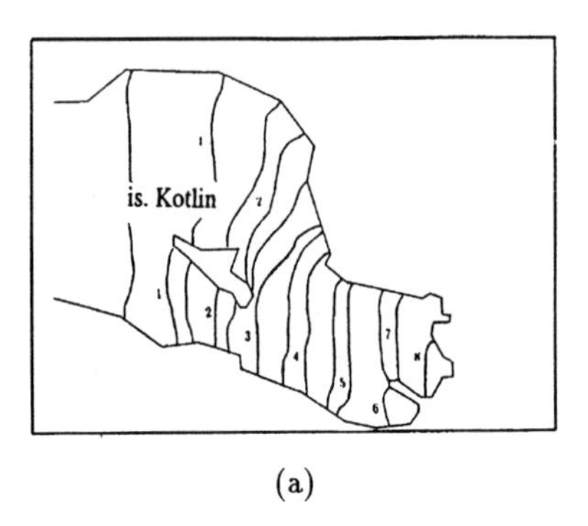

(a)

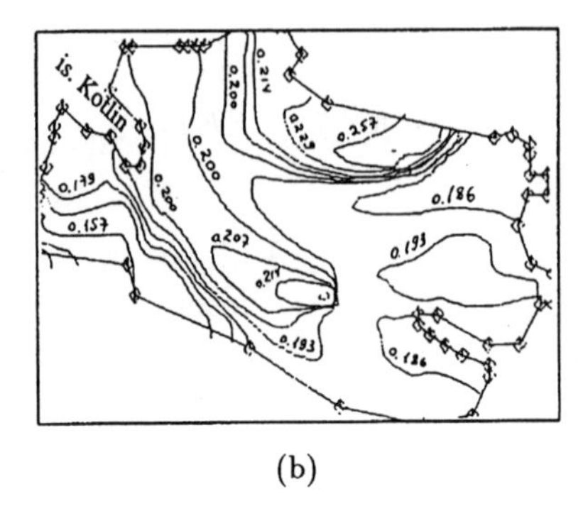

(b)

Figure 12.2: (a) Difference in average and actual water levels in cm without dam and (b) streamlines with dam.

for all $\tilde{\varphi} \in B_h$, where

$$\begin{aligned} L_\Omega(\varphi, \psi) &= \int_\Omega \frac{gkH^2}{k^2 + \ell^2 H^2} \nabla\varphi \cdot \nabla\psi \, dx \, dy \\ &+ \int_\Omega \frac{g\ell H^3}{k^2 + \ell^2 H^2} \left(\frac{\partial\varphi}{\partial y}\frac{\partial\psi}{\partial x} - \frac{\partial\varphi}{\partial x}\frac{\partial\psi}{\partial y} \right) dx \, dy \end{aligned} \tag{12.24}$$

with $\varphi, \psi \in W_2^1(\Omega)$.

If $\xi \in W_2^2(\Omega)$, we obtain the error estimates [4, 5, 11, 17, 18]

$$\int_\Omega \frac{gkH^2}{k^2 + \ell^2 H^2} |\nabla(\xi - \tilde{\xi})|^2 \, dx \, dy + h^2 \int_{Q\setminus\Omega} |\nabla(\xi - \tilde{\xi})|^2 \, dx \, dy \le ch^2$$

$$\left\{ \int_\Omega |\xi - \tilde{\xi}|^2 \, dx \, dy \right\}^{1/2} \le ch^2 \tag{12.25}$$

Remark: We can add a diffusion equation for transport of a pollutant,

$$\frac{\partial C}{\partial t} + U\frac{\partial C}{\partial x} + V\frac{\partial C}{\partial y} = k\Delta C + Gx, \tag{12.26}$$

where C is the concentration of the pollutant, k is a turbulent diffusion coefficient and G is a source. This equation is solved in the same way as the previous hydrodynamic models, except for small k which require upwind schemes [25].

In the next section, results of practical numerical calculations are presented: Figures 12.2 and 12.3 show the structure of a flow and pollution

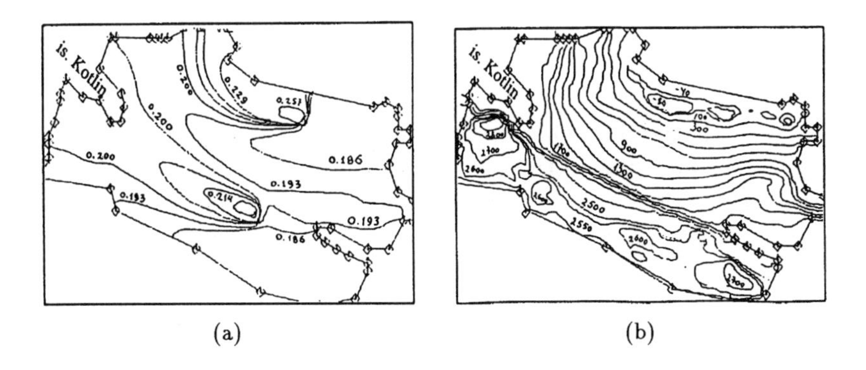

(a) (b)

Figure 12.3: Pollutant concentration from 3 sources.

near the mouth of the river Neva and the beginning of the Finnish Gulf for cases corresponding to the presence and absence, respectively, of a dam across island Kotlin. A uniform grid, with 100×80 elements, was used ($h_x = 500\,\mathrm{m}$, $h_y = 250\,\mathrm{m}$). The difference between the average level and the actual level of water in *cm* is given in Figure 12.2 (a) (without dam) for a prescribed wind velocity of 5 *m/sec*. In Figure 12.2 (b) the streamlines for the calculation with a dam are given. Figure 12.3 shows the pollutant concentration in the river Neva from three sources of pollution.

12.3 Lake Circulation and Transport

The increase of anthropogenic loads on lakes, which appear frequently as main sources of drinking water, enforces the processes of their eutrophication. We give a mathematical model which simulates the water circulation in basins and the admixture of water with biogens, passive organic substance and pollutants.

Suppose that the cartesian coordinate system is chosen so that xy-plane coincides with the unperturbed fluid surface, while the z-axis is directed upward. Let Ω be a three-dimensional basin region. The two-dimensional region corresponding to the unperturbed basin surface is denoted by S_0, the vertical lateral boundary by S_1, and the basin bottom, i.e. the surface $z = -H(x, y)$ by S_2. Let $\partial\Omega = S_0 \cup S_1 \cup S_2$.

The equations of motion are

$$\frac{\partial u}{\partial t} - \ell v = \frac{\partial}{\partial z}\left(k_z \frac{\partial u}{\partial z}\right) - g\frac{\partial \xi}{\partial x} - \frac{g}{\rho_w}\int_z^0 \frac{\partial \rho}{\partial x}\, dz', \qquad (12.27)$$

$$\frac{\partial v}{\partial t} - \ell u = \frac{\partial}{\partial z}\left(k_z \frac{\partial v}{\partial z}\right) - g\frac{\partial \xi}{\partial y} - \frac{g}{\rho_w}\int_z^0 \frac{\partial \rho}{\partial y}\, dz', \quad (12.28)$$

$$\nabla \cdot \boldsymbol{u} = 0 \quad (12.29)$$

where $\rho(t)$ is the density of unsalted water, ρ_w is the average density of water, $\xi(x, y, t)$ is the deviation of the liquid surface from the equilibrium state, and g is the acceleration due to gravity. Here ℓ is the Coriolis parameter which is assumed, for simplicity, to be constant. The heat transfer equation has the form

$$\frac{\partial T}{\partial t} + u\frac{\partial T}{\partial x} + v\frac{\partial T}{\partial y} + w\frac{\partial T}{\partial z} = \frac{\partial}{\partial x}\left(v_x \frac{\partial T}{\partial x}\right) + \frac{\partial}{\partial y}\left(v_y \frac{\partial T}{\partial y}\right) + \frac{\partial}{\partial z}\left(v_z \frac{\partial T}{\partial z}\right) \quad (12.30)$$

The boundary conditions for $\boldsymbol{u} = (u, v, w)$ on the fluid surface $z = 0$ are

$$k_z \frac{\partial u}{\partial z}\bigg|_{S_0} = \frac{\tau_x}{\rho_w}, \qquad k_z \frac{\partial v}{\partial z}\bigg|_{S_0} = \frac{\tau_y}{\rho_w}, \quad (12.31)$$

$$w|_{S_0} = \alpha \frac{\partial \xi}{\partial t}, \qquad \alpha = 0, 1. \quad (12.32)$$

where τ_x, τ_y define the vector of wind tangential stress and $\alpha = 0$ corresponds to the rigid-lid approximation [14]. The boundary conditions at the bottom $z = -h(x, y)$ are

$$k_z \frac{\partial u}{\partial z}\bigg|_{S_2} = k_2 U\sqrt{U^2 + V^2}, \quad k_z \frac{\partial v}{\partial z}\bigg|_{S_2} = k_2 V\sqrt{U^2 + V^2}, \quad (12.33)$$

$$w = -u\frac{\partial H}{\partial y} - v\frac{\partial H}{\partial x} \quad (12.34)$$

where (U, V) is the vector of integral velocity, defined as

$$U \equiv \frac{1}{H}\int_{-H}^0 u(x, y, z, t)\, dz, \quad V \equiv \frac{1}{H}\int_{-H}^0 v(x, y, z, t)\, dz. \quad (12.35)$$

Finally, on the vertical boundary

$$u_n|_{S_1} = v_0, \quad (12.36)$$

where u_n is the projection in the direction of the outward normal n to $\partial\Omega$. Here v_0 is equal to 0 at a solid boundary and v_0 is equal to the known flow values at inflowing river boundaries (R_{in}) and outflowing river boundaries (R_{out}).

Boundary conditions for temperature T on the fluid surface are

$$\frac{\partial T}{\partial n} \equiv \sum_{i=1}^{3} v_i \frac{\partial T}{\partial x_i} \cos(n, x_i) = \frac{1}{c_p^v \rho_w} Q, \quad x_1 = x;\ x_2 = y;\ x_3 = z \tag{12.37}$$

where Q denotes the heat flow and c_p^v is the specific heat of water.

On the entire solid boundary and on outflowing river boundaries

$$\frac{\partial T}{\partial n} = 0, \tag{12.38}$$

whereas on inflowing river boundaries

$$\frac{\partial T}{\partial n} = v_0(T - T_r) \tag{12.39}$$

where T_r is the water temperature of the river.

The solution $T \in W_1^2(\Omega)$ satisfies

$$\begin{aligned}\int_\Omega \frac{\partial T}{\partial t}\phi\, dx\, dy = &-\int_\Omega \left(v_x \frac{\partial T}{\partial x}\frac{\partial \phi}{\partial x} + v_y \frac{\partial T}{\partial y}\frac{\partial \phi}{\partial y} + v_z \frac{\partial T}{\partial z}\frac{\partial \phi}{\partial z}\right) dx\, dy \\ &- \int_\Omega \left(u\frac{\partial T}{\partial x} + v\frac{\partial T}{\partial y} + w\frac{\partial T}{\partial z}\right)\phi\, dx\, dy + \frac{1}{c_p^v \rho_w}\int_{S_0} Q\phi\, ds \\ &+ \int_{R_{\text{in}}} v_0(T - T_r)\phi\, ds,\end{aligned} \tag{12.40}$$

for all $\phi \in W_1^2(\Omega)$.

Setting $\phi \equiv 1$ in (12.40) and integrating by parts the convective terms

$$\begin{aligned}\frac{\partial}{\partial t} c_p^v \rho_w \int_\Omega T\, dx\, dy &= -c_p^v \rho_w \int_{R_{\text{out}}} \boldsymbol{u}_n T\, ds - c_p^v \rho_w \int_{R_{\text{in}}} \boldsymbol{u}_n T_r\, ds \\ &\quad + \int_{S_0} Q\, ds - \alpha c_p^v \rho_w \int_{S_0} \frac{\partial \xi}{\partial t} T\, ds.\end{aligned} \tag{12.41}$$

This equality is referred to as the "heat variation law". We can construct similar integral statements for u and v.

The numerical schemes studied here are based on a modification of the discrete Galerkin method [17–19] on rectangular grids. A significant feature of these schemes is the fact that discrete analogs of conservative laws for mass, heat, and kinetic energy are also valid for these schemes: A rectangular nonuniform mesh is defined by the planes $x = x_i$, $y = y_j$, $z = z_k$ with mesh increments $h_i(x) = x_{i+1} - x_i$, $h_j(y) = y_{j+1} - y_j$, $h_k(z) = z_{k+1} - z_k$. The maximum union of the mesh cells is denoted $\Pi_{ijk} = \{x_i \le x \le x_{i+1},\ y_j \le y \le y_{j+1},\ z_k \le z \le z_{k+1}\}$.

The discrete scheme is constructed as follows: Integrals in (12.40) over Ω are presented as sums of integrals over the mesh cells; Integrals over cells are approximated using a quadrature formula. This implies that

$$\int_{\Pi_{ijk}} \frac{\partial u}{\partial t}\phi\, dx\, dy \approx \frac{1}{8}\sum_{\alpha,\beta,\gamma=0}^{1}\left(\frac{\widehat{T}-T}{h_t}\right)_{i+\alpha,j+\beta,k+\gamma}\phi_{i+\alpha,j+\beta,k+\gamma}\delta_{ijk}, \quad (12.42)$$

where h_t is the time step, $\widehat{T}_{ijk} \equiv T(x_i, y_j, z_k, t+h_t)$, $\delta_{ijk} = \mathrm{mes}(\Pi_{ijk}\cap\Omega)$;

$$\int_{\Pi_{ijk}} V_x \frac{\partial T}{\partial x}\frac{\partial \phi}{\partial x}\, dx\, dy \approx \frac{1}{4}\sum_{\beta,\gamma=0}^{1}(T_x\phi_x)_{i,j+\beta,k+\gamma}(v_x)_{i+1/2,j+1/2,k+1/2}\delta_{ijk}, \quad (12.43)$$

where $(T_x)_{i,j+\beta,k+\gamma} \equiv (T_{i+1,j+\beta,k+\gamma} - T_{i,j+\beta,k+\gamma})/h_i(x)$

$$\int_{\Pi_{ijk}} u\frac{\partial T}{\partial x}\phi\, dx\, dy$$
$$\approx \left[\frac{1}{4}\sum_{\beta,\gamma=1}^{1}(T_x)_{i,j+\beta,k+\gamma}\phi_{i,j+\beta,k+\gamma}\right]\left(\frac{u-|u|}{2}\right)_{i+1/2,j+1/2,k+1/2}\delta_{ijk} \quad (12.44)$$
$$+\left[\frac{1}{4}\sum_{\beta,\gamma=1}^{1}(T_x)_{i,j+\beta,k+\gamma}\phi_{i+1,j+\beta,k+\gamma}\right]\left(\frac{u+|u|}{2}\right)_{i+1/2,j+1/2,k+1/2}\delta_{ijk},$$

and so on with the half-integer indices corresponding to midside and central nodes in the rectangular elements.

From these equations together with the approximation

$$(\nabla\cdot u_h)_{ijk} = -\frac{1}{4}(v_0)_{ijk}\int_{\omega_{ijk}} ds - \delta_k^0\alpha\left(\frac{\hat{\xi}-\xi}{h_t}\right)_{ij}\delta_{ij}^a, \quad (x_i, y_j, z_k)\in\overline{\Omega} \quad (12.45)$$

we get a discrete scheme for T

$$c_p^v\rho_w \sum_{(x_i,y_j,z_k)\in\overline{\Omega}}\left(\frac{\widehat{T}-T}{h_t}\right)_{i,j,k}\delta_{ijk}^a$$
$$= \sum_{(x_i,y_j,0)\in\overline{S}_0} Q_{ij}\delta_{ij}^a - c_p^v\rho_w \sum_{(x_i,y_j,z_k)\in R_{\text{in}}}\frac{1}{4}(v_0T_r)_{ijk}\int_{(R_{\text{in}})_{ijk}} ds$$

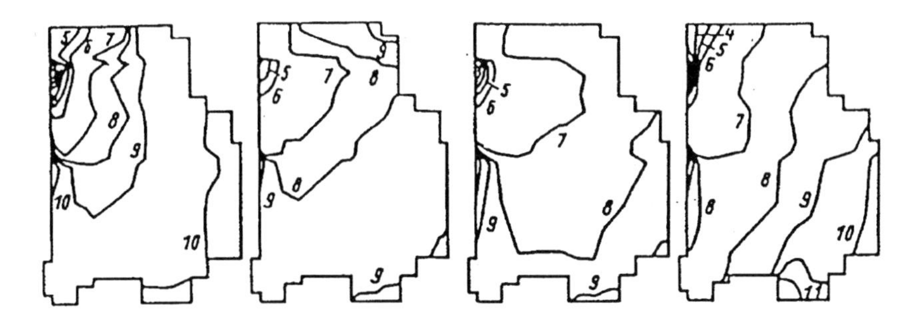

Figure 12.4: The concentration of the pollutant on the surface of Ladoga lake: (a) summer, (b) autumn, (c) winter, (d) spring.

$$-c_p^v \rho_w \sum_{(x_i,y_j,z_k)\in R_{\text{out}}} \frac{1}{4}(v_0\widehat{T})_{ijk} \int_{(R_{\text{out}})_{ijk}} ds$$

$$-c_p^v \rho_w \alpha \sum_{(x_i,y_j,0)\in \overline{S}_0} \widehat{T}_{ij0} \left(\frac{\hat{\xi}-\xi}{h_t}\right)_{ij} \delta_{ij}^a. \tag{12.46}$$

The mesh continuity equation (12.45) plays an important role in this discrete model. Note that for nodes on the surface of Ω, equation (12.45) provides the mesh counterpart of boundary conditions (12.32), (12.34), (12.36).

Discrete systems for the equations of motion are constructed similarly by element quadrature. However, equation (12.45) for u and v usually is used for the determination of mesh function w. This function satisfies two boundary conditions: one on the surface S_0 and the second on the bottom.

Figure 12.4 shows the seasonal concentration of pollutant on the surface of Ladoga lake. The spring (source) of pollution is at the point A at a depth of $6m$ (due to the outlet of the Priosersk paper factory). The values for the contour labels are as follows: '1' $\equiv 1mg/l$; '2' $\equiv 0.8mg/l$; '3' $\equiv 0.6mg/l$; '4' $\equiv 0.5mg/l$; '5' $\equiv 0.4mg/l$; '6' $\equiv 0.3mg/l$; '7' $\equiv 0.2mg/l$; '8' $\equiv 0.19mg/l$; '9' $\equiv 0.18mg/l$; '10' $\equiv 0.16mg/l$; '11' $\equiv 0.008mg/l$. Related calculations are given in [1,2,6].

On the basis of this three-dimensional model and algorithms, it is also possible to calculate the production of biogenic and organic matter in lakes. The most difficult aspect of this modeling is to define the source parameters and coefficients of interactions between zooplankton and phytoplankton. Using experiments, these coefficients are given in the model presented in [3]. In Figure 12.5, the concentration of zooplankton and phytoplankton in mg/l

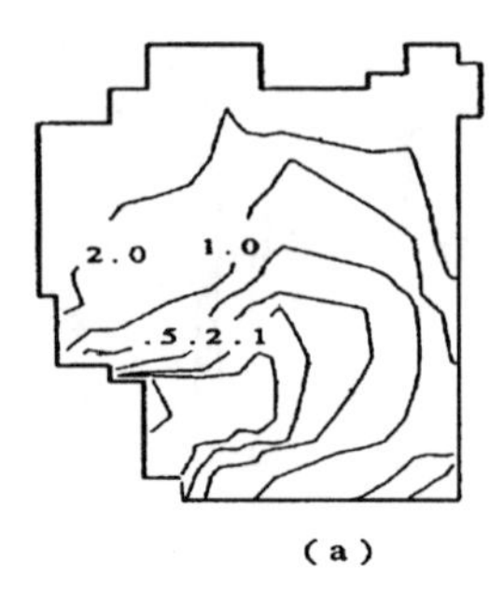

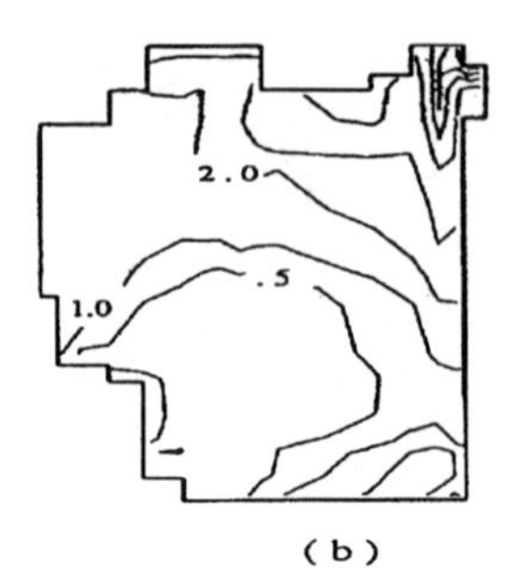

Figure 12.5: The concentration of zooplankton (a) and phytoplankton (b).

are shown.

12.4 Flow and Pollution in a Deep Hollow

Here we consider a nonstationary problem with flow and diffusion of a pollutant around a hollow in the plane. The formulation permits arbitrary reliefs, velocity profiles at the entrance region, initial distribution of a concentration of a pollutant and sources acting on the boundaries. The mathematical model involves the nonstationary Navier–Stokes equations, the continuity equation for incompressible fluid flow and the transport equation for a pollutant.

The nondimensional parameters are: $Re_T = L_B u_\infty / k_\infty = 160$, the turbulent Reynolds number; $Pr_T = k/k_\infty = 1$, the diffusion Prandtl number which characterizes the relation between the turbulent viscosity coefficient and the diffusion coefficient R; $u_\infty = 4\,\mathrm{m/s}$ and $k_\infty = 5\ \mathrm{m^2/s}$ (see Figure 12.6).

The initial domain Ω of integration is bounded by the solid oblique wall, inlet and outlet vertical cross sections and the upper horizontal boundary. A typical control volume in which one can observe the solution development with respect to time is shown in Figure 12.6. The height of the integration domain, the characteristic longitudinal scale value of the quarry and its depth are denoted correspondingly by H, L_B and h_{Γ_1}. At the domain entrance $x = 0$, $f_u(y)$ and $f_c(y,t)$ are the distribution of wind velocity, u and a concentration of a pollutant, c with respect to the altitude parameter. On the upper boundary $y = H$, $u = f_u(H)$, $w = 0$ is applied and $c = 0$. At the outlet cross section $x = X_{\text{exit}}$, $\partial u/\partial x = \partial v/\partial x = \partial c/\partial x = 0$ apply. Fi-

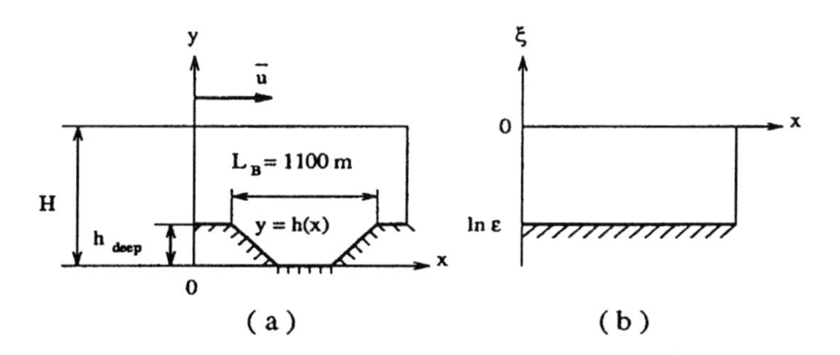

Figure 12.6: (a) The domain of integration in physical coordinates x, y and (b) its transformation to rectangular form in oblique coordinates x, ξ.

nally, on the solid surface $y = h(x)$ we set the no slip condition for velocities $u = w = 0$ and

$$(-1/Pr_d)\frac{\partial c}{\partial n} = L_B Q/R_\infty D R \varphi_1(x)\varphi_2(t), \tag{12.47}$$

where n is the normal to the surface, and the value of Q and functions $\varphi_1(x)$ and $\varphi_2(t)$ characterize correspondingly the power of the boundary sources, nonregularity of their distribution on the quarry surface and their variation in time.

At $t = 0$ we specify the initial conditions

$$u = v = p = 0, \qquad c = c_0(x, z). \tag{12.48}$$

The vertical Cartesian coordinate is transformed to a new variable ξ according to

$$\xi = \ln\left(\frac{y(1-\varepsilon) + H\varepsilon - h(x)}{H(x) - h(x)}\right) \tag{12.49}$$

where $h(x)$ is an arbitrary function describing the relief and ε is the tension parameter. This change of variable maps the domain to a rectangular region. The grid is graded toward the solid boundary where larger solution gradients are known to occur. The simplification in the domain is offset by the increased complexity of the equations under the stated transformation.

For the numerical solution we used Chorin's two-step method [20]. On the first step we solve the systems using derivatives of p from the previous step, etc. For the transport equation we use upwind finite elements and variational difference schemes.

Figure 12.7 illustrates dissipation from the sources. The momentary overlap of the steady stream upon the initial concentration field is due to

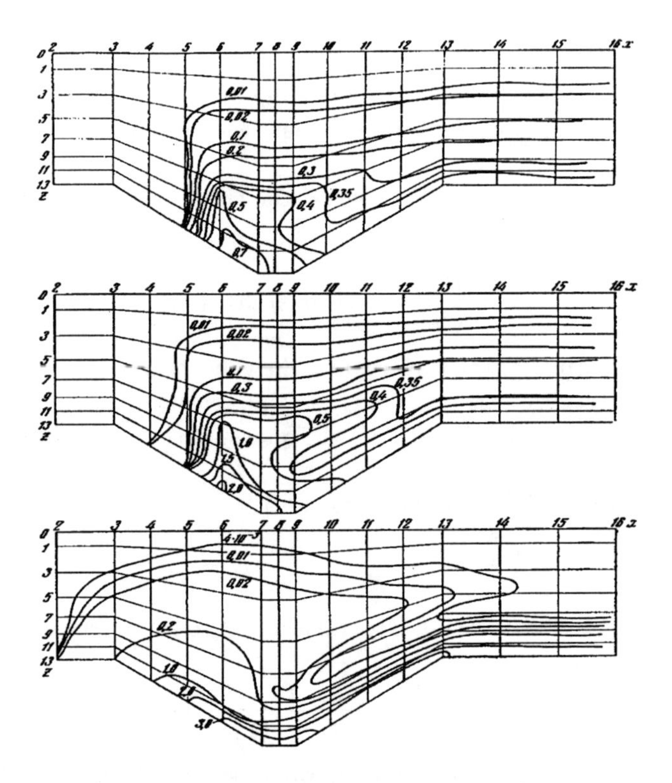

Figure 12.7: The concentration $\overline{c} = c/0.7 \cdot 10^{-6}\,\mathrm{kg/m^3}$ of a pollutant in time: $t = 0.01$; $t = 0.5$; $t = 0.9$. $\mathrm{Re}_T = 160$, $\mathrm{Pr}_T = 1$.

rotational movement. The trajectories of this motion are closed curves and thereby confine material. Nevertheless, the turbulent layer near the line of separation between the free shift layer and recirculation zone transports material into the mixing layer from which it moves to the main stream.

12.5 Conclusions

The models and numerical methods discussed here may be used for many other environmental problems (for example, for ocean problems, the ventilation of quarries; see [21] for further details).

References

[1] Astrakhantsev, G. P., N. B. Egorova and L. A. Rukhovets, "Numerical modelling of the year circulation of deep lakes", *Dokl. Akad. USSR*, 296, 1987, (In Russian).

[2] Astrakhantsev, G. P., N. B. Egorova and L. A. Rukhovets, "Modelling of the streams and thermal regime of Ladoga Lake", Preprint of Institute of Limnology, Russian Academy of Sciences, Leningrad, 1988, (In Russian).

[3] Astrakhantsev, G. P., V. Menshutkin, I. Pisulin and L. A. Rukhovets, "Mathematical model for a research of a reaction ecosystem of the Ladoga lake", Preprint, St. Petersburg, 1993, (In Russian).

[4] Astrakhantsev, G. P. and L. A. Rukhovets, "A numerical method for solving the stratified basin dynamics problem I", *Sov. J. Numer. Anal. Math. Modelling*, 41, 1–18, 1989.

[5] Astrakhantsev, G. P. and L. A. Rukhovets, "A numerical method for solving the stratified basin dynamics problem II", *Sov. J. Numer. Anal. Math. Modelling*, 42, 87–98, 1989.

[6] Astrakhantsev, G. P., L. A. Rukhovets and N. B. Egorova, "Mathematical modelling of distribution of pollution in basins", *Journal Meteorology and Hydrology*, 6, 71–79, 1988, (In Russian).

[7] Glowinski, R., T. W. Pan, J. Periaux and M. Ravachol, "A fictitous domain method for the incompressible Navier–Stokes equations", in E. Quabe, J. Periaux and A. Samuelson (eds.), *Finite element methods in the 90's*, 410–417, Springer-Verlag, Berlin, 1991.

[8] Gordeev, R., B. Kagan, G. Marchuk and V. Rivkind, "A numerical method for the solution of tidal dynamics equations and the results of its applications", *J. Computational Phys.*, 13, 15–34, 1973.

[9] Gordeev, R., B. Kagan and V. Rivkind, "Numerical methods for solutions of dynamics equations of tides in the world ocean", *Dokl. Akad. USSR*, 2092, 340–344, 1973, (In Russian).

[10] Kamenkovich, V. M., *The Frame of Ocean Dynamics*, Gidrometeoizdat, 1973, (In Russian).

[11] Kříźek, M. and P. Neittaanmäki, "Finite element approximation of variational problems and applications", in *Pitman Monographs and Surveys in Pure and Applied Mathematics 50*, Longman Scientific & Technical, Harlow (Copubl. Wiley, New York), 1990.

[12] Ladyzhenskaya, O. A., *Mixed Problem for a Hyperbolic Equation*, Gostekhizdat, Moscow, 1953, (In Russian).

[13] Lewantowich, R. and L. A. Rukhovets, "About one approximate method to calculate the streams in nonconnected domains", *Journal of Oceanology*, 281, 35–41, 1988, (In Russian).

[14] Marchuk, G. I., B. P. Dymnikov and V. B. Zalesskyi, *Mathematical Models in Geophysical Hydrodynamics and Numerical Method of Their Realization*, Gidrometeoizdat, Leningrad, 1987, (In Russian).

[15] Marchuk, G. I., V. P. Kockergin, A. S. Sarkisyan *et al.*, *Mathematical Modelling of the Ocean Circulation*, Nauka, Novosibirsk, 1984, (In Russian).

[16] Mushelischvili, N. I., *Singular Integral Equations*, Nauka, Moscow, 1968, (In Russian).

[17] Oganesyan, L. A., V. J. Rivkind and L. A. Rukhovets, *Variational-Difference Method for Solving Elliptic Equations. I, II*, 5,8 of *Differential equations and their applications*, Vilnius, Lithuania, 1973, 1974, (In Russian).

[18] Oganesyan, L. A. and L. A. Rukhovets, *Variation-Difference Method for Solving Elliptic Equations*, Arm. SSR, Acad. Sci., Erevan, 1979, (In Russian).

[19] Rivkind, V., "Estimates of the rate of convergence of solutions difference schemes to exact solutions elliptical equations with discontinuous coefficients and about one fiction domain method of solving the Dirichlet problem", *Dokl. Akad. Nauk SSSR*, 1496, 1264–1268, 1963, English transl. in *J. Soviet Math* 4, 1964, pp. 546–550.

[20] Rivkind, V. and B. Epstein, "Projection difference schemes for solving Navier–Stokes equations in orthogonal curvilinear coordinate systems", *Vestnik Leningrad Univ.*, 3, 56–63, 1974, English transl. Vestnik Leningrad Univ. 7, 1979 pp. 269–277.

[21] Rivkind, V., V. Markov and E. Potashnik, "Two dimensional mathematical model of process of natural quarry ventilation", *J. Physics of Atmosphere and Ocean XIV*, 5, 500–509, 1978, (In Russian).

[22] Rukhovets, L. A., "To the question of construction the variation-difference schemes", *Journal of Computational Mathematics and Mathematical Physics*, 123, 781–785, 1972, (In Russian).

[23] Rukhovets, L. A., "The fictitious domain method for the problem of established winding currents", *Numerical Methods of Mechanics of Medium*, 123, 98–116, 1981, (In Russian).

[24] Rukhovets, L. A., "Mathematical modelling of the water exchange and distribution of pollution in the Neva-Gulf", *Meteorology and Hydrology*, 7, 78–87, 1982, (In Russian).

[25] Rukhovets, L. A., *Numerical Modelling of the Climatic Circulation of Stratified Lakes*, Doctor of sciences dissertation, Institute Computer Center of Sibrian Branch of Russian Academy of Sciences, Novosibirsk, 1990, (In Russian).

Chapter 13

Radionuclide Release, Ground-Water Transport and Geochemical Processes

R.J. MacKinnon[1] *and T.M. Sullivan*[2]

13.1 Introduction

During the past several years Brookhaven National Laboratory has been developing a family of computer models that predict the release and groundwater transport of radionuclides from low-level nuclear waste (LLW) disposal facilities. These models are capable of predicting waste container degradation, waste form leaching, and radionuclide migration with retardation and first-order decay in unsaturated or saturated porous media. Two of these models, DUST (Disposal Unit Source Term) [14] and BLT (Breach,Leach,Transport) [17] are single-species (single-solute) codes. DUST is a one-dimensional finite-difference model which was developed to permit simulation of a large number of simple cases, yet flexible enough to allow simulation of a wide range of conditions. The DUST code, because of its flexibility and ability to compute release rates quickly, is ideally suited for screening studies to determine, for example, the radionuclides released at the highest rate and upper bounds to release rates. BLT is a two-dimensional finite-element model designed to permit more detailed analyses that take

[1]Daniel B. Stephens & Associates, Inc. Albuquerque, NM, 87109

[2]Brookhaven National Laboratory, Upton, NY 11973

into account the effects of geometry and material anisotropy due to different facility designs and hydrogeologic conditions.

Although DUST and BLT are applicable to a wide range of LLW performance assessment problems, they can only calculate the release and aqueous phase transport of one chemical species at a time. In some applications, this limitation precludes accurate assessment of potentially important processes of radionuclide decay and geochemical interactions between the various chemicals and host porous media. In particular, radionuclide sorption and solubility properties are dependent on spatial and temporal variations in the chemical environment, both in the disposal unit and along the migration path. As a consequence, modeling radionuclide transport without considering solution chemistry can result in unrealistic predictions of long-term radionuclide migration. Another important situation involves the capability for adequately accounting for the chain-decay process. This capability is essential when decay products pose potentially greater health risks and exhibit substantially different mobility characteristics than the parent. In such cases, satisfactory representations of chain decay may require a multispecies modeling capability; that is, one transport equation for each parent and daughter species may be required to adequately predict total radiological exposures.

To better address the issues of chain decay and the impact of the chemical environment on facility performance, we have developed two extended versions of BLT, BLT-MS (Breach, Leach, Transport-Multiple Species) and BLT-EC (Breach, Leach, Transport- Equilibrium Chemistry). Both of these codes simulate waste container degradation, waste form leaching, and multiple-species transport in two dimensions. BLT-MS will simulate both sequential and branched decay wherein each parent radionuclide may decay to two daughter nuclides. BLT-MS is discussed fully in a separate report [13]. BLT-EC is designed to simulate important chemical processes and their impact on radionuclide transport in addition to sequential and branched radioactive decay.

In this chapter, we discuss and demonstrate some current capabilities of BLT-EC. We begin by providing a simple conceptualization of a hypothetical low-level waste disposal facility and associated radionuclide release and transport processes. We next summarize the governing mathematical equations for flow, radionuclide release, transport, and chemical processes. The numerical models, basic code structure, and code-user implementation are then briefly described. Finally, we demonstrate the application of BLT-EC on an example problem.

13.2 Important Processes and Existing Models

A low-level waste disposal facility is a complex open chemical system. The chemical composition and physical characteristics of a facility's contents, engineered barriers (if any), and surrounding soil all contribute to the chemical environment within and in the vicinity of a facility. The chemical environment within the facility influences, and may even control, releases from the waste forms. The chemical environment in the facility, together with the site geochemistry and hydrogeology, determine the transport rates of radionuclides beyond the facility.

Figure 13.1 depicts a contamination scenario associated with the release of radionuclides and their subsequent migration into a ground water system comprised of unsaturated and saturated zones. A disposal unit may contain a multi-layered cover to divert water away from the waste, an engineered barrier to further reduce water flow to the wastes (for trench disposal there is no engineered barrier), and metallic, concrete, or high density polyethylene waste containers. The waste exists in several forms, a partial list of which includes: wastes solidified by one of several processes (e.g., cement, VES, bitumen), dewatered resins, activated metals, and dry active solids (e.g. contaminated paper, cloth, rubber, plastic, glass, etc.). Unsaturated zones are typically shallow and range to depths as shallow as a few meters, to as deep as several hundred meters in arid environments. The extent of saturated zones is determined in the vertical direction by geologic stratigraphy and in the horizontal direction by geologic features such as faults or the location of man-made receptors. Horizontal distances of interest may range up to thousands of meters.

The first step in the ground-water contaminant transport process begins when water infiltrates into the disposal unit and comes in contact with the waste containers. This contact initiates container corrosion, which eventually results in container breach and communication between water and waste form. As a result, contaminants transfer from the waste form into the water and are subsequently transported into adjacent backfill material and soil exterior to the disposal facility. As depicted in Figure 13.1, the bulk of the contaminants escaping the disposal facility are migrating downward with the infiltrating water. In time, some of the aqueous leachate may eventually reach the saturated zone, mix, and spread laterally in the flowing ground water. Dissolved contaminant concentrations in the migrating leachate are primarily determined by waste, waste form, and container characteristics and limited by sorption, ion-exchange, and precipitation reactions. These reactions are strongly affected by contaminant speciation, which in turn is influ-

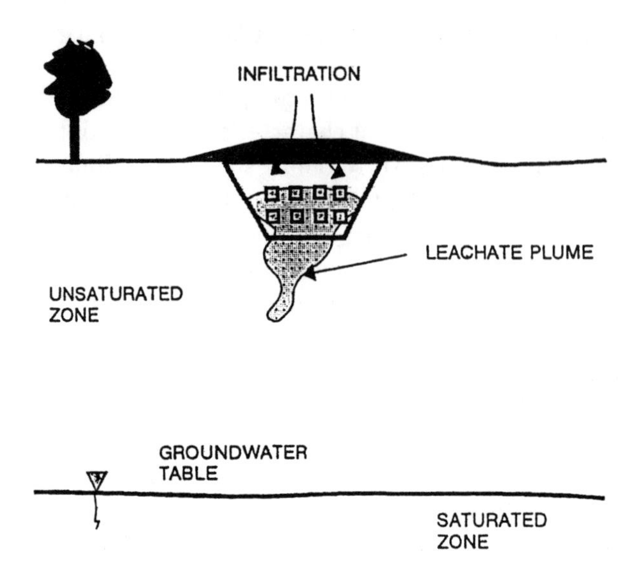

Figure 13.1: Schematic diagram of low-level waste leachate infiltration into the subsurface.

enced by ground water and leachate composition, pH, and redox state of the leachate. Additional mechanisms including microbially-mediated reactions, colloid facilitated transport, container and engineered barrier degradation, and radioactive decay also combine to influence dissolved concentrations of contaminants.

Based on the foregoing conceptualization, modeling the release of contaminants from a LLW disposal unit can be divided into five processes: (1) water flow in the unit and surrounding subsurface, (2) container degradation to the point that water can contact the waste form (breach), (3) release of the contaminants from the waste form to the water (leach), (4) transport of the contaminants within and beyond the disposal unit, and (5) geochemical interactions between the contaminants, natural materials in the ground water, and the host soil. These processes are not independent of one another and therefore proper simulation requires coupling between their corresponding models.

Several models have been developed in the past decade to assess the role of coupled transport and chemical processes on the subsurface migration of reactive solutes, notably [2–6, 9, 12], and [21]. The most comprehensive of these models are in [5, 6, 21]. Liu and Narasimhan [5] developed the model, DYNAMIX, to simulate redox-controlled, multispecies, reactive chemical transport in two dimensions. This model represents both kinetic and equilibrium chemical interactions between aqueous and solid phases. Yeh and Tripathi [21] developed the two-dimensional, finite-element model, HYDROGEOCHEM, for simulating transport of reactive multi- species solutes in two dimensions under the assumption of chemical equilibrium. This model employs a two-step solution approach wherein the transport and chemistry equation systems are solved separately and sequentially in iterative fashion. The model is designed for application to heterogeneous, anisotropic, variably saturated porous media under transient or steady state flow conditions, and simulates the chemical processes of dissolution-precipitation, complexation, sorption, ion exchange, redox and acid-base reactions.

Although existing models treat many of the physical and chemical phenomena identified to be influential in radionuclide transport, these models do not account for container degradation, leaching of radionuclides from waste forms, engineered barrier degradation, and radioactive decay processes. An additional shortcoming with many existing reactive transport models, as with HYDROGEOCHEM for example, is that their implementation requires the user to identify relevant reactions *a priori* and construct an input file containing the corresponding stoichiometry and thermodynamic data. This process requires fairly extensive knowledge of the pertinent geochemistry and quickly becomes cumbersome and time consuming for most practical problems involving several chemical reactions. Moreover, this approach is prone to misapplication through improper identification of important reactions. BLT-EC is designed to alleviate many of the aforementioned shortcomings by incorporating realistic models for waste form release processes and a thermodynamic database which permits automatic selection of pertinent reactions and thermodynamic data [8].

13.3 Governing Equations

Consider a variably water-saturated plane porous region occupying domain Ω and enclosed by boundary Γ. The mathematical description of important processes occurring in this region is based upon flow and transport equations governing the movement and conservation of ground water and chemical species, mass transfer equations governing the release of radionuclides from waste forms, and reaction equations governing the interaction of the various chemical species with each other, as well as with the host porous media. These equations are presented below.

Water Flow in Variably Saturated Porous Media

Advection of radionuclides with the ground-water flow field will often be the dominant process controlling the release and migration of radionuclides from a LLW facility. The ground water flow model is therefore a crucial component of disposal facility analyses because of the need to define the ground-water flow field, both in unsaturated and saturated (variably saturated) porous media. The governing equation for variably saturated flow presented here is similar to that used in previous formulations [23].

The flow field in region Ω can be described by the well known multidimensional form of Richard's equation:

$$\frac{\partial \theta_a}{\partial t} = F\frac{\partial h}{\partial t} = -\nabla \cdot \boldsymbol{v}_a + Q_a \tag{13.1}$$

with

$$\boldsymbol{v}_a = -\boldsymbol{K}^c \cdot \nabla H \tag{13.2}$$

where the subscript a is used to denote the aqueous or ground-water phase. In (13.1), the hydraulic conductivity tensor $\boldsymbol{K}^c[cm{\cdot}s^{-1}]$, coefficient $F[cm^{-1}]$, total pressure head $H[cm]$, and source-sink term $Q_a[cm^3 \cdot cm^{-3} \cdot s^{-1}]$ are given, respectively, by

$$\boldsymbol{K}^c = \frac{\boldsymbol{K} r_{ra}\rho_a g}{\mu_a}, \quad H = h + z = \frac{P_a}{\rho_a g} + z, \quad F = \frac{d\theta_a}{dh}, \quad Q_a = \frac{q_a}{\xi_a} \tag{13.3}$$

where z is the potential head (cm), h is the pressure head (cm), q_a is the molar source/sink term $[mol \cdot cm^{-3} \cdot s^{-1}]$, θ_a is the volume fraction of the aqueous phase $[cm^3 \cdot cm^{-3}]$, ξ_a is the molar density of the aqueous phase $[mol \cdot cm^{-3}]$, μ_a is the aqueous phase viscosity $[g \cdot s^{-1} \cdot cm^{-1}]$, p_a is the phase pressure $[cm^{-1} \cdot g \cdot s^{-2}]$, ρ_a is the mass density $[g \cdot cm^{-3}]$, and g is gravitational acceleration $[cm \cdot s^{-2}]$.

The initial distribution of pressure head

$$h = h_I(x, y, z) \tag{13.4}$$

in region Ω is required for the solution of (13.1).

The boundary conditions include:

Dirichlet conditions (specified head)

$$h = h_D(x, y, z, t), \quad (x, y, z) \in \Gamma_D, \quad t > 0 \tag{13.5}$$

Neumann conditions (specified dispersive flux)

$$-\boldsymbol{K}^c \cdot \nabla H \cdot \boldsymbol{n} = q_N \cdot \boldsymbol{n} \quad (x, y, z) \in \Gamma_N, \quad t > 0 \tag{13.6}$$

Cauchy (mixed) conditions

$$\nabla H \cdot \boldsymbol{n} + \beta H = f_c \cdot \boldsymbol{n} \quad (x, y, z) \in \Gamma_C, \quad t > 0 \tag{13.7}$$

where H_D and q_N are the prescribed total head $[cm]$ and Neumann flux $[cm^3 \cdot cm^{-2} \cdot s^{-1}]$, β and f_c are known functions, and Γ_D, Γ_N, and Γ_C are segments of boundary Γ on which the corresponding boundary conditions apply.

Multispecies Aqueous-Phase Transport with Chemistry

In this section, we present equations that couple transport and chemical reactions under the assumption that all chemical reactions are sufficiently fast that the principle of local equilibrium is applicable. Other major assumptions include:

1. Volume changes in the precipitated phase are negligible.

2. The system temperature is invariant.

3. Microbially mediated reactions are ignored.

4. Colloid facilitated transport is negligible.

5. Radioactive decay reactions are homogeneous and occur only in the aqueous, precipitated, and adsorbed phases.

6. Intraphase transport occurs in the aqueous phase only.

To facilitate the equation presentation, it will be helpful to first define some terminology: The ground water, or aqueous phase, may contain a number of chemical elements and free electrons, each of which are referred to hereafter as "elements". "Elements" are independent basis entities that linearly combine to produce all "species" in the system [19]. For example, two chemical "elements" can combine to form a chemical "species". In similar fashion, a chemical"element" can combine with an electron "element" to form a reduced chemical "species." A more detailed discussion of this notion can be found in [1] and [20].

Now, consider a multiphase, permeable system composed of N_e elements. These elements combine to form N_s distinct chemical species which participate in N_{cr} chemical reactions and N_{dr} radioactive decay reactions. These reactions take place within and between a single mobile fluid phase (aqueous) and two stationary phases (soil and precipitate). Within the aqueous phase, species transport can occur due to bulk phase advection, as well as molecular diffusion and mechanical dispersion effects. Interphase transfer of mass can occur due to the heterogeneous reactions of precipitation-dissolution and sorption. Additional transformations due to radioactive decay can also occur. In the following, we denote phases as follows: the aqueous phase, $\alpha = a$, precipitated phase $\alpha = p$, and sorbed phase $\alpha = s$. The starting point for our presentation is the description of equilibrium chemical reactions.

Equilibrium Chemical Reactions

The various chemical species include radioactive and nonradioactive materials introduced into the subsurface by waste disposal and those naturally present in the subsurface. The molecular formula, $B_{j\alpha}$, of each species j in phase α can be represented by

$$B_{j\alpha} = \sum_{k=1}^{N_e} \epsilon_{jk}^{\alpha} e_k \tag{13.8}$$

for $j = 1, 2, ..., N_s$, $\alpha = a, p, s$, where $B_{j\alpha}$ is a symbol vector for the species (e.g., $B_{1a} = H_2^{16}O, B_{2a} = {}^{238}U^{16}O_2$, etc.), e_k is the symbol vector for elements (e.g. ${}^3H^+$, ${}^{14}C^{2+}$, etc.), and ϵ_{jk}^{α} is the formula matrix for phase α, with each row vector denoting the quantity of each element in the corresponding species. Note that by definition, the set of elements e_k is comprised of the minimum number of elements necessary to define all species present in the system.

Key reactions involving species $B_{j\alpha}$ include complexation, acid-base, oxidation-reduction, precipitation-dissolution, sorption, and ion-exchange. These reactions may be written in symbolic form as

$$0 \longleftrightarrow \sum_{\alpha} \sum_{j=1}^{N_s} \nu_{jr}^{\alpha} B_{j\alpha} \tag{13.9}$$

for $r = 1, 2, ..., N_{cr}$, where the notation ($\longleftrightarrow$) is used to indicate a reaction and ν_{jr}^{α} and B_{jr}^{α} are stoichiometric reaction matrices whose members are negative for reactant species and positive for product species.

Equation (13.9) can be rewritten in symbolic matrix form as

$$0 \longleftrightarrow \boldsymbol{NB} \tag{13.10}$$

As a consequence of the assumption of thermodynamic equilibrium, if the rank of reaction matrix $\boldsymbol{N}$ is M, it follows that the collection of N_s chemical species can be partitioned into M independent reactant species, referred to herein as "components", and $(N_s - M)$ dependent or "product" species [1,19] that is, $\{c_{1\alpha}, ..., c_{M\alpha} : c_{(M+1)\alpha}, ..., c_{N_s\alpha}\}$. The total number of components, M, is further divided into two sets $\{c_{1\alpha}, ..., c_{M_a\alpha} : c_{(M_a+1)\alpha}, ..., c_{M\alpha}\}$ consisting of M_a aqueous components and $M_s = M - M_a$ adsorbent components (including ion-exchange sites). In addition, each aqueous component can exist in each phase present (i.e., aqueous, precipitate, sorbed) in the system. Moreover, each product species can be expressed as a linear combination of the components, and no component can be expressed as linear combinations of the other components. Thus, $(N_s - M)$ linear combinations of M components are chemical reactions which form $(N_s - M)$ product species. These reactions can be separated into aqueous complexation, dissolution-precipitation, and sorption reactions as follows:
Aqueous Phase Reactions

$$\sum_{k=1}^{M} \nu_{ki}^{a} B_{ka} \longleftrightarrow B_{ia} \quad \forall \text{ product species } i \in \boldsymbol{A} \tag{13.11}$$

Dissolution-Precipitation Reactions

$$\sum_{k=1}^{M} \nu_{kl}^{a} B_{ka} \longleftrightarrow B_{lp} \quad \forall \text{ product species } l \in \boldsymbol{P} \tag{13.12}$$

Sorption Reactions

$$\sum_{k=1}^{M} \nu_{km}^{a} B_{ka} + \sum_{k=1}^{M} \nu_{ks}^{s} B_{ks} \longleftrightarrow B_{ms} \quad \forall \text{ product species } m \in \boldsymbol{S} \tag{13.13}$$

where $\boldsymbol{A}, \boldsymbol{P}$, and $\boldsymbol{S}$denote the sets of species that may be present in the aqueous, precipitated, and sorbed phases, respectively.

To simplify the representation of radioactive decay reactions, we make the assumption that radionuclides are component species. This assumption, in conjunction with assumption (13.3), permits the following simplified representation:

Radioactive Decay Reactions

$$\sum_{k=1}^{M} \omega_{kr}^{\alpha} B_{ka} \longrightarrow 0 \tag{13.14}$$

for $\alpha = a, p, s, r = 1, ..., N_{dr}$, and where ω_{kr}^{α} is the component species stoichiometric radioactive decay matrix.

Under the assumption of chemical equilibrium, the mass action relations corresponding to reactions (13.11)-(13.13) are:
Aqueous Complexation

$$Y_{ia} = K_{ia} \prod_{k=1}^{M} X_{ka}^{\nu_{ki}^{a}} \tag{13.15}$$

Precipitation-Dissolution

$$Y_{lp} = K_{lp} \prod_{k=1}^{M} X_{ka}^{\nu_{kl}^{a}} \tag{13.16}$$

Sorption

$$Y_{ms} = K_{ms} \prod_{k=1}^{M} X_{ka}^{\nu_{km}^{a}} \prod_{k=1}^{M} X_{ks}^{\nu_{km}^{s}} \tag{13.17}$$

where $\prod$ indicates the product over all component species $k, K_{j\alpha}$ is the equilibrium constant for the formation of species j in phase α, and $X_{k\alpha}$ and $Y_{i\alpha}$ are thermodynamic activities of component species and product species $[mol \cdot cm^{-3}]$, respectively.

The net rate of radioactive decay associated with each decay reaction in equation (13.14) is given by

$$r_{k\alpha}^{d} = \sum_{k=1}^{N_{dr}} \omega_{kr}^{\alpha} f_{r}^{d} = -\lambda_{j} \theta_{a} c_{ja} + \sum_{p} \lambda_{pj} c_{pa} \tag{13.18}$$

where λ_{pk} is the partial decay constant for the parent component species of daughter component species $k[s^{-1}]$, the sum is over all parents of species k, and λ_k is the total radioactive decay constant for component species $k[s^{-1}]$.

The thermodynamic activities are approximated by the relations

$$Y_{i\alpha} = \gamma_{i\alpha} c_{i\alpha} \tag{13.19}$$

and

$$X_{k\alpha} = \gamma_{k\alpha} c_{k\alpha} \tag{13.20}$$

where $\gamma_{k\alpha}$ and $\gamma_{i\alpha}$ are activity coefficients of the component and product species.

In general, equilibrium constants in equations (13.15)-(13.17) and activity coefficients in (13.19) and (13.20) are functions of temperature and ionic strength. Often-used approximations for these parameters can be found in [1].

Transport with Chemical Equilibrium Reactions

The species transport equations can be written symbolically as

$$L_{j\alpha}(c) = \Gamma_{j\alpha} \quad \alpha = a, p, s \tag{13.21}$$

where the operator $L_{j\alpha}$ is given by

$$\begin{aligned} L_{j\alpha}(c) &= \frac{\partial}{\partial t}(\theta_\alpha c_{j\alpha}) + \nabla \cdot (c_{j\alpha} v_\alpha - \theta_\alpha D_{j\alpha} \cdot \nabla c_{j\alpha}), \quad \alpha = a \\ L_{j\alpha}(c) &= \frac{\partial}{\partial t}(\theta_\alpha c_{j\alpha}) = 0, \quad \alpha = p, s \end{aligned} \tag{13.22}$$

Here $\Gamma_{j\alpha}$ represents the net rate at which the jth species is added to phase α due to chemical ($r^c_{j\alpha}$) and radioactive decay ($r^d_{j\alpha}$) reactions, waste form leaching ($s^w_{j\alpha}$), and injection/extraction ($q_{j\alpha}$). That is,

$$\Gamma_{j\alpha} = r^c_{j\alpha} + r^d_{j\alpha} + s^w_{j\alpha} + q_{j\alpha} \tag{13.23}$$

Since we make use of the assumption of local chemical equilibrium, we may eliminate the chemical reaction rate terms, $r^c_{j\alpha}$ $[mol \cdot cm^{-3} \cdot s^{-1}]$, as variables from transport equations (13.21). To accomplish this task, we make use of the conservation relation

$$R^c_k = \sum_i \nu^a_{ki} r^c_{ia} + \sum_l \nu^a_{kl} r^c_{lp} + \sum_m \nu^a_{km} r^c_{ms} \tag{13.24}$$

which states that the net change in total mass of aqueous component k, by chemical tranformation, is zero. Solving (13.21) for r^c_{ia}, r^c_{lp} and r^c_{ms}, respectively, and introducing the results into (13.24) we obtain

$$\sum_{i \in A} \nu_{ki}^{a} L_{ia}(c) + \sum_{l \in P} \nu_{kl}^{a} L_{lp}(c) + \sum_{m \in S} \nu_{km}^{a} L_{ms}(c) =$$
$$\theta_a \frac{\partial T_k}{\partial t} + (C_{ks} + C_{kp}) \frac{\partial \theta_a}{\partial t} + \boldsymbol{v}_a \cdot \nabla C_{ka} - \nabla \cdot \theta_a \boldsymbol{D} \cdot \nabla C_{ka}$$
$$+ C_{ka} \left(\frac{\partial \theta_a}{\partial t} + \nabla \cdot \boldsymbol{v}_a \right) = R_k^d + S_k^w + Q_k \quad (13.25)$$

The last term in brackets can be replaced by Q_a using the aqueous phase continuity equation to obtain

$$\theta_a \frac{\partial T_k}{\partial t} + T_k \frac{\partial \theta_a}{\partial t} + \boldsymbol{v}_a \cdot \nabla T_{ka} - \nabla \cdot \theta_a \boldsymbol{D} \cdot \nabla T_k =$$
$$C_{ka} \frac{\partial \theta_a}{\partial t} + \boldsymbol{v}_a \cdot \nabla (C_{ks} + C_{kp}) - \nabla \cdot \theta_a \boldsymbol{D} \cdot \nabla (C_{ks} + C_{kp})$$
$$+ R_k^d + S_k^w + Q_k - Q_a C_{ka} \quad (13.26)$$

where

$$T_k = C_{ka} + C_{kp} + C_{ks} \quad (13.27)$$

Here, C_{ka}, C_{kp} and C_{ks} are total concentrations of component k in moles of k per unit volume of the aqueous phase $[mol \cdot cm^{-3} \cdot s^{-1}]$ in the aqueous, precipitated, and sorbed phases, respectively,

$$C_{kp} = \frac{V_p}{V_a} \sum_{i \in P} \nu_{ki}^{a} c_{ip} \quad (13.28)$$
$$C_{ka} = c_{ka} + \sum_{i \in A} \nu_{ki}^{a} c_{ia} \quad (13.29)$$
$$C_{ks} = \frac{V_s}{V_a} \sum_{i \in S} \nu_{ki}^{a} c_{is} \quad (13.30)$$

The total radioactive decay rate, R_k^d and total source terms, S_k^w and Q_k, for component k, are

$$R_k^d = \sum_i \nu_{ki}^{a} r_{ia}^{d} + \sum_l \nu_{kl}^{a} r_{lp}^{d} + \sum_m \nu_{km}^{a} r_{ms}^{d}$$
$$= r_{ka}^{d} + r_{kp}^{d} + r_{ks}^{d} \quad (13.31)$$

and

$$S_k^w = \sum_i \nu_{ki}^a s_{ia}^w \tag{13.32}$$

$$Q_k^w = \sum_i \nu_{ki}^a q_{ia} \tag{13.33}$$

Equations governing conservation of adsorbent and ion- exchange site components are required to complete the system of transport equations. From (13.21)

$$\frac{\partial(\theta_s A_{ks})}{\partial t} = 0 \tag{13.34}$$

with

$$A_{ks} = \frac{V_s}{V_a} \sum_k \bar{\nu}_{ki}^s c_{is} \tag{13.35}$$

for k ranging over the adsorbent and ion-exchange site components only and where A_{ks} is the total concentration of adsorbent component k.

The solution of aqueous phase transport equation (13.26) and sorption/ion exchange conservation equation (13.34) requires that the initial total analytical concentrations of all components $[mol \cdot cm^{-3}]$, including aqueous phase and adsorbent components and number of equivalents of ion exchange sites $[mol \cdot gm^{-1}]$, be specified in the flow domain Ω; that is

$$T_k(x,y,z,0) = T_k^I(x,y,z) \tag{13.36}$$

and

$$C_{ms}(x,y,z,0) = C_{ms}^I(x,y,z) \tag{13.37}$$

where k ranges over the aqueous components, m ranges over the adsorbent components and ion-exchange sites, and the superscript I denotes initial values.

Three types of boundary conditions may be applied on the boundary Γ of flow domain Ω. Dirichlet conditions prescribe analytical concentrations on boundary segment Γ_D:

$$T_k(x,y,z,t) = T_k^D(x,y,z,t), \quad (x,y,z) \in \Gamma_D, \ t > 0 \tag{13.38}$$

Note that boundary conditions are not required for adsorbent components and ion exchange sites because these quantities are not transported and are properties of the porous media. The second and third types of boundary conditions are the Neumann and Cauchy conditions, respectively,

$$-\theta_a \boldsymbol{D} \cdot \nabla(C_{ka}) \cdot \boldsymbol{n} = \boldsymbol{q}_{ka}^N \cdot \boldsymbol{n}, \quad (x,y,z) \in \Gamma_N,\ t > 0 \qquad (13.39)$$

and

$$-\theta_a \boldsymbol{D} \cdot \nabla(C_{ka}) \cdot \boldsymbol{n} + \boldsymbol{v}_a C_{ka} \cdot \boldsymbol{n} = \boldsymbol{q}_{ka}^C \cdot \boldsymbol{n}, \quad (x,y,z) \in \Gamma_C,\ t > 0 \quad (13.40)$$

where k ranges over the aqueous components, $\boldsymbol{n}$ is the outward unit normal vector to Γ, q_{ja}^N is the prescribed outward Neumann (dispersive) flux across boundary segment $\Gamma_N [mol \cdot cm^{-2} \cdot s^{-1}]$, and q_{ja}^c is the outward Cauchy (total) flux vector across $\Gamma_C [mol \cdot cm^{-2} \cdot s^{-1}]$.

Release of Radionuclides

In this section we formulate the radionuclide release term S_k^w. Recall that this term (see (13.26) and (13.32)) represents the mass release rate of radionuclides $(mol \cdot cm^{-3} \cdot t^{-1})$ from waste forms and waste containers.

A schematic representation of radionuclide release from a breached waste container is depicted in Figure 13.2. The following processes are illustrated: 1) container degradation, which, upon container breach, allows water to enter the container and contact the waste forms; 2) the transfer of radionuclides from the waste forms into the contacting water; and 3) the flow of the radionuclide-laden water (leachate) out of the container and into the adjacent backfill. Mathematical representations for these processes and how they are combined to yield S_k^w are described below.

Container Degradation

Before water can contact a waste form and mediate the release of radioactivity, the container surrounding the waste form must be breached. Therefore, to predict total release it is essential to know the time at which breach occurs for each container in the system, the number of breached containers at any given time, the area breached per container, and the rate at which the breached area increases. Because of low cost and relative durability, carbon steel containers are the most commonly used low-level waste package containers. Containers made of various stainless steels, concrete, and high density polyethylene (HDPE) are used to a lesser extent. Carbon steel containers are subject to chemical attack (corrosion) which eventually leads

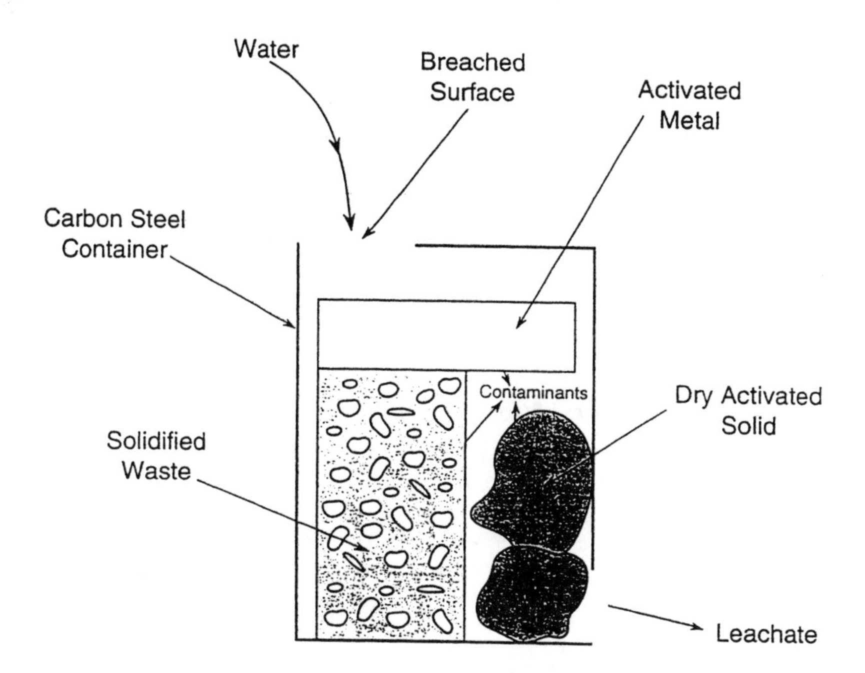

Figure 13.2: Idealized representation of radionuclide release from a breached container.

to breach. These containers are susceptible to general as well as localized corrosion in soil environments. In the present treatment, both localized corrosion, as represented by pitting corrosion, and generalized corrosion are considered.

Localized Corrosion

We represent the localized corrosion process by empirical correlations for pitting depth and area breached that are based on data obtained by the National Bureau of Standards (currently the National Institute of Standards and Technology) [11].

The maximum pit depth takes the form:

$$h = kt^n \left(\frac{A}{372}\right)^a \tag{13.41}$$

where h is the maximum pit depth in cm, k is the pitting parameter in cm/yr^n, t is the time in years, n is the pitting exponent which depends on soil properties, A is the surface area of the container in cm^2, the constant

372 cm^2 is a scaling factor, and a is an experimentally derived correlation coefficient. Values of a depend on the material and soil. Extensive studies [7] indicated that, for wrought irons and carbon steels, a ranged form 0.08 to 0.32 with a mean value of 0.15.

Values of n strongly depend on soil aeration; in practice n is often selected as 0.26, 0.39, 0.44, or 0.59 for good, fair, poor, and very poor aeration, respectively. These values are averages determined from the NBS study for their respective soil aeration. If the clay content is known, n may be estimated from

$$n = n_0 \theta_1 (1 - CL)^{0.4} \tag{13.42}$$

where $n_0 = 1$, 1.5, 2.0, or 2.5 for good, fair, poor, and very poor aeration, respectively, where CL is the clay fraction. Values of pitting parameter k are determined from the following relationships [17]

$$\begin{aligned} k &= 0.01458(10 - pH), \quad pH < 6.8 \\ k &= 0.0457, \quad 6.8 < pH < 7.3 \\ k &= 0.0256(pH - 5.13), \quad 7.3 < pH \end{aligned} \tag{13.43}$$

If the pit depth, h, given by equation (13.41) does not exceed the container thickness, the container is unbreached and water cannot access the waste form. When h does exceed the container thickness, the area breached is represented by the relationship

$$A_b = N_p \pi \left(h^2 - MT^2\right) \tag{13.44}$$

where A_b is the area breached in cm^2; N_p is the number of penetrating pits per container (estimates of this value range from 1000 to 10000 for the surface area of a 55 gallon drum, 21,000 cm^2 [16]), and MT is the thickness of the metal container (cm). Equation (13.44) arises from the assumption that the pits are hemi-spherical in shape and continue to grow at the same rate once the metal has been penetrated.

Generalized Corrosion - Time to Failure

The general corrosion of metal is calculated assuming that the corrosion rate is constant and independent of time. This approach is likely to be conservative because the NBS general corrosion data indicate that the rate decreases in time. For a constant corrosion rate, the thickness of metal corroded, $d(cm)$, is simply

$$d = gt \tag{13.45}$$

where g is the general corrosion rate in $(cm \cdot sec^{-1})$ and t is the time in seconds. If d exceeds the container thickness, the entire surface area of the container is assumed to be corroded away. At this time, the container does not provide any barrier for water access to the container.

The general corrosion model can be viewed as a time to failure model. Time to failure models are commonly used in low-level waste performance assessment codes [14]. In the absence of site-specific corrosion rate data, analysts often assume that all containers of a certain type fail at a fixed time. This can be accomplished in the BLT-EC model through appropriate choice of the corrosion rate, container thickness, and time of failure.

Release from Waste Forms

The waste form is the physical form of the waste in the disposal container. A wide range of waste forms are used in LLW disposal. A review of the compilation of data from the commercial shipping manifests [10] indicates that there are over 22 categories of waste streams. These waste streams may be placed untreated into the container, or they may be treated with sorbents to absorb free liquids, solidification agents such as Portland cement, modified sulfur cement, vinyl-ester styrene, or bitumen, compacted to reduce volume, or surrounded with sand to minimize void space in the container. Knowledge of the waste form is crucial in developing the conceptual models for release from the waste package.

In general each waste form may release radionuclides by a combination of release mechanisms. These mechanisms include:

1. Diffusional transport of material through a porous solidified waste form to the waste form surface;

2. The release of materials from bulk solids by dissolution of the matrix or solid phase; and

3. The release of surface residing materials by surface rinse.

Recall the conceptual picture of the breached container shown in Figure 13.2. In the following development, we treat the environment within the container as a mixing bath; that is, transport processes outside the waste forms are fast enough to maintain a uniform concentration within the container environment. We also make the following important assumptions:

1. The volumetric flow rate of water through the mixing bath occurs at a rate determined by the product of the Darcy velocity in the neighborhood of the container and one half the breached area of the container (the factor of one half is based on the assumption that water enters a container through half of the breached area and exits through the other half);

2. Radionuclides are delivered to the mixing bath from the waste forms at a rate determined by one or more of the three release mechanisms noted above; and

3. The impact of chemical processes on radionuclide transport within the waste forms may be ignored.

Performing a mass balance on the mixing bath gives

$$\rho_a^{mb}\theta_a^{mb}\frac{d\left(c_{ja}^{mb}\right)}{dt} = \theta_a^{mb}\nu^{wf}\left(c_{ja}^{I}-c_{ja}^{mb}\right)+S_{ja}^{dif}+S_{ja}^{dis}+S_{ja}^{r} -\lambda_j\theta_a^{mb}c_{ja}^{mb}+\theta_a^{mb}\sum_p \lambda_{pj}c_{pa}^{mb} \quad (13.46)$$

where j ranges over the number of radionuclide species, ρ_a^{mb} and θ_a^{mb} represent average aqueous phase density and moisture content values in the mixing bath, c_j^{mb} is the average concentration of species j in the mixing bath, c_{ja}^{I} is the average concentration of species j entering the mixing bath, and s_{ja}^{dif}, s_{ja}^{dis}, and s_{ja}^{r} are release rates of radionuclides from waste forms by diffusion, dissolution, and surface rinse, respectively. The last two terms on the rhs of equation (13.46) represent radioactive decay and ingrowth.

The term, ν^{wf}, in (13.46) represents the leachant renewal frequency (s^{-1}) and is a measure of how fast water within the mixing bath is replenished. ν^{wf} is the ratio of volumetric flow rate into the mixing bath divided by the volume of water in the mixing bath; that is,

$$\nu^{wf} = \frac{0.5v_a A_b}{\theta_1^{mb}V^{mb}} \quad (13.47)$$

where v_a is the Darcy velocity of the aqueous phase given by equation (13.3) and V^{mb} is the volume of the mixing bath.

The source term s_{ja}^{r} is due to mass release via surface rinse and takes the form

$$s_{ja}^{r} = \Gamma_j^{r}\left(1-\frac{c_{ja}^{mb}}{c_{ja}^{sat}}\right) \quad (13.48)$$

where c_{ja}^{sat} is the saturation concentration (solubility limit) of species j in the mixing bath solution and Γ_j^r is given by [15]

$$\Gamma_j^r = \frac{\delta \left[M_j^a(t) - \left(\frac{K_j^p}{\beta} \right) M_j^s(t) \right]}{V^{mb} \theta_1^{mb} \left(1 + \frac{K_j^p}{\beta} \right)} \tag{13.49}$$

where δ is an empirical constant having units of s^{-1}, $M_j^a(t)$ is the mass of species j available for rinse release at time t, $M_j^s(t)$ is the mass of species j in the mixing bath at time t, K_j^p is the partition coefficient for species j, and $\beta = \theta_a V^{mb} / \rho V^{wf}$. Equation (13.49) is based on the assumption that release is governed by equilibrium between the solid waste form and the aqueous phase. As material is moved out of the mixing bath, more material is released to maintain equilibrium.

We next consider the release of radionuclide species j from a solid, porous, dissolving waste form. A one-dimensional representation of this waste form undergoing dissolution is shown in Figure 13.3. As the outer surface of the waste form dissolves, radionuclides present in the removed region enter the contacting water. Simultaneously, radionuclides diffuse from within the porous waste form outwards towards the moving surface where they exit according to Fick's Law. Under these conditions the release of radionuclides can be written

$$s_{ja}^{dis} + s_{ja}^{dif} = \frac{SA}{V^{mb}} \boldsymbol{u} \left[\theta_a^{wf} c_{ja}^{wf} - (1 - \eta) F_j \right] - \frac{SA}{V^{mb}} \theta_a^p D_j \frac{\partial c_{ja}^p}{\partial x} \tag{13.50}$$

where the first and second terms on the right of (13.50) are contributions due to dissolution and diffusion, respectively, $\boldsymbol{y}$ is the dissolution velocity, η is the porosity of the waste form, SA is the surface area of the waste form, D_j is the species diffusion coefficient for the waste form, F_j is the immobile matrix concentration of species j, c_{ja}^{wf} is the aqueous phase concentration in the waste form, and θ_a^{wf} is the volumetric moisture content in the waste form. Note that, the dissolution contribution in equation (13.50) consists of two parts, radionuclides present in the aqueous phase and radionuclides associated with the solid matrix.

The second term on the right in equation (13.46)(or the last term on the rhs of (13.50)) requires the concentration field of species j, c_{ja}^{wf}, in the waste form. In a dissolving region, this concentration field is described by [15]

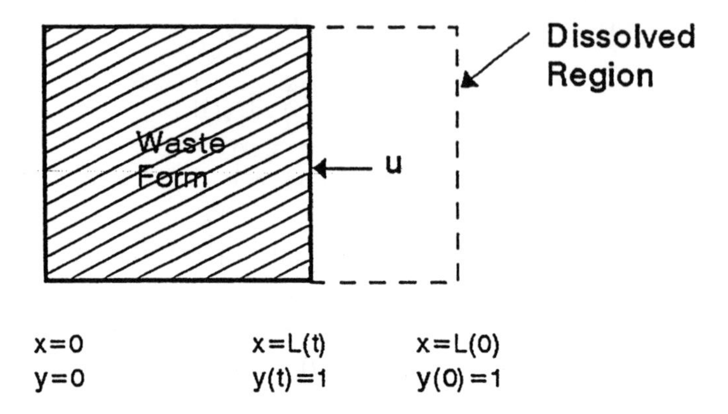

Figure 13.3: One-dimensional representation of a waste form that undergoes dissolution at a velocity $\boldsymbol{u}$. Here x is the distance variable and y is a normalized distance variable that always has the dissolution front at $y = 1$.

$$\theta_a^{wf}\frac{\partial c_{ja}^{wf}}{\partial t} = \frac{\theta_a^{wf} D}{L^2}\frac{\partial^2 c_{ja}^{wf}}{\partial y^2} + \frac{(v_a^{wf} + \theta_a^{wf} uy)}{L}\frac{\partial c_{ja}^{wf}}{\partial y} + s_{ja}^{wf}$$
$$+C_{yl}\left(\frac{-\theta_a^{wf} D}{yL^2}\frac{\partial c_{ja}^{wf}}{\partial y} + \frac{v_a^{wf} c_{ja}^{wf}}{yL}\right) - \theta_a \lambda_j c_{ja}^{wf} + \theta_a^p \sum_p \lambda_{pj} c_{pa}^{wf} \quad (13.51)$$

where v_a^{wf} is the advection velocity through the waste form, λ_j is the species decay constant, s_{ja}^{wf} is the source/sink term for the diffusive species (it may be a function of concentration, solubility limit, immobile species concentration or other parameters), C_{yl} takes the value of 0 for plane geometry and 1 for cylindrical geometry, and L is the length of the modeled region at time t. Finally, y is a dimensionless distance variable given by $y = \frac{x}{L(t)}$. Some species are attached to the structure of the waste form, these species are not transported in the solid phase. In this case, equation (13.51) simplifies to

$$(1-\eta)\frac{\partial F_j}{\partial t} = \frac{(1-\eta)uy}{L}\frac{\partial F_j}{\partial y} - (1-\eta)\left(\lambda_j F_j + \sum_p \lambda_p F_p\right) \quad (13.52)$$

where F_j and η are the immobile phase concentration and waste form porosity.

The associated boundary and initial conditions then are

$$c_{j1} = c_{j1}^{mb}, \quad \text{on } \Gamma_{wf} \quad (13.53)$$

and

$$c_{j1} = c_{j1}^0, \quad t = 0, \ (x, y) \in \Omega^{wf} \tag{13.54}$$

where Ω^{wf} and Γ_{wf} are body and boundary of the waste form.

Equations (13.46)-(13.54) form the basis for calculating the concentration of species j, c_{ja}^{mb}, in the mixing bath. The release of chemical species j from the mixing bath for transport, source term s_{ja}^w, is

$$s_{ja}^w = \nu_{wf} \left(c_{ja}^{mb} - c_{ja}^I \right) \tag{13.55}$$

The total source term for component k, S_k^w, is obtained by introducing equation (13.55) into equation (13.32) so that

$$S_k^w = \nu_{wf} \sum_j \bar{\nu}_{kj}^a \left(c_{ja}^{mb} - c_{ja}^I \right) \tag{13.56}$$

The release models presented above, by equations (13.46)-(13.54), describe the release of different radionuclide species (UO_2OH^+ vs. $UO_2(OH)_2$). Note that the chemical processes that determine the solubility and sorption properties of radionuclides in waste forms are species dependent. This point is important because species-specific data parameters D_j, M_{ja}, K_{jp}, and c_{ja}^{sat} are typically not available or measured during waste form leaching studies. Rather, these parameters are typically available as homogenized values; that is, they are determined experimentally by measuring total radionuclide release rates without specification of the chemical form of the radionuclides. The implications of this issue on characterization and modeling of waste form leaching processes need to be examined further. From the standpoint of implementing the source models in BLT-EC, however, this issue poses no difficulty as long as radionuclides are chosen by the user to be "components".

13.4 Description of Models

Water Flow

The flow of water in unsaturated and saturated porous media is described by nonlinear partial differential equation (13.1) and constitutive relations for $\boldsymbol{K}^c$ and F (see (13.3)) that relate water content to hydraulic conductivity and pressure head. We presently solve the water flow problem using a modified version of the two-dimensional finite-element code FEMWATER [22]. This code discretizes the flow equations in space with linear finite elements and in time with a variable two-point finite-difference scheme. The final linearized matrix equation is solved using direct elimination with Picard iteration to iterate the nonlinearity. The flow solution is postprocessed to provide the velocity vector defined by (13.2) required by the discrete advection-dispersion chemical transport equations in BLT-EC. Although we use FEMWATER in our applications other codes may be easily adapted to BLT-EC calculations.

Chemical Processes: EC Module

Several chemical processes potentially play a role in determining the sorption and solubility properties and hence mobilities of radionuclides in solution. The EC module accounts for several of these processes including complexation, dissolution-precipitation, reduction-oxidation, sorption, and ion exchange. These processes, under the assumption of chemical equilibrium, are represented by the set of mass action equations (13.15)-(13.17) that describe how the various chemical constituents are distributed among the aqueous, precipitated, and adsorbed phases present in the pore space. EC has the ability, for a specified solution composition, to automatically select the relevant reactions and thermodynamic data from its associated thermodynamic data base. In addition, the EC module can compute pH and Eh as solution composition changes in response to complexation and precipitation-dissolution processes. The user also has the option of specifying the pH and Eh values. Adsorption models in EC include activity K_d, Langmuir, Freundlich, ion exchange, constant capacitance, triple-layer, and diffusive-layer models.

The geochemical model used to solve the chemical equilibrium equations in BLT-EC is a modified version of the geochemical computer code MINTEQA2 [1]. MINTEQA2 uses the Newton-Raphson method to solve the governing equations for a mixture of specified composition. Further details of the computational algorithms can be found in [1]. In a BLT-EC simulation, mixture compositions at each node and time step are provided by the transport module. The pre-solution routines in MINTEQA2 have been modified extensively to interface efficiently with the transport module.

Transport Processes: Transport Module

The transport module in the BLT-EC computer code is based on a modified version of the hydrological transport module contained in the finite-element code HYDROGEOCHEM [21]. This module approximates the governing transport equations (13.26) and (13.34) with bilinear finite elements for the spatial discretization, a variable two-point finite-difference scheme for time integration, and either direct or pointwise iteration methods for solution of the matrix equations. The code user can also select the following options: (a) an upstream weighting finite-element approximation for advection-dominated flows, (b) lumping of the mass matrix, and (c) tetrahedral or quadrilateral elements (although the BREACH and LEACH models require quadrilateral elements for their implementation).

Container Degradation: BREACH Module

The BREACH module in BLT-EC computes the following quantities:

1. The time at which breach occurs for each container in the system;

2. The number of breached containers at any given time;

3. The area breached per container;

4. The rate at which the breached area increases.

These quantities are computed for both localized as well as generalized corrosion (recall (13.41)-(13.45)). General failure is modeled through a user-specified time of failure. The time to failure may be estimated, for example, as the corrosion allowance of the container divided by the time-averaged corrosion rate, that is, $t = d/g$.(recall equation (13.45)). Corrosion rates should be obtained from site specific data whenever possible. If such data are not available, the data base generated by the National Bureau of Standards (NBS) [11] for stainless steels may be used for these materials. Localized corrosion is modeled by empirical correlations, (13.41)-13.44), which are based on the NBS data base. The parameters in this correlation depend on soil-water pH, degree of soil aeration, moisture content, and clay content.

Release from Waste Forms: LEACH Module

Radionuclide release from the waste form commences upon container failure. Several different waste forms may be present in a LLW facility. In general, each waste form may release radionuclides by a combination of different mechanisms. To cover a wide range of conditions, four release mechanisms are modeled in the LEACH model: surface rinse limited by partitioning (see (13.48) and (13.49)), uniform release (13.50) and diffusion through solidified waste forms (13.51). All models include the effects of solubility limits and radioactive decay. Daughter ingrowth in the waste forms is currently not accounted for in the present version of BLT-EC. This capability will be implemented in the future. These leaching models and their implementation are briefly described below.

The most complex of the waste form release models considers all four release mechanisms simultaneously. A partial differential equation is used to represent diffusion, waste form dissolution, radioactive decay, and surface rinse with partitioning (see 13.46). The solution concentration is supplied from calculations which model the transport of radionuclides in the ground water after they have been released from the waste form (13.26). Release from waste forms is directly coupled to solution concentration. The equation representing movement within the waste form (see (13.51)) is solved using the method of finite differences.

A simpler approach, which is frequently used, is to consider each of the four release mechanisms as an independent process. In this case, total release from the waste form is the sum of the rinse, diffusion, and dissolution release models with the constraint that the release will not cause user-specified solubility limits to be exceeded. This approach is a useful approximation because, for many waste forms, one release mechanism will dominate over the others.

13.5 Implementation

The coupling between the five process models as implemented in BLT-EC is depicted schematically in Figure 13.4. In this figure, it can be seen that the water flow is assumed to be independent of all other processes. The independence of water flow is assumed because the dependence of water flow parameters such as hydraulic conductivity on container degradation and waste form leaching are expected to be small. All other processes, however, are coupled to water flow. Container degradation (breach) modeling involves the prediction of pitting and general corrosion rates. These processes are, in general, functions of the moisture content, pH, and redox conditions in the backfill adjacent to the containers. Waste form release (leach) modeling requires information from all four other process models. This information includes the breached area as a function of time from the container degradation model; the concentration of contaminants in solution adjacent to the waste form from the transport-chemical process models; and the inlet and outlet flow rates and moisture content from the water flow model. Chemical process modeling requires, besides thermodynamic data, concentrations of

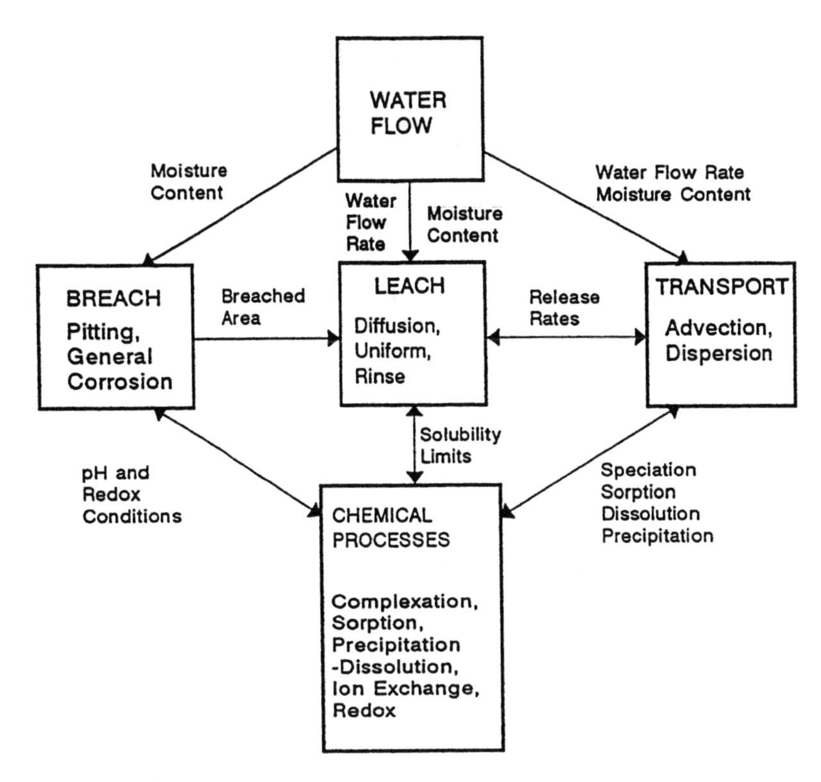

Figure 13.4: Coupling between FLOW, BREACH, LEACH, TRANSPORT and CHEMICAL PROCESS models.

all chemical components in the system; these are provided by the transport model. Transport modeling requires the release rate from the leaching models, and the water velocity field and moisture content distribution in the soil.

Substantial effort has been directed towards making the BLT-EC code modular, user friendly, and transportable between UNIX and DOS based platforms. To help the user create an input file, menu driven preprocessors that guide the user through the necessary steps of creating or modifying input files have been developed. We are also currently developing a menu-driven postprocessor to facilitate graphical display of one- and two-dimensional output data.

The modular structure of the complete BLT-EC code package is illustrated in Figure 13.5. This package consists of six modules: (1) an optional preprocessor that assists the user in preparing source, transport, and chem-

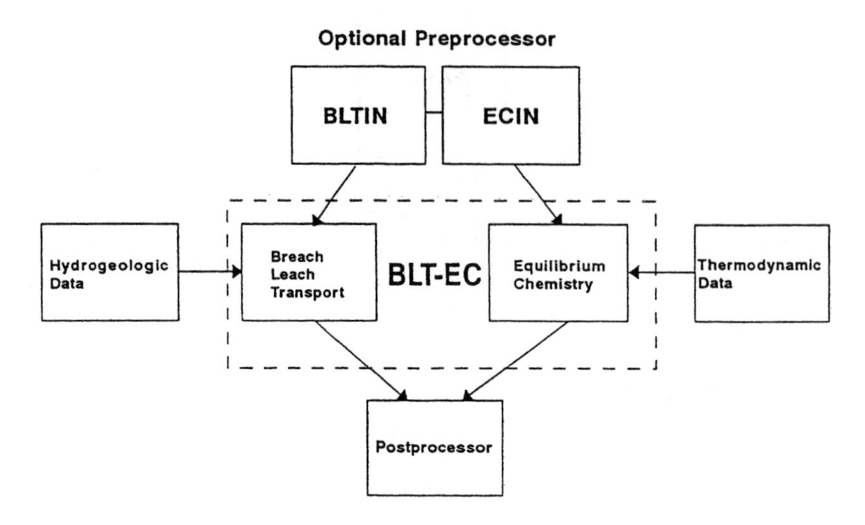

Figure 13.5: Top-level flow chart of the BLT-EC code.

ical input data; (2) the BLT module that simulates radionuclide release and migration; (3) the EC module that simulates the chemical reactions; (4) the hydrogeologic data module that transfers data from FEMWATER's finite-element mesh to BLT-EC's finite-element mesh for use by the BLT module; (5) the thermodynamic data base that provides the EC module with the pertinent reactions and data; (6) the postprocessing module that provides tabular and graphical displays of output.

13.6 Example Problem

We present an example application that considers the release of radionuclides from a hypothetical shallow land burial trench. This example is non-site specific and is designed to demonstrate BLT-EC's ability to simulate important processes associated with the release of radionuclides from LLW shallow land disposal facilities. Interactions between transport, complexation, precipitation/dissolution, and adsorption of uranium are considered.

The problem domain is taken to be a two-dimensional vertical cross-section perpendicular to the longitudinal axis of a disposal trench as shown in Figure 13.6. It is assumed that the length of the trench in the longitudinal direction is much longer than the width of the trench. Therefore, a two-dimensional cross-section provides a reasonable representation for simulating radionuclide migration near the central portion of the trench. Symmetry

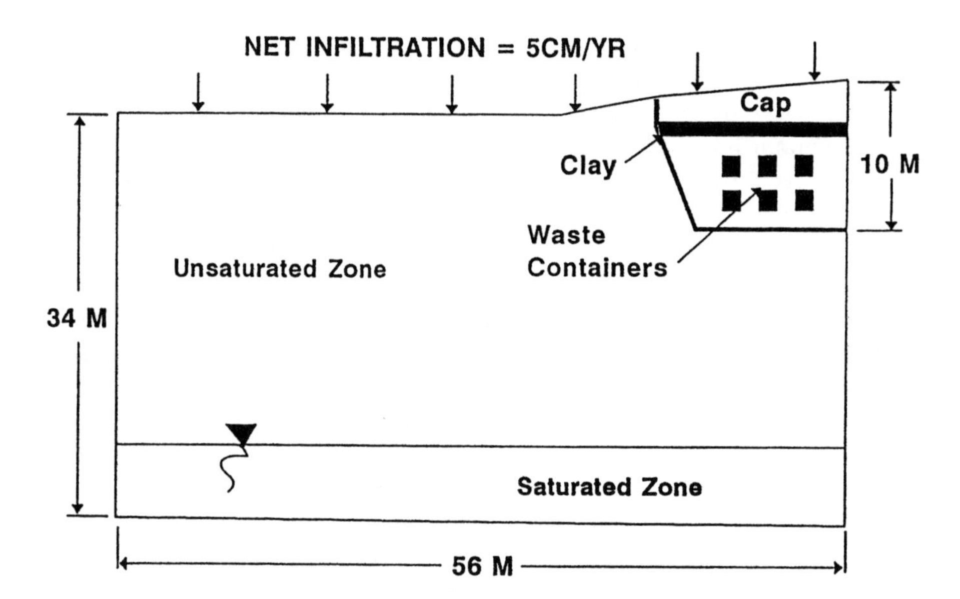

Figure 13.6: Schematic diagram of example field-scale problem.

within the cross section is further assumed, thus requiring only half the problem domain to be modeled. The water table is located approximately 30 meters below the ground surface. The waste containing portion of the trench is taken to be 7 meters deep and 28 meters wide, with the side walls slanting at an angle approximately 12 degrees from vertical. The trench contains 12 waste containers (6 in the left half are shown as shaded regions) that are surrounded by backfill. Above the waste region is a 1 meter thick clay layer with a low hydraulic conductivity to minimize water intrusion from above. The clay layer is covered by a high conductivity cap layer, which is 2 meters thick and slants off towards the edge of the trench. The soil properties are assumed to be uniform in each trench region and in the underlying unsaturated zone; values of these parameters are given in [8].

The example application involves two simulations, a steady-state water-flow simulation and a radionuclide transport simulation. The water-flow problem is simulated over the entire cross section, which is divided into 624 bilinear finite elements and 675 nodes. The steady-state flow problem is solved once and the hydrogeologic data (moisture contents and flow velocities) are used for both transport problems. The radionuclide transport problem is solved over a smaller region comprised of 405 elements and 450 nodes.

The boundary conditions for the flow simulation are as follows. The vertical left and right boundaries are homogeneous Neumann or no-flow boundaries. The bottom horizontal boundary is a Dirichlet boundary with a prescribed 1000 cm hydraulic head. The top boundary (ground surface) is a Neumann boundary with a prescribed rainfall infiltration of 5 cm/yr. Boundary conditions for the radionuclide migration problem are described separately below.

The transport problem considers a soil-surface component, SOH, and five aqueous components including hydronium, carbonate, calcium, sulfate, and uranium in the form of uranium oxide. These components form 19 aqueous and 3 surface complexes and 11 possible minerals (see Table 13.1). The initial composition of the water (mole/l) was $CO_3^-=2.0\times10^{-5}$, $SO_4^-=5.0\times10^{-5}$, $Ca^{+2}=5.0\times10^{-4}$, $SOH=2.0\times10^{-5}$, and $UO_2 = 0.0$. The component concentrations in the trench (in mole/l) are similar except for the following: $CO_3^-=2.0\times10^{-5}$ and $SO_4^- = 5.0\times10^{-5}$. At the top infiltration boundary the infiltrating rain water is assumed to contain Ca^{+2} and CO_3^- at concentrations of 1.0×10^{-6} and 1.0×10^{-5} moles/l, respectively. The vertical and bottom boundaries are no-flow boundaries. The pH of the ground water is held fixed at 7.5 throughout the problem domain. Redox reactions were not considered.

The source concentration in each 1.0 m^3 container was 0.2 mole/m^3 of UO_2. Localized and general corrosion results in gradual and complete container breach by 20 years. Release from the waste form occurs by rinse release during the first 20 simulation years.

Simulation results are presented for 30 and 200 years. The distributions of total uranium in Figure 13.7 show that low concentrations of uranium have migrated from the disposal unit. Most of the uranium, however, has precipitated in the trench as the mineral schoepite. Distributions of uranium precipitation are shown in Figure 13.8. Results show the dissolution of the uranium over time, with the left portion of the precipitation zone dissolving at a much faster rate. This is a consequence of the higher infiltration rates in the outer region of the trench.

Figures 13.9 and 13.10 illustrate the evolution of the dissolved uranium plume and associated adsorbed uranium zone. The adsorbate uranium species and surface reactions are specified in Table 13.1. Adsorption was modeled using the activity Langmuir model. The fraction of adsorption sites occupied by uranium never exceeded 50 per cent of those available at any given location. Observe that concentrations of dissolved uranium are substantially less than adsorbed concentrations on a mole per liter of solution basis. However, because of the low moisture content in the unsaturated region (0.025 over most of the unsaturated region) the adsorbed

Table 13.1: Hypothetical field-scale problem.

Species	Log K	Components and Stoichiometry					
		H^+	CO_3^{2-}	Ca^{2+}	SO_4^{2-}	UO_2^{2+}	SOH
Aqueous Species							
H^+	0.00	1	0	0	0	0	0
CO_3^{2-}	0.00	0	1	0	0	0	0
Ca^{2+}	0.00	0	0	1	0	0	0
SO_4^{2-}	0.00	0	0	0	1	0	0
UO_2^{2+}	0.00	0	0	0	0	1	0
$CaOH^+$	-12.59	-1	0	1	0	0	0
$CaHCO_3^+$	11.33	0	1	1	0	1	0
$CaCO_3$	3.15	0	1	1	0	0	0
$CaSO_4$	2.31	0	0	1	1	0	0
HCO_3^-	10.33	1	1	0	0	0	0
HSO_4^-	1.99	1	0	0	1	0	0
H_2CO_3	16.68	2	1	0	0	0	0
UO_2OH^+	-5.09	-1	0	0	0	1	0
$(UO_2)_2(OH)_2^{2+}$	-5.64	-2	0	0	0	2	0
$(UO_2)_3(OH)_5^+$	-15.59	-5	0	0	0	1	0
UO_2CO_3	10.07	0	1	0	0	1	0
$UO_2(CO_3)_2^{-2}$	17.01	0	2	0	0	1	0
$UO_2(CO_3)_3^{-4}$	21.38	0	3	0	0	1	0
UO_2SO_4	2.71	0	0	0	1	1	0
Surface Species							
SOH	0.00	0	0	0	0	0	1
SO^-	-7.30	-1	0	0	0	0	1
SOH_2^+	-1.40	1	0	0	0	0	1
$(SO^-)(UO_2OH^+)$	-5.10	-2	0	0	0	1	1
Minerals							
Schoepite	-5.40	-2	0	0	0	1	0
UO_3	-7.72	-2	0	0	0	1	0
Gummite	-10.40	-2	0	0	0	1	0
B-$UO_2(OH)_2$	-5.54	-2	0	0	0	1	0
Rutherfordin	14.44	0	1	0	0	1	0
Anhydrite	4.64	0	0	1	1	0	0
Aragonite	8.36	0	1	1	0	0	0
Calcite	8.48	0	1	1	0	0	0
Gypsum	4.85	0	0	1	1	0	0
Lime	-32.80	-2	0	1	0	0	0
Portlandite	-22.68	-2	0	1	0	0	0

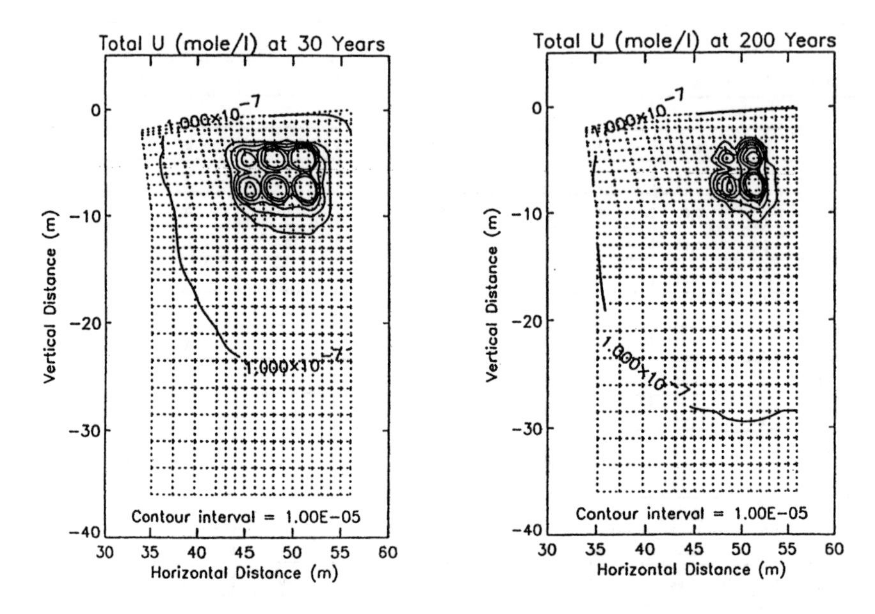

Figure 13.7: Distribution of total uranium at 30 and 200 years.

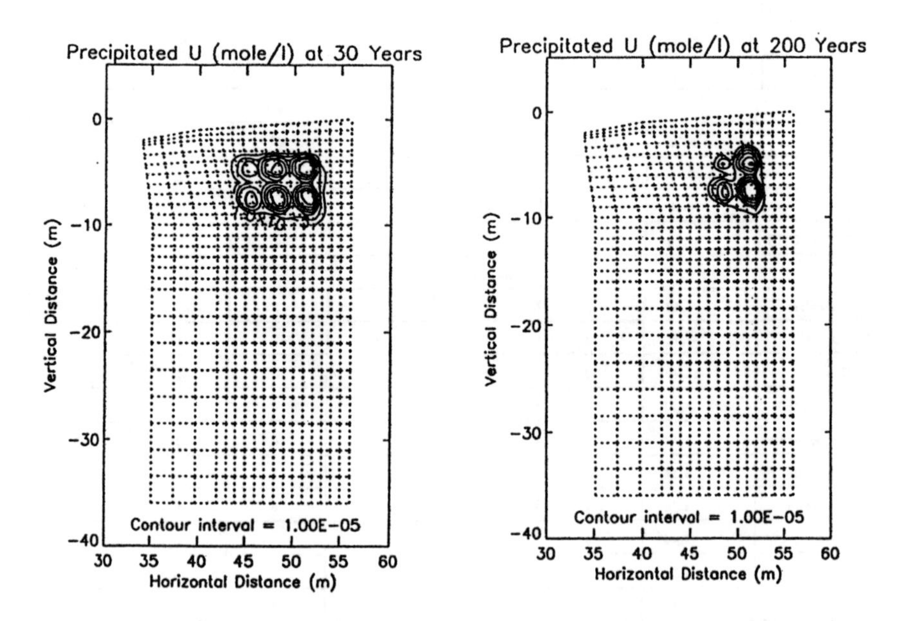

Figure 13.8: Distribution of precipitated uranium at 30 and 200 years.

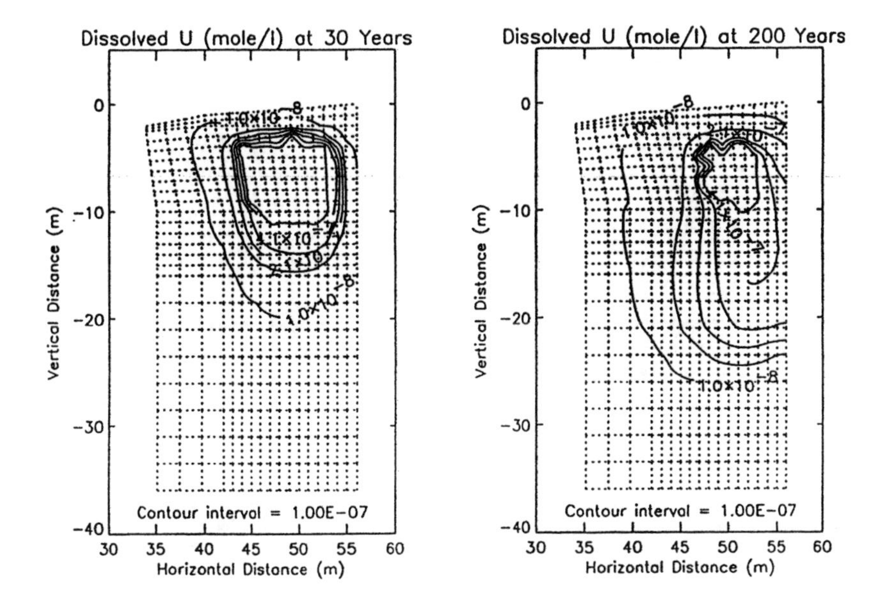

Figure 13.9: Distribution of dissolved uranium at 30 and 200 years.

concentrations translate into a relatively low K_d.

The K_d distributions are shown in Figure 13.11. Note that K_d varies spatially and temporally as well. These results illustrate the importance of speciation. Recall Table 13.1 and note the various species of uranium considered. In this hypothetical example the only species allowed to adsorb was UO_2OH^+. At the higher uranium concentrations, however, the fraction of dissolved uranium present in the form of other hydroxides and carbonates increased. As a result, a smaller fraction of the dissolved uranium was available for adsorption. Note that if the species UO_2OH^+ was absent completely, adsorption would not occur and K_d would be zero except in the region of precipitation.

Although the example presented here is somewhat simplified it demonstrates some important capabilities of BLT-EC. It also hints at both the limitations of the constant K_d approach and the potential importance of speciation on retardation. We refer readers to [8] for additional details and discussion of this problem.

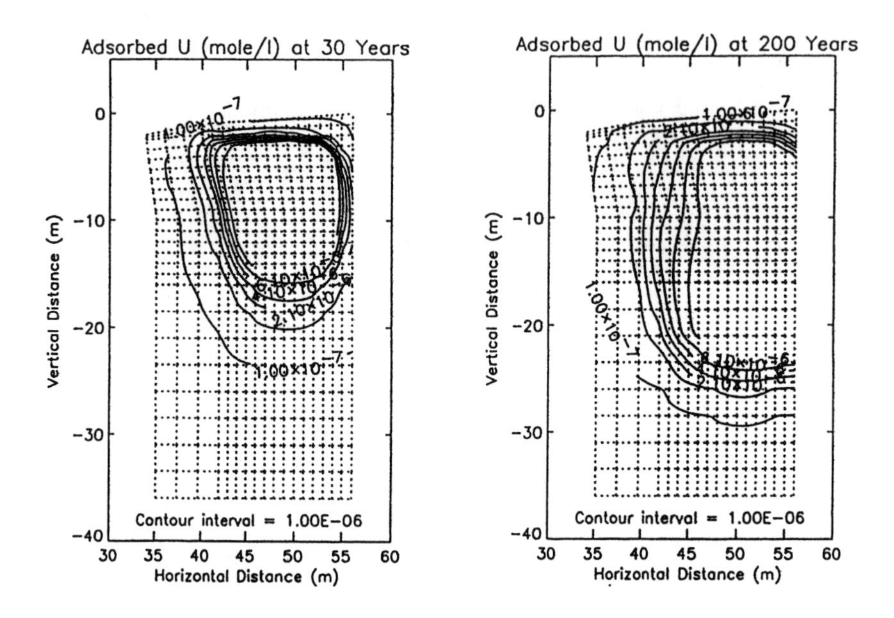

Figure 13.10: Distribution of adsorbed uranium at 30 and 200 years.

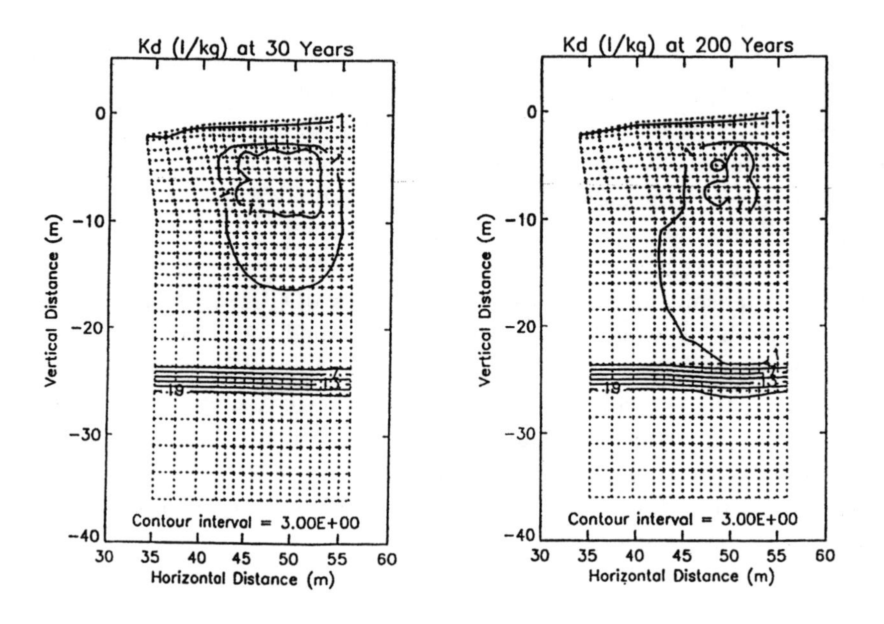

Figure 13.11: Distribution of K_d at 30 and 200 years.

13.7 Conclusion

In this paper we have described a new finite-element model, BLT-EC, for predicting the release and ground-water transport of radionuclides from LLW shallow land burial facilities. This model is capable of predicting container degradation, waste-form leaching, and advective-dispersive multispecies solute transport with retardation in saturated or unsaturated porous media. BLT-EC accounts for retardation directly by modeling the chemical processes of complexation, sorption, dissolution-precipitation, ion exchange, and oxidation-reduction. The ability of BLT-EC to simulate interactions between chemical processes and transport was demonstrated on the example field-scale problem. BLT-EC is presently undergoing further development and will be released for public use in the near future.

References

[1] Allison, J.D., D.S. Brown, and K.J. Novo-Gradac, "MINTEQA2/PRODEFA2, A Geochemical Assessment Model For Environment System: Version 3.0 User's Manual," EPA/600/3- 91/021, U.S. Environmental Protection Agency, 1991.

[2] Cederberg, G.A., R.L. Street, and J.O. Leckie, "A groundwater mass transport and equilibrium chemistry model for multicomponent systems," *Water Resources Research*, 21, 1095-1104, 1985.

[3] Kirkner, D.J., T.L. Theis, and A.A. Jennings, "Multicomponent solute transport with sorption and soluble complexation," *Adv. Water Resources*, 7, 120-125, 1984.

[4] Lichtner, P.C., "Continuum model for simultaneous chemical reactions and mass transport in hydrothermal systems," *Geochim. Cosmochim. Acta*, 49, 779-800, 1985.

[5] Liu, W. C. and T. N. Narasimhan, "Redox-controlled multiple-species reactive chemical transport 1. Model development," *Water Resources Research*, 25, 5, 869-882, 1989.

[6] Liu, W.C., and T.N. Narasimhan, "Redox-Controlled multiple-species reactive chemical transport. 2. Verification and application," *Water Resources Research*, 25, 883-910, 1989.

[7] Logan, K.H., "Engineering significance of National Bureau of Standards soil-corrosion data," *Journal of Research of the National Bureau of Standards*, 22, 1939.

[8] MacKinnon, R.J., T.M. Sullivan, C.J. Suen, and S.A. Simonson, "BLT-EC (Breach, Leach, Transport, and Equilibrium Chemistry): A Geochemical Ground-Water Transport Model for Assessing the Release of Radionuclides from Low-Level Waste Disposal Units," (in press), BNL-NUREG 1994.

[9] Matsunaga, T., G. Karametaxas, H.R. Von Gunten and P.L. Lichtner, "Redox chemistry of iron and manganese minerals in river-recharging aquifers: A model interpretation of a column experiment." *Geochim. et Cosmochim. Acta*, 57, 1691- 1704, 1993.

[10] Roles, G.W., "Characteristic of Low-level Radioactive Waste Disposed during 1987 through 1989," NUREG-1418, U.S. Nuclear Regulatory Commission, Washington, D.C., 1990.

[11] Romanoff, M., "Underground Corrosion," *National Bureau of Standards Circular*, 579, 1957.

[12] Sevougian, S.D., R.S. Schechter, and L.W. Lake, "Effect of partial local equilibrium on the propagation of precipitation/dissolution waves," *Ind. Eng. Chem. Res.*, 32, 10, 281-2304, 1993.

[13] Sullivan, T.M., "Data Input Guide For BLT-MS" (Breach, Leach, Transport-Multispecies), (NUREG in preparation), 1994.

[14] Sullivan, T.M., "Disposal Unit Source Term (DUST) Data Input Guide," NUREG/CR-6041, BNL-NUREG-52375, Brookhaven National Laboratory, 1993.

[15] Sullivan, T.M., "Selection of Models to Calculate the LLW Source Term," NUREG/CR-5773, BNL-NUREG-52295, Brookhaven National Laboratory, 1991.

[16] Sullivan, T.M., C.R. Kempf, C.J. Suen, and S.F. Mughbahab, "Low-level Radioactive Waste Source Term Model Development and Testing," NUREG/CR-5204, BNL-NUREG-52160, Brookhaven National Laboratory, 1988.

[17] Sullivan, T.M., and C.J. Suen, "Low-level Waste Shallow Land Disposal Source Term Model: Data Input Guides," NUREG/CR-5387, BNL-NUREG-52206, Brookhaven National Laboratory, 1989.

[18] Sullivan, T.M. and C.J. Suen, "Low-level Waste Source Term Model Development and Testing," NUREG/CR-5681, BNL-NUREG-52280, Brookhaven National Laboratory, 1991.

[19] Van Zeggeren, F. and S.H. Storey, *The Computation of Chemical Equilibria*, Cambridge University Press, Cambridge, U.K., 1970.

[20] Walsh, M.P., *Geochemical Flow Modeling, Ph.D. Dissertation*, The University of Texas at Austin, Austin, TX, 1983.

[21] Yeh, G.T., and V.S. Tripathi, "A Model for Simulating Transport of Reactive Multispecies Components: Model development and demonstration," *Water Resources Research*, 27, 12, 3075-3094, 1991.

[22] Yeh, G.T., and D.S. Ward, "FEMWASTE: A Finite Element Model of Waste Transport Through Saturated-Unsaturated Porous Media," ORNL - 5601, Oak Ridge National Laboratory, 1981.

[23] Yeh, G.T. and D.S. Ward, "FEMWATER: A Finite Element Model of Water Flow Through Saturated-Unsaturated Porous Media," ORNL-5567, Oak Ridge National Laboratory, 1980.

Chapter 14
Contaminant Transport with Nonlinear, Nonequilibrium Adsorption Kinetics

C.N. Dawson[1]

14.1 Introduction

Solutes in groundwater may undergo several types of chemical reactions during transport. Some of these reactions, such as biodegradation, involve other species and microorganisms in solution. Other reactions, such as sorption, involve interaction between the solute and the porous skeleton. Adsorption/desorption is a retardation/release reaction between the solute and a thin layer of molecules on the pore surface, and in effect, represents mass transfer between the fluid and the solid phases. Knowledge of the influence of sorption on the transport of the solutes is of fundamental importance in understanding how pollutants spread through the soil.(See [16] for a fundamental discussion of sorption processes and their mathematical description.)

Contaminant transport with sorption can be described by the following mathematical model. For simplicity, we consider the case of steady, single-phase flow and a single chemical species. We neglect sources and sinks and incorporate only those chemical reactions due to sorption. We also assume all sorption sites are described by a single kinetic model. Let $c(x,t)$

[1]Department of Computational and Applied Mathematics, Rice University, Houston, Texas

(μg/cm^3) ≥ 0 denote the concentration of the solute in the groundwater, and let $s(x,t)$ (μg/g porous medium) denote the mass of contaminant adsorbed on the solid matrix per unit mass of solid. Then c and s satisfy [6]

$$\phi c_t + \rho_b s_t + \nabla \cdot (\boldsymbol{u}c - \boldsymbol{D}\boldsymbol{\nabla} c) = 0, \quad \text{on } \Omega \times (0,T], \tag{14.1}$$

$$\rho_b s_t = \rho_b(k_a \psi(c) - k_s s), \tag{14.2}$$

In (14.1), ϕ represents porosity, ρ_b (g porous medium/cm^3) > 0 the bulk mass density of the soil, $\boldsymbol{u}$ (cm/s) the Darcy velocity, and $\boldsymbol{D}$ (cm^2/s) the hydrodynamic diffusion/dispersion tensor. We assume Ω is a polygonal domain in $\mathbb{R}^d$, with boundary $\partial\Omega = \Gamma_1 \cup \Gamma_2$, for Γ_1 an inflow boundary, where $\boldsymbol{u} \cdot \boldsymbol{n} < 0$, and Γ_2 an outflow or noflow boundary, where $\boldsymbol{u} \cdot \boldsymbol{n} \geq 0$; here $\boldsymbol{n}$ is the unit outward normal to $\partial\Omega$. We supplement (14.1) with the initial and boundary conditions

$$c(\boldsymbol{x},0) = c^0(\boldsymbol{x}), \qquad \boldsymbol{x} \in \Omega, \tag{14.3}$$

$$(\boldsymbol{u}c - \boldsymbol{D}\boldsymbol{\nabla} c) \cdot \boldsymbol{n} = (\boldsymbol{u}c_1) \cdot \boldsymbol{n}, \quad \text{on } \Gamma_1 \times (0,T], \tag{14.4}$$

$$(\boldsymbol{D}\boldsymbol{\nabla} c) \cdot \boldsymbol{n} = 0, \quad \text{on } \Gamma_2 \times (0,T] \tag{14.5}$$

where c^0 and c_1 are specified.

In (14.2), $\psi(c)$ is a sorption isotherm. Two common isotherms are the Langmuir isotherm, given by

$$\psi(c) = \frac{K_1 c}{1 + K_2 c}, \quad K_1, K_2 \geq 0 \tag{14.6}$$

and the Freundlich isotherm, given by

$$\psi(c) = K_3 c^p, \quad K_3 \geq 0 \tag{14.7}$$

where the exponent $p > 0$, and in most cases satisfies $0 < p \leq 1$. The constants $k_a(1/s)$ and $k_s(1/s)$ in (14.2) are reaction rates, with $k_a, k_s < \infty$ in the case of non-equilibrium sorption. In this case we supplement (14.2) with the initial condition

$$s(\boldsymbol{x},0) = s^0(\boldsymbol{x}), \quad \boldsymbol{x} \in \Omega \tag{14.8}$$

When $k_a, k_s \to \infty$, (14.2) reduces to the algebraic equation $s = \psi(c)$, and (14.1) becomes

$$\phi c_t + \rho_b \psi(c)_t + \nabla \cdot (\boldsymbol{u}c - \boldsymbol{D}\boldsymbol{\nabla} c) = 0 \tag{14.9}$$

Note that the Langmuir isotherm is nonlinear whenever $K_2 \neq 0$ and the Freundlich isotherm is nonlinear whenever $K_3 > 0$ and $p < 1$. In the latter case, assuming equilibrium kinetics and one-dimensional flow, (14.9) reduces to

$$\phi c_t + \rho_b K_3 (c^p)_t + u c_x - D c_{xx} = 0 \tag{14.10}$$

Making the change of variables $\gamma = c + \rho_b K_3 c^p / \phi$ and $\eta(\gamma) = c$, (14.10) can be rewritten as

$$\phi \gamma_t + u \eta(\gamma)_x - D \eta(\gamma)_{xx} = 0 \tag{14.11}$$

which is a nonlinear parabolic equation in γ. For $p < 1$, $\eta'(\gamma) = 0$ whenever $c = 0$; thus (14.11) is degenerate parabolic for $c \geq 0$ but uniformly parabolic for c bounded away from zero. The result is that classical, smooth solutions to (14.10) may not exist, even for smooth initial data.(For more on the behavior of solutions to these equations, see [10].) In the case of nonequilibrium adsorption, the smoothness of the solution improves, however, full regularity of the solution cannot be expected when $p < 1$ [11].

The numerical approximation to the model given by (14.1)-(14.8) using upwind-mixed finite element techniques has been described and analyzed in previous papers [4–6]. The method is an extension of the Godunov-mixed methods developed in [2,7,8]. Numerical results based on these ideas appear in these papers and in [9,12], where the technique is used to study the asymptotic behavior of solutions to one- and two-dimensional equations of the form (14.9). Much of our numerical work and analysis has focused on the case of equilibrium adsorption, assuming the Freundlich isotherm with various values of the exponent p. This case is the most difficult to handle numerically and to analyze mathematically.

In the next section, we describe our numerical approach to solving (14.1)-(14.8), and in Section 14.3, some numerical results for nonequilibrium sorption are presented.

14.2 The Numerical Model

In this section, we outline the upwind-mixed finite element approximation to the model given by (14.1)-(14.8). For the purposes of discussion we assume $k_a, k_s < \infty$ in (14.2), though the scheme easily handles the case where both are infinite; that is, where the model reduces to (14.9).

In (14.1), denote $\boldsymbol{u}c$ by $\boldsymbol{a}$ and $\boldsymbol{D}\nabla c$ by $\boldsymbol{z}$. Let $(\cdot, \cdot)$ represent the $L^2(\Omega)$ inner product, and let $H(div; \Omega) = \{\boldsymbol{v} \in (L^2(\Omega))^d : \nabla \cdot \boldsymbol{v} \in L^2(\Omega)\}$. Let $W = L^2(\Omega)$, $\boldsymbol{V} = H(div; \Omega)$, and $\boldsymbol{V}^0 = \boldsymbol{V} \cap \{\boldsymbol{v} : \boldsymbol{v} \cdot \boldsymbol{n} = 0\}$.

The mixed weak form of (14.1)-(14.2) is given by

$$(\phi c_t + \rho_b s_t + \nabla \cdot (\boldsymbol{a} + \boldsymbol{z}), w) = 0, \quad w \in W, \tag{14.12}$$

$$(\boldsymbol{D}^{-1}\boldsymbol{z}, v) = (c, \nabla \cdot \boldsymbol{v}), \quad \boldsymbol{v} \in \boldsymbol{V}^0, \tag{14.13}$$

$$(\rho_b s_t, w) = (\rho_b(k_a \psi(c) - k_s s), w), \quad w \in W \tag{14.14}$$

Let $W_h \subset W$ and $\boldsymbol{V}_h \subset \boldsymbol{V}$ denote the lowest-order Raviart-Thomas [16] approximating spaces defined on a triangulation $\mathcal{T}_h$ of Ω into elements with maximum diameter $h > 0$. Let $\boldsymbol{V}_h^0 = \boldsymbol{V}_h \cap \{\boldsymbol{v} : \boldsymbol{v} \cdot \boldsymbol{n} = 0\}$. We note that W_h consists of functions which are piecewise constant on the triangulation $\mathcal{T}_h$. The space $\boldsymbol{V}_h$ is more complicated, and is discussed below. Let $\Delta t = T/N$ for some positive integer N, and $t^n = n\Delta t$, $n = 0, \ldots, N$. For a time dependent function $g(t)$, let $g^n = g(t^n)$ and let $\partial_t g^n = (g^n - g^{n-1})/\Delta t$.

Approximate $c(\cdot, t^n)$, $s(\cdot, t^n)$ by C^n, $S^n \in W_h$, respectively, and approximate $\boldsymbol{a}(\cdot, t^n)$, $\boldsymbol{z}(\cdot, t^n)$ by $\boldsymbol{A}^n \in \boldsymbol{V}_h$, $\boldsymbol{Z}^n \in \boldsymbol{V}_h^0$, respectively. For $n = 0$, set

$$(C^0, w) = (c^0, w), \quad w \in W_h, \tag{14.15}$$
$$(S^0, w) = (s^0, w), \quad w \in W_h \tag{14.16}$$

Given C^{n-1}, S^{n-1}, determine C^n, S^n, and $\boldsymbol{Z}^n$ for $n = 1, \ldots, N$ by

$$(\phi \partial_t C^n + \rho_b \partial_t S^n + \nabla \cdot (\boldsymbol{A}^{n-1} + \boldsymbol{Z}^n), w) = 0, \quad w \in W_h, \tag{14.17}$$

$$(D^{-1}\boldsymbol{Z}^n, \boldsymbol{v}) = (C^n, \nabla \cdot \boldsymbol{v}), \quad \boldsymbol{v} \in \boldsymbol{V}_h^0, \tag{14.18}$$

$$(\rho_b \partial_t S^n, w) = (\rho_b(k_a \psi(C^n) - k_s S^n), w), \quad w \in W_h \tag{14.19}$$

The advective flux $\boldsymbol{A}^{n-1} \approx (\boldsymbol{u}c)(\cdot, t^{n-1})$ is the only undetermined quantity in (14.17). In our method $\boldsymbol{A}^{n-1} \in \boldsymbol{V}_h$ is calculated explicitly from the solution given at t^{n-1}. In the lowest-order spaces, a function $\boldsymbol{v} \in \boldsymbol{V}_h$ is completely determined by the values of $\boldsymbol{v} \cdot \boldsymbol{n}_T$, where $\boldsymbol{n}_T$ is the normal vector to the boundary of an element T. For example, suppose $\Omega \subset \mathbb{R}^2$ is divided into triangles, and suppose the triangle edges are numbered $k = 1, \ldots, N_e$. Let $\boldsymbol{n}_k$ denote a unit vector normal to edge k. We associate with each edge a basis function $\boldsymbol{v}_k \in \boldsymbol{V}_h$ which satisfies (i) each component of $\boldsymbol{v}_k$ is linear on each triangle, and (ii) $\boldsymbol{v}_k \cdot \boldsymbol{n}_l = \delta_{kl}$, where δ_{kl} is one if $k = l$ and zero otherwise. Then, $\boldsymbol{v} \in \boldsymbol{V}_h$ can be written as

$$\boldsymbol{v}(\boldsymbol{x}) = \sum_{k=1}^{N_e} (\boldsymbol{v} \cdot \boldsymbol{n}_k)\boldsymbol{v}_k(\boldsymbol{x}) \tag{14.20}$$

We remark that $\boldsymbol{v}_k$ is nonzero only on the triangles which share edge k.

On inflow boundary edges, corresponding to the region Γ_1, we set $\boldsymbol{A}^{n-1} \cdot \boldsymbol{n}_k = (\boldsymbol{u}c_1) \cdot \boldsymbol{n}_k$. Since $\boldsymbol{Z}^{n-1} \cdot \boldsymbol{n}_k = 0$, the inflow boundary condition (14.4) is satisfied. On internal and outflow element edges, there are many ways to calculate $\boldsymbol{A}^{n-1} \cdot \boldsymbol{n}_k$. One way is by simple one-point upwinding. Let C_L^{n-1} denote the value of C^{n-1} on the element to the "left" of edge k (relative to the direction of the normal vector $\boldsymbol{n}_k$), and C_R^{n-1} the value to the "right" of edge k. Then define

$$\boldsymbol{A}^{n-1} \cdot \boldsymbol{n}_k = (\boldsymbol{u}(\boldsymbol{x}_k) \cdot \boldsymbol{n}_k) \begin{cases} C_L^{n-1}, & \text{if } \boldsymbol{u}(\boldsymbol{x}_k) \cdot \boldsymbol{n}_k > 0, \\ C_R^{n-1}, & \text{if } \boldsymbol{u}(\boldsymbol{x}_k) \cdot \boldsymbol{n}_k < 0, \end{cases} \tag{14.21}$$

where $\boldsymbol{x}_k$ is the midpoint of edge k. This simple type of upwinding is usually insufficiently accurate and results in solutions which are numerically diffusive. Accuracy can be improved by calculating better approximations to the left and right "states" C_L and C_R. For rectangular elements, such approximations are discussed in [2,8]. These higher-order approximations are constructed by a two-step process. First, the constant values C^{n-1} are postprocessed to obtain a linear function on each element. This improves the spatial accuracy. To improve the temporal accuracy, characteristic tracing can be used to predict the advective flux across edge k from time t^{n-1} to time t^n. Using these ideas, the formal accuracy of the advective flux can be improved to second-order in space and time,(see [7]). Similar types of advective flux approximations for triangular elements can be found in [3,13].

The system (14.17)-(14.19) is a nonlinear system of equations in C^n, S^n, and $\boldsymbol{Z}^n$. Using (14.18) and (14.19), $\boldsymbol{Z}^n$ and S^n can be eliminated, giving a system in C^n only. The existence and uniqueness of C^n follows from the fact that $\psi(c)$ is a continuous, monotone function of c, by invoking standard theory for continuous, monotone operators [14].

14.3 Numerical Results

In this section, we present numerical results for the nonequilibrium case described above. The equilibrium case $k_a = k_s = \infty$ has been studied previously in [4–6, 9, 12]. We consider the one-dimensional equation

$$\phi c_t + \rho_b s_t + uc_x - Dc_{xx} = 0, \quad 0 < x < L, \quad 0 < t \leq T, \tag{14.22}$$

$$s_t = k_a \phi(c) - k_s s, \tag{14.23}$$

and take

$$\phi(c) = K_3 c^p, \tag{14.24}$$

for $p \in (0, 1]$.

In the simulations presented here we have assumed $\phi = .5$, $\rho_b K_3 = 1.5$, $u = 3$ cm/hour, $D = .005$ cm^2/hour, $L = 100$ cm, $T = 30$ hours, and we vary k_a, k_s, and p. At the inflow boundary $x = 0$, $c_1 = 1$ μg/cm^3, and the initial conditions are $c(x, 0) = s(x, 0) \equiv 0$. Thus, the simulations below represent an inflow of contaminant into a previously clean aquifer, and the retardation of the contaminant due to adsorption.

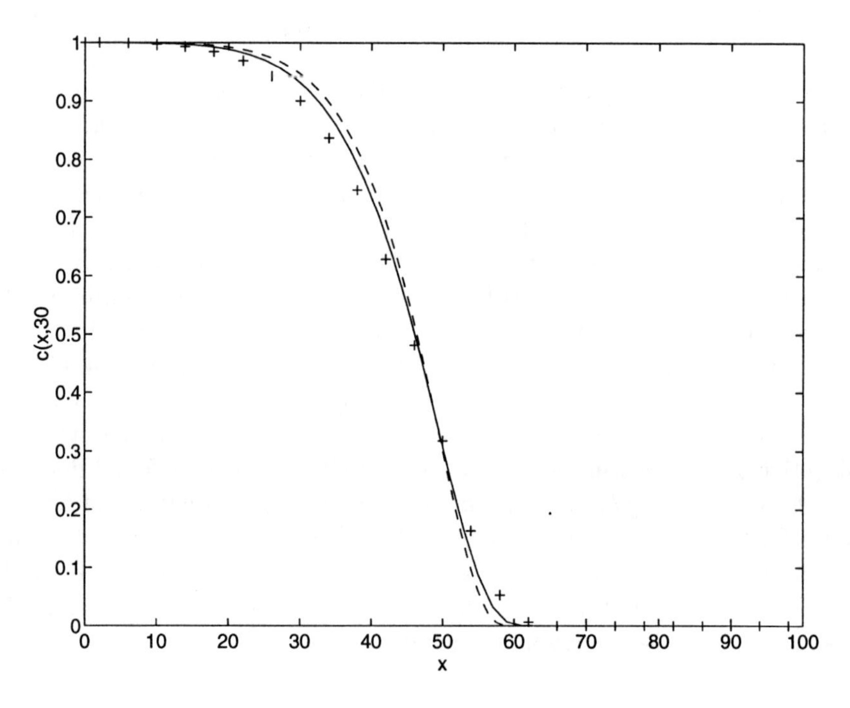

Figure 14.1: c at $T = 30$ hours for $\Delta x = 4(+)$, $2(- -)$, and $1(-)$ cm.

First, we choose $k_a = k_s = .5$/hour, $p = .6$, and vary Δx; Δt is always chosen so that the CFL constraint $\Delta t = \phi \Delta x / u$ is satisfied. In Figure 14.1, solutions for $\Delta x = 1$ cm, 2 cm, and 4 cm at time $T = 30$ are given. As seen in the figure, the solutions are very similar, with the difference between $\Delta x = 2$ and $\Delta x = 1$ small.

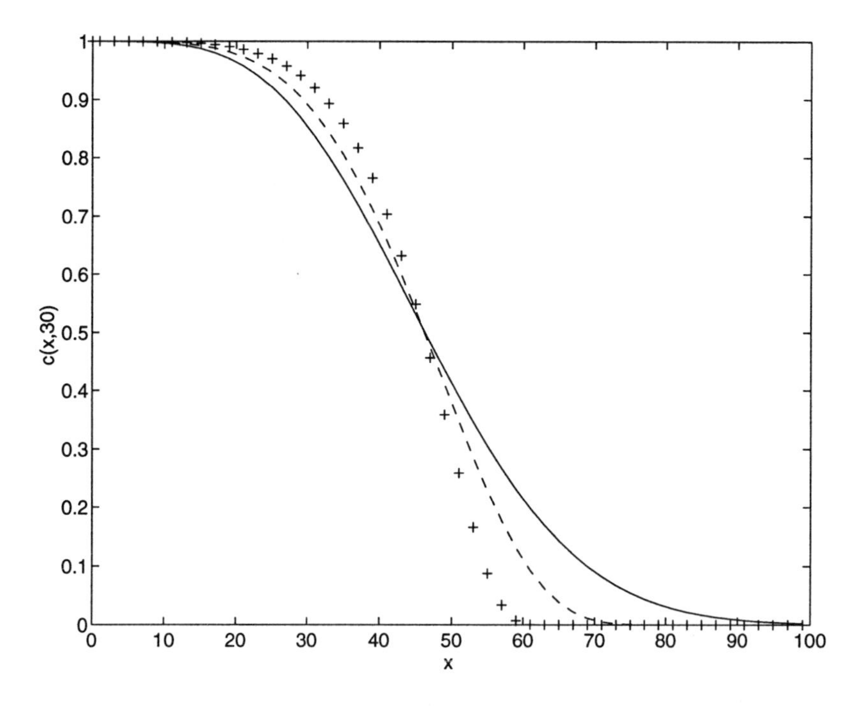

Figure 14.2: Comparison of solutions for $p = 1(-)$, $.8(- \, -)$, and $.6(+)$.

Next, we choose $\Delta x = 2$, and vary p, keeping k_a and k_s as above. In Figure 14.2, a comparison of solutions for $p = 1$, .8, and .6 are given. This figure shows that the effect of the exponent p on the solution is very dramatic, with $p = .6$ resulting in substantial adsorption (hence retardation) of the contaminant. This is to be expected upon examination of (14.23), since for $c \leq 1$, $c^{.6} \geq c^{.8} \geq c$. Notice also that the primary differences in the solutions are in the region where $c \to 0$.

Next, we compare solutions for different values of k_a and k_s, keeping $p = .6$ and $\Delta x = 2$. We remark that, as can be seen in (14.23), the effect of the first term on the right hand side involving k_a is to adsorb contaminant, retarding the front and transferring mass from the fluid to the solid phase. The higher k_a is, the more pronounced this effect should be. The effect of the second term involving k_s is to desorb contaminant, which balances adorption and in some cases, depending on the relative values of c, s, k_a,

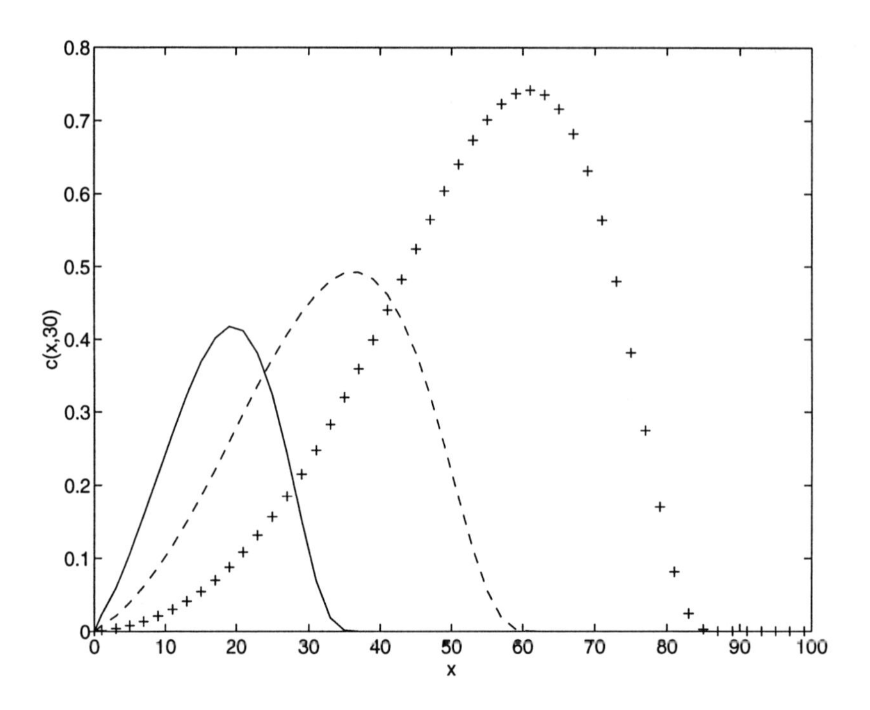

Figure 14.3: Comparison of solutions for $k_a = 1, k_s = .5(-)$, $k_a = k_s = .5(\text{- -})$, and $k_a = .5, k_s = 1(+)$.

and k_s, can transfer mass from the solid to the fluid phase.

In these simulations, we set the inflow concentration to one for $t \leq 15$ hours, then shut-off the source of contaminant. In Figure 14.3, we compare solutions at $T = 30$ hours for three scenarios: $k_a = 1$ and $k_s = .5$, $k_a = k_s = .5$, and $k_a = .5$ and $k_s = 1$. The first case, $k_a = 1$, $k_s = .5$, represents stronger adsorption and, as can be observed in the figure, substantial retardation of the front and loss of mass occur. The second case $k_a = k_s = .5$ represents an equal amount of adsorption/desorption, resulting in a less retarded front and less mass loss. The third case $k_a = .5$ and $k_s = 1$ is biased toward desorption and results in much less mass loss and retardation of the contaminant front. These results agree with the expected behavior as determined by (14.22) and (14.23).

To examine the effects of no desorption, we compared the case $k_a = 1$, $k_s = .5$ described above with the case $k_a = 1$, $k_s = 0$. For the latter set of parameters, the solution $c(x, 30) \equiv 0$, which means all of the contaminant was adsorbed.

Acknowledgments: This work was supported by grants from the U.S. Department of Energy.

References

[1] Bear, J., *Dynamics of fluids in porous media*, Elsevier, New York, 1972.

[2] Bell, J. B., C. N. Dawson, and G. R. Shubin, "An unsplit higher-order Godunov scheme for scalar conservation laws in two dimensions", *J. Comput. Phys.*, 74, 1-24, 1988.

[3] Cockburn, B., S. Hou, and C.-W. Shu, "The Runge-Kutta local projection discontinuous Galerkin finite element method for conservation laws IV: The multidimensional case", *Math. Comp.*, 54, 545-581, 1990.

[4] Dawson, C. N., "Simulation of nonlinear contaminant transport in groundwater by a higher order Godunov-mixed finite element method", in *Applications of supercomputers in engineering II*, C. A. Brebbia, D. Howard, and A. Peters, eds., Computational Mechanics Publications, Southampton, UK, 419-433, 1991.

[5] Dawson, C. N., "Analysis of Upwind-Mixed Methods for Nonlinear Contaminant Transport Equations", Technical Report TR93-57, Dept. of Comp. and Appl. Math., Rice University, Houston, TX, and to appear.

[6] Dawson, C. N., "Modeling of nonlinear adsorption in contaminant transport", to appear in *Proceedings, Computational Methods in Water Resources '94*.

[7] Dawson, C. N., "Godunov-mixed methods for advective flow problems in one space dimension", *SIAM J. Numer. Anal.*, 28, 1282-1309, 1991.

[8] Dawson, C. N., "Godunov-mixed methods for advection-diffusion equations in multidimensions", *SIAM J. Numer. Anal.*, 30, 1315-1332, 1993.

[9] Dawson, C. N., C. J. van Duijn, and R. E. Grundy, "Asymptotic profiles with finite mass in two-dimensional contaminant transport through porous media: the fast reaction case", submitted 1994.

[10] van Duijn, C. J. and P. Knabner, "Solute transport in porous media with equilibrium and non-equilibrium multiple-site adsorption: Travelling waves", *J. Reine Angewandte Math.*, 415, 1-49, 1991.

[11] van Duijn, C. J. and P. Knabner, "Solute transport through porous media with slow adsorption", in *Free boundary problems: Theory and applications, Vol. 1*, K. H. Hoffman and J. Sprekels, eds., Longman, New York, 1990, 375-388.

[12] Grundy, R. E., C. J. van Duijn, and C. N. Dawson, "Asymptotic profiles with finite mass in one-dimensional contaminant transport through porous media: the fast reaction case", *Quarterly Journal of Mechanics and Applied Math*, 47, 69-106, 1994.

[13] Harten, A. and S. R. Chakravarthy, "Multi-Dimensional ENO Schemes for General Geometries", ICASE Report No. 91-76, 1991.

[14] Ortega, J. M. and W. C. Rheinboldt, *Iterative solution of nonlinear equations in several variables*, Academic Press, New York, 1970.

[15] Raviart, P. A. and J. M. Thomas, "A mixed finite element method for 2nd order elliptic problems", in *Mathematical aspects of the finite element method*, Rome 1975, Lecture Notes in Mathematics, Springer-Verlag, Berlin, 1977.

[16] Weber, Jr., W. A., P. M. McGinley, and L. E. Katz, "Sorption phenomena in subsurface systems: Concepts, models and effects on contaminant fate and transport", *Wat. Res.*, 25, 499-528, 1991.

Chapter 15

Mixed Methods for Flow and Transport Problems on General Geometry

T. Arbogast[1]

15.1 Introduction

The mixed finite element method was developed by Raviart and Thomas [13], and we restrict our attention to their lowest-order method. It was first used for subsurface problems by Douglas *et al* [6], though Russell and Wheeler [14] pointed out that the commonly used cell-centered finite difference method on rectangular grids for problems with diagonal coefficients is actually a mixed method with appropriate quadrature rules applied to the integrals. The problem with mixed methods is that they can be difficult to implement directly, especially if the aquifer domain Ω is not rectangular.

This chapter describes some recent work on the development of numerical schemes for groundwater flow and transport problems with tensor coefficients on a geometrically general domain. The schemes are cell-centered finite difference approximations of a mixed finite element method; thus, they are easy to implement, have only one unknown per element, and are locally conservative. The elements can be either deformed rectangles or bricks, or they can be triangles. The work on rectangular elements is joint with Mary

[1]Department of Computational and Applied Mathematics, Rice University, Houston, TX

F. Wheeler and Ivan Yotov [1–3], and the work on triangles is joint with Clint N. Dawson and Philip T. Keenan [1].

Our main requirement is that there be a smooth mapping F of a reference, computational domain $\hat{\Omega}$ onto the aquifer domain Ω in $d = 2$ or 3 dimensions. The Jacobian matrix of F is DF, $(DF)_{ij} = \partial F_i / \partial \hat{x}_j$, and the Jacobian of the mapping is $J = |\det(DF)|$. The reference domain is divided into standard elements (rectangles, bricks, or equilateral triangles) of maximal diameter $\hat{h}$ to form a regular partition $\hat{\mathcal{T}}_{\hat{h}}$ of $\hat{\Omega}$. Then F defines a smooth, curvilinear grid $\mathcal{T}_h$ on Ω with elements of maximal diameter h. (There are grid generation codes available for creating F and its Jacobian matrix.)

We consider the two main problems of subsurface flow and transport, in somewhat simplified form. In the flow problem, we solve for the pressure p and the velocity $\boldsymbol{u}$ satisfying

$$\nabla \cdot \boldsymbol{u} = f, \quad \boldsymbol{x} \in \Omega, \tag{15.1}$$

$$\boldsymbol{u} = -K\nabla p, \quad \boldsymbol{x} \in \Omega, \tag{15.2}$$

$$p = g, \quad \boldsymbol{x} \in \partial\Omega, \tag{15.3}$$

where K is the hydraulic conductivity tensor, f is a source term, and g gives the boundary condition. In the transport problem, we solve for the concentration c such that

$$\frac{\partial(\theta c)}{\partial t} + \nabla \cdot (\boldsymbol{u}c - D(\boldsymbol{u})\nabla c) = f_c(c), \tag{15.4}$$

where θ is the water content and D is the dispersion tensor. The dispersion part of the transport problem, $-\nabla \cdot D(\boldsymbol{u})\nabla c$, is of the same form as the flow problem; therefore, we concentrate on (15.1)–(15.3), a second order elliptic equation with, for simplicity of exposition, Dirichlet boundary conditions.

In the next two sections, before we begin the development of our new numerical schemes, we define the mixed method finite element spaces and review the standard mixed method.

15.2 The Raviart-Thomas Mixed Spaces

Let $L^2(\Omega)$ denote the set of square integrable functions, and let $H(\Omega; \mathrm{div})$ denote the space of vector functions that have a divergence; that is,

$$L^2(\Omega) = \left\{ w(\boldsymbol{x}) : \int_\Omega |w|^2 \, dx < \infty \right\}, \tag{15.5}$$

$$H(\Omega; \mathrm{div}) = \{ \boldsymbol{v}(\boldsymbol{x}) : \boldsymbol{v} \in (L^2(\Omega))^d \text{ and } \nabla \cdot \boldsymbol{v} \in L^2(\Omega) \}. \tag{15.6}$$

The mixed method requires a vector approximating space $V_h \subset H(\Omega; \text{div})$ for the velocity $\boldsymbol{u}$, as well as a scalar space $W_h \subset L^2(\Omega)$ for the pressure p. In the lowest order Raviart-Thomas mixed spaces [11,13], the pressure is approximated by a constant on each element; that is,

$$W_h = \{w \in L^2(\Omega) : w \text{ is constant on each element } E \in \mathcal{T}_h\}.$$

The nodal degrees of freedom can be considered as the function values at the centers of the elements.

The elemental space of vectors $\hat{V}(\hat{E})$ can be described on a standard reference element $\hat{E}$. If $\hat{T}$ is a rectangle or brick, then

$$\hat{V}(\hat{E}) = \{\boldsymbol{v} : v_i = \alpha_i x_i + \beta_i \text{ for any real numbers } \alpha_i \text{ and } \beta_i,\ i = 1, 2(, 3)\};$$

that is, the ith component of $\boldsymbol{v}$ is linear in the ith coordinate direction and constant in the other direction(s). If $\hat{T}$ is a triangle, then

$$\hat{V}(\hat{E}) = \{\boldsymbol{v} : v_i = \alpha x_i + \beta_i \text{ for any real numbers } \alpha \text{ and } \beta_i,\ i = 1, 2\}.$$

In either case, the important fact is that if ν denotes the outward unit normal vector (to the element), then $\boldsymbol{v} \cdot \nu$ is a constant; therefore, the nodal degrees of freedom can be considered as the values of $\boldsymbol{v} \cdot \nu$ at the centers of the element edges or faces. By matching these degrees of freedom across elements, we define $\hat{V}_{\hat{h}} \subset H(\hat{\Omega}; \text{div})$ on the reference domain:

$$\begin{aligned} \hat{V}_{\hat{h}} &= \{\boldsymbol{v} : \boldsymbol{v}|_{\hat{E}} \in \hat{V}(\hat{E}) \text{ for all } \hat{E} \in \hat{\mathcal{T}}_{\hat{h}} \text{ and the normal} \\ &\qquad \text{fluxes } \boldsymbol{v} \cdot \nu \text{ agree across element boundaries}\}. \end{aligned}$$

On general elements, there is no simple polynomial with the requisite properties. Instead, the Piola transform is used to define the space V_h [5,15]. The construction is

$$V(E) = \{\boldsymbol{v} : \boldsymbol{v}(\boldsymbol{x}) = J^{-1} DF\, \hat{\boldsymbol{v}}(F^{-1}(\boldsymbol{x})) \text{ for some } \hat{\boldsymbol{v}} \in \hat{V}(\hat{E})\},$$

where $E = F(\hat{E})$. The Piola transform has the property that normal fluxes are preserved (in an average sense); therefore, local mass conservation is preserved under this mapping. Moreover, since $\boldsymbol{v} \cdot \nu$ is a constant on each edge or face, the nodal degrees of freedom are as for reference elements and V_h is defined as above. Denote by $\{w_i\}$ a basis for W_h, and $\{\boldsymbol{v}_i\}$ a basis for V_h.

15.3 The Standard Mixed Method

We rewrite (15.1)–(15.3) in variational form as follows. Note that (15.2) can be written as

$$K^{-1}\boldsymbol{u} = -\nabla p, \tag{15.7}$$

and then the mixed variational form of the equations is

$$\int_\Omega \nabla \cdot \boldsymbol{u}\, w\, dx = \int_\Omega f\, w\, dx, \quad w \in L^2(\Omega), \tag{15.8}$$

$$\int_\Omega K^{-1}\boldsymbol{u} \cdot \boldsymbol{v}\, dx = \int_\Omega p \nabla \cdot \boldsymbol{v}\, dx - \int_{\partial\Omega} g\, \boldsymbol{v} \cdot \nu\, ds, \quad \boldsymbol{v} \in H(\Omega; \mathrm{div}) \tag{15.9}$$

after integrating the pressure gradient term by parts.

Restricting $\boldsymbol{u}$ and $\boldsymbol{v}$ to V_h and p and w to W_h gives the standard mixed method for (15.1)–(15.3). It is a saddle point linear system; in matrix form it is

$$\begin{pmatrix} A & -B^T \\ -B & 0 \end{pmatrix} \begin{pmatrix} U \\ P \end{pmatrix} = \begin{pmatrix} -G \\ -F \end{pmatrix}, \tag{15.10}$$

where U represents $\boldsymbol{u}$ and P represents p in the bases, and

$$A_{ij} = \int_\Omega K^{-1}\boldsymbol{v}_i \cdot \boldsymbol{v}_j\, dx, \qquad B_{ij} = \int_\Omega w_i \nabla \cdot \boldsymbol{v}_j\, dx, \tag{15.11}$$

$$F_i = \int_\Omega f\, w_i\, dx, \qquad G_i = \int_{\partial\Omega} g\, \boldsymbol{v}_i \cdot \nu\, ds. \tag{15.12}$$

To reduce the size of this linear system, we can solve for the Schur complement system by eliminating U to obtain

$$(BA^{-1}B^T)P = F + BA^{-1}G. \tag{15.13}$$

This system is symmetric and positive definite and has small size. Unfortunately, although A is sparse, A^{-1} is in general full. Iterative solution will require the following steps for the application of the matrix: a matrix vector multiply $B^T P$; the solution of the system $Ay = B^T P$; and another matrix vector multiply By. Thus, we need inner iterations within our overall iterative solution, which can become somewhat expensive.

Arnold and Brezzi [4] defined the hybrid form of the mixed method to obtain a sparse, symmetric, and positive definite linear system. This formulation incorporates additional unknowns approximating the pressure on the edges or faces of the elements. These additional unknowns are often referred to as the "Lagrange multipliers". These approximations are piecewise constant; thus, the nodal degrees of freedom are the values at the centers of the edges or faces of the elements. The Schur complement system is defined in terms of these Lagrange multipliers; unfortunately, there are more unknowns than the number of elements. In two dimensions, if rectangles or triangles are used, there are two or 1.5 times as many unknowns, respectively. In three dimensions, brick elements give three times as many unknowns.

As mentioned in the introduction, Russell and Wheeler [14] defined an approximation to the mixed method in the case of a diagonal conductivity or permeability tensor and a rectangular grid. They used a nodal quadrature rule on the integral $\int_\Omega K^{-1}\boldsymbol{u}\cdot\boldsymbol{v}\,dx$ above, so that A becomes diagonal. This scheme is identical to the cell-centered finite difference method used in the petroleum industry (e.g., see [12]). The Schur complement system is then sparse, having a 5 point stencil for the pressure unknown in 2 dimensions, and 7 points in 3 dimensions. Unfortunately, if K is not diagonal, the interaction of the basis functions is problem and space dependent; thus, A cannot be diagonalized.

15.4 The Expanded Mixed Method

In this and the next three sections, we generalize the approach in [14] to define our numerical methods for the full case of tensor K and general geometry. Through a modification of the standard mixed method, we are able to simplify the interaction of the basis functions when K is not diagonal. If the elements are restricted to standard shapes, appropriate quadrature rules can be applied to approximate the Schur complement system by a sparse, symmetric, and positive definite system for the pressure unknowns. By mapping to the computational domain, geometrically general domains can be handled easily.

Let M be a tensor defined by the mapping F as

$$M = J((DF)^{-1})^T(DF)^{-1}; \tag{15.14}$$

then $\boldsymbol{M}$ is symmetric and positive definite. Its purpose will become clear in the next section. We introduce the additional unknown $\tilde{\boldsymbol{u}}$ defined by

$$M\tilde{\boldsymbol{u}} = -\boldsymbol{\nabla}p, \qquad \boldsymbol{u} = KM\tilde{\boldsymbol{u}}. \tag{15.15}$$

Then as before, the variational form is

$$\int_\Omega \nabla\cdot\boldsymbol{u}\,w\,dx = \int_\Omega f\,w\,dx, \quad w\in L^2(\Omega), \tag{15.16}$$

$$\int_\Omega M\tilde{\boldsymbol{u}}\cdot\boldsymbol{v}\,dx = \int_\Omega p\,\nabla\cdot\boldsymbol{v}\,dx - \int_{\partial\Omega} g\,\boldsymbol{v}\cdot\nu\,ds, \ \boldsymbol{v}\in H(\Omega;\mathrm{div}), \tag{15.17}$$

$$\int_\Omega M\boldsymbol{u}\cdot\tilde{\boldsymbol{v}}\,dx = \int_\Omega MKM\tilde{\boldsymbol{u}}\cdot\tilde{\boldsymbol{v}}\,dx,\ \tilde{\boldsymbol{v}}\in(L^2(\Omega))^d \tag{15.18}$$

Restricting $\boldsymbol{u}$, $\boldsymbol{v}$, $\tilde{\boldsymbol{u}}$, and $\tilde{\boldsymbol{v}}$ to V_h and p and w to W_h gives the expanded mixed method. As we will see, the linear system is again a saddle point problem, but first let us transform back to the computational domain.

A shift back to the reference domain considerably simplifies the form of the equations (15.16)–(15.18). Let $\boldsymbol{x} = F(\hat{\boldsymbol{x}})$ for $\hat{\boldsymbol{x}}\in\hat{\Omega}$. Then vector and scalar functions transform according to

$$\boldsymbol{v}(\boldsymbol{x}) = \frac{1}{J}DF\,\hat{\boldsymbol{v}}(\hat{\boldsymbol{x}}), \tag{15.19}$$

$$w(\boldsymbol{x}) = \hat{w}(\hat{\boldsymbol{x}}). \tag{15.20}$$

If one notes that the Piola transform satisfies the property [5, 15]

$$\nabla\cdot\boldsymbol{v} = \frac{1}{J}\hat{\nabla}\cdot\hat{\boldsymbol{v}}, \tag{15.21}$$

then the change of variables in the integrals of (15.16)–(15.18) is straightforward. By our choice of M,

$$\int_{\hat{\Omega}}\hat{\nabla}\cdot\hat{\boldsymbol{u}}\,\hat{w}\,d\hat{x} = \int_{\hat{\Omega}}\hat{f}\,\hat{w}\,J\,d\hat{x},\quad \hat{w}\in\hat{W}_{\hat{h}}, \tag{15.22}$$

$$\int_{\hat{\Omega}}\hat{\tilde{\boldsymbol{u}}}\cdot\hat{\boldsymbol{v}}\,d\hat{x} = \int_{\hat{\Omega}}\hat{p}\,\hat{\nabla}\cdot\hat{\boldsymbol{v}}\,d\hat{x} - \int_{\partial\hat{\Omega}}\hat{g}\,\hat{\boldsymbol{v}}\cdot\hat{\nu}\,d\hat{s},\quad \boldsymbol{v}\in\hat{V}_{\hat{h}}, \tag{15.23}$$

$$\int_{\hat{\Omega}}\hat{\boldsymbol{u}}\cdot\hat{\tilde{\boldsymbol{v}}}\,d\hat{x} = \int_{\hat{\Omega}}\kappa\hat{\tilde{\boldsymbol{u}}}\cdot\hat{\tilde{\boldsymbol{v}}}\,d\hat{x},\quad \hat{\tilde{\boldsymbol{v}}}\in\hat{V}_{\hat{h}}, \tag{15.24}$$

where

$$\kappa = J(DF)^{-1}K((DF)^{-1})^T. \tag{15.25}$$

This is an approximation to problem (15.1)–(15.3) on $\hat{\Omega}$, except that the conductivity tensor has been modified. The true solution is given by mapping back to Ω according to (15.19)–(15.20); one would normally map only the nodal degrees of freedom.

In matrix form, the method is

$$\begin{pmatrix} \mathcal{A} & -C & 0\\ -C & 0 & B^T\\ 0 & B & 0\end{pmatrix}\begin{pmatrix}\tilde{U}\\ U\\ P\end{pmatrix} = \begin{pmatrix}0\\ G\\ F\end{pmatrix}, \tag{15.26}$$

where $\tilde{U}$, U, and P represent $\hat{\tilde{\boldsymbol{u}}}$, $\hat{\boldsymbol{u}}$, and $\hat{p}$ in the bases, respectively, B, G, and F are as before (except now defined on the computational domain), and

$$\mathcal{A}_{ij} = \int_{\hat{\Omega}}\kappa\hat{\boldsymbol{v}}_i\cdot\hat{\boldsymbol{v}}_j\,d\hat{x},\qquad C_{ij} = \int_{\hat{\Omega}}\hat{\boldsymbol{v}}_i\cdot\hat{\boldsymbol{v}}_j\,d\hat{x} \tag{15.27}$$

(note that $\mathcal{A}$ and C are symmetric).

Again we can construct the Schur complement system by eliminating U and $\tilde{U}$ to obtain

$$(BC^{-1}\mathcal{A}C^{-1}B^T)P = F + (BC^{-1}\mathcal{A}C^{-1})G, \tag{15.28}$$

but C^{-1} is in general full. Depending on the type of grid used, we can introduce quadrature rules to diagonalize C.

15.5 Approximation by Cell-Centered Finite Differences

In this section, assume that $\hat{\Omega}$ and the computational grid is rectangular in d dimensions. We denote the trapezoidal quadrature rule on the reference domain for a function $\psi(\hat{\boldsymbol{x}})$ by

$$Q(\psi) = \sum_{\hat{E}\in\hat{\mathcal{T}}_{\hat{h}}} v\frac{1}{2^d}\sum_{i=1}^{2^d} \psi(\hat{\boldsymbol{x}}_i(\hat{E}))\,\mathrm{vol}(\hat{E}), \tag{15.29}$$

where the $\hat{\boldsymbol{x}}_i(\hat{E})$ go through the vertices of $\hat{E}$ and $\mathrm{vol}(\hat{E})$ is its area or volume. We then approximate (15.1)–(15.3) by

$$\int_{\hat{\Omega}} \hat{\nabla}\cdot\hat{\boldsymbol{u}}\,\hat{w}\,d\hat{x} = \int_{\hat{\Omega}} \hat{f}\,\hat{w}\,J\,d\hat{x}, \quad \hat{w}\in\hat{W}_{\hat{h}}, \tag{15.30}$$

$$Q(\hat{\hat{\boldsymbol{u}}}\cdot\hat{\boldsymbol{v}}) = \int_{\hat{\Omega}} \hat{p}\,\hat{\nabla}\cdot\hat{\boldsymbol{v}}\,d\hat{x} - \int_{\partial\hat{\Omega}} \hat{g}\,\hat{\boldsymbol{v}}\cdot\hat{\nu}\,d\hat{s}, \quad \boldsymbol{v}\in\hat{V}_{\hat{h}}, \tag{15.31}$$

$$Q(\hat{\boldsymbol{u}}\cdot\hat{\hat{\boldsymbol{v}}}) = Q(\kappa\hat{\hat{\boldsymbol{u}}}\cdot\hat{\boldsymbol{v}}), \quad \hat{\hat{\boldsymbol{v}}}\in\hat{V}_{\hat{h}}. \tag{15.32}$$

Indeed, now

$$\mathcal{A}_{ij} = Q(\kappa\hat{\boldsymbol{v}}_i\cdot\hat{\boldsymbol{v}}_j) \quad\text{and}\quad C_{ij} = Q(\hat{\boldsymbol{v}}_i\cdot\hat{\boldsymbol{v}}_j), \tag{15.33}$$

and so C is diagonal.

It is easy to unravel the procedure in terms of the nodal degrees of freedom of the $\hat{\boldsymbol{u}}$, $\hat{\hat{\boldsymbol{u}}}$, and p. Consider an element $\hat{E}$ (not adjacent to the outer boundary). Equation (15.30) requires that the divergence of $\hat{\boldsymbol{u}}$ be set equal to the source term $\hat{f}$. This involves differences of the velocities that live on the four edges or six faces of the element. Equation (15.32) relates the velocities to the gradients of pressure. The velocity $\hat{\boldsymbol{u}}$ on a given edge or face is related to the gradient $\hat{\hat{\boldsymbol{u}}}$ that "lives" on the given edge or face and to those that live on the adjacent but perpendicular four edges or eight faces (if κ is not diagonal). Finally, (15.31) relates the $\hat{\tilde{\boldsymbol{u}}}$ on an edge or face to the difference of the adjacent pressures. Combining this together, we get a 9 point stencil for the pressure on $\hat{E}$ if $d = 2$, and 19 points if $d = 3$.

15.6 Triangular Finite Difference Approximation

In this section, we assume that the computational grid is composed of equilateral triangles. We need to define a very special quadrature rule on equilateral triangles to diagonalize C. For a function $\psi(\hat{\boldsymbol{x}})$, let

$$Q(\psi) = \sum_{\hat{E}\in\hat{\mathcal{T}}_{\hat{h}}} \frac{\text{area}(\hat{E})}{6}\left[\sum_{i=1}^{3}\psi(\hat{\boldsymbol{x}}_i(\hat{E})) + 3\psi(\hat{\boldsymbol{x}}_c(\hat{E}))\right], \tag{15.34}$$

where again $\hat{\boldsymbol{x}}_i(\hat{E})$ goes through the vertices of $\hat{E}$, $\hat{\boldsymbol{x}}_c(\hat{E})$ is the centroid of $\hat{E}$, and area$(\hat{E})$ is its area. The usual formula with 3 replaced by 9 and 6 replaced by 12 is third order accurate. Our formula is accurate only to order two; however, if

$$C_{ij} = Q(\hat{\boldsymbol{v}}_i \cdot \hat{\boldsymbol{v}}_j) \tag{15.35}$$

and each $\hat{E}$ is equilateral, then C is diagonal.

We approximate (15.1)–(15.3) by (15.30)- (15.31), and

$$Q(\hat{\boldsymbol{u}} \cdot \hat{\tilde{\boldsymbol{v}}}) = \int_{\hat{\Omega}} \kappa\hat{\tilde{\boldsymbol{u}}} \cdot \hat{\tilde{\boldsymbol{v}}}\, d\hat{x}, \quad \hat{\tilde{\boldsymbol{v}}} \in \hat{V}_{\hat{h}}. \tag{15.36}$$

A 10-point stencil for the pressure results. For an element $\hat{E}$ (not adjacent to the outer boundary), (15.30) involves differences of the velocities that live on the three edges. Equation (15.36) relates the velocities to the $\hat{\tilde{\boldsymbol{u}}}$ that live on the nine edges of the three adjacent triangles. Finally, (15.31) relates the $\hat{\tilde{\boldsymbol{u}}}$ living on an edge to the difference of the adjacent pressures.

15.7 Convergence Results

Because of the way (15.30) is approximated in mixed methods, mass is conserved cell-by-cell; that is, the test function $w \in W_h$ is a piecewise constant, and so the average divergence is set equal to the average mass source over each cell. Such is not the case in, say, the more straightforward Galerkin finite element method. Moreover, as we describe below, the standard and expanded mixed finite element methods give accurate approximations of both the velocity and pressure.

The convergence theory for mixed methods is fairly well established (see, e.g., [7,9]). Below, we summarize the main theoretical results (which hold under appropriate hypotheses). For a scalar function ψ, let

$$\|\psi\| = \left\{ \int_\Omega |\psi(\boldsymbol{x})|^2 \, dx \right\}^{1/2} \tag{15.37}$$

denote its $L^2(\Omega)$-norm; if ψ is a vector function, this defines its $(L^2(\Omega))^d$-norm. Let C denote a generic positive constant that depends on the solution, the domain, K, f, g, and the regularity of the partition $\mathcal{T}_h$, but not on the element size h.

As we refine the grid $\mathcal{T}_h$, both the standard [5,7,15,16] and expanded [1,2] mixed methods are first order accurate in the velocities and pressures; that is, the errors satisfy

$$\|\boldsymbol{u}_{\text{approx}} - \boldsymbol{u}_{\text{true}}\| + \|p_{\text{approx}} - p_{\text{true}}\| \leq C\,h. \tag{15.38}$$

This convergence rate is to be expected, since both V_h and W_h contain the piecewise constant functions.

It should be noted that at certain discrete points, one can observe computationally and prove theoretically that the pressure approximation is second order accurate. This phenomenon is referred to as superconvergence. Let the midpoint rule be used to approximate the $L^2(\Omega)$-norm:

$$|||\psi|||_M = \left\{ \sum_{E \in \mathcal{T}_h} |\psi(\boldsymbol{x}_c(E))|^2 \, \text{vol}(E) \right\}^{1/2}, \tag{15.39}$$

where $\boldsymbol{x}_c(E)$ denotes the centroid of E and vol(E) is the area or volume of E. Then in fact [1,2,5,7,15]

$$|||p_{\text{approx}} - p_{\text{true}}|||_M \leq C\,h^2. \tag{15.40}$$

This result is not surprising, since $\boldsymbol{u}$, essentially a derivative of p, is first order accurate.

A more interesting result is that superconvergence is sometimes observed for the velocity approximation. If the standard mixed method is used, and if the grid is rectangular and K is diagonal, then we can define the discrete approximation to the $(L^2(\Omega))^d$-norm for a vector ψ by

$$|||\psi|||_{TM} = \left\{ \sum_{E \in \mathcal{T}_h} \frac{1}{2d} \sum_{i=1}^{d} [|\psi_i(\boldsymbol{x}_i^1(E))|^2 + |\psi_i(\boldsymbol{x}_i^2(E))|^2] \, \text{vol}(E) \right\}^{1/2}, \tag{15.41}$$

where $\boldsymbol{x}_i^1(E)$ and $\boldsymbol{x}_i^2(E)$ are the centers of the two edges or faces perpendicular to the ith coordinate direction. That is, we construct the discrete norm by looking at the nodal points of V_h and applying the trapezoidal or midpoint rule as appropriate in each coordinate direction. Then the theoretical result [8,10] is

$$|||\boldsymbol{u}_{\text{approx}} - \boldsymbol{u}_{\text{true}}|||_{TM} \leq C\,h^2. \tag{15.42}$$

In fact, if homogeneous Neumann boundary conditions are used, the above result holds for the cell-centered finite difference approximation of the standard mixed method [16]; computationally speaking, other boundary conditions also exhibit some superconvergence, but perhaps not to order 2.

The cell-centered finite difference approximation of the expanded mixed method gives somewhat better velocities. On rectangular grids, but with a tensor K, we have the same superconvergence of the velocity (15.42) in the interior of the domain [2]. Convergence is degraded near the boundary of the domain, so the overall superconvergence rate is of order 1.5. We observe computationally that the velocity exhibits superconvergence of order in the range 1.5 to 2, depending on whether Dirichlet or Neumann conditions are used, and on the eigenvalues and eigenvectors of K.

If Ω is a general domain that can be constructed as the image by F of a rectangular computational domain $\hat{\Omega}$, then (15.42) holds on the computational domain (at least in the interior). Provided that F is smooth, (15.42) induces a superconvergent discrete norm on Ω. This induced norm is not easily described; however, the midpoint approximation of the $(L^2(\Omega))^d$-norm is also superconvergent. Therefore, the velocity approximation is superconvergent to the true velocity on rectangular or merely logically rectangular grids, to order 1.5 in general and to order 2 in the strict interior of the domain.

15.8 Conclusions

We have presented a cell-centered finite difference approximation to an expanded mixed finite element method, suitable for groundwater flow and transport problems on general geometry. It has only as many unknowns as elements. If the elements are deformed rectangles, the stencil for the pressure has only 9 points, involving the central point and its 8 nearest neighbors. If the elements are deformed bricks, the stencil has 19 points (not 27). If the elements are triangles, the stencil has 10 points.

Because these are mixed methods, mass is conserved locally cell-by-cell. The rate of convergence of the velocities is order 1 in general; however, on logically rectangular grids, superconvergence is seen in the solution. We do not require orthogonal grids, we merely require that the grid be smooth.

Acknowledgments: This research was supported in part by the Department of Energy, the State of Texas Governor's Energy Office, and the National Science Foundation.

References

[1] Arbogast, T., C. N. Dawson, P. T. Keenan, M. F. Wheeler, and I. Yotov, "Implementation of mixed finite element methods for elliptic equations on general geometry", to appear.

[2] Arbogast, T., M. F. Wheeler, and I. Yotov, "Mixed finite elements for elliptic problems with tensor coefficients as cell-centered finite differences", to appear.

[3] Arbogast, T., M. F. Wheeler, and I. Yotov, "Logically rectangular mixed methods for groundwater flow and transport on general geometry," in *Computational methods in water resources X*, A. Peters *et al.*, eds., Kluwer Academic Publishers, Dordrecht, The Netherlands, 149-156, 1994.

[4] Arnold, N. D. and F. Brezzi, "Mixed and nonconforming finite element methods: implementation, postprocessing and error estimates", *R.A.I.R.O. Modél. Math. Anal. Numér.*, 19, 7–32, 1985.

[5] Brezzi, F. and M. Fortin, *Mixed and Hybrid Finite Elements*, Springer Series in Computational Mathematics, Vol. 15, Springer-Verlag, Berlin, 1991.

[6] Douglas, Jr., J., R. E. Ewing, and M. F. Wheeler, "Approximation of the pressure by a mixed method in the simulation of miscible displacement", *R.A.I.R.O. Modél. Math. Anal. Numér.*, 17, 17-33, 1983.

[7] Douglas, Jr., J. and J. E. Roberts, "Global estimates for mixed methods for second order elliptic equations", *Math. Comp.*, 44, 39–52, 1985.

[8] Ewing, R. E., R. D. Lazarov, and J. Wang, "Superconvergence of the velocity along the Gauss lines in mixed finite element methods", *SIAM J. Numer. Anal.*, 28, 1015-1029, 1991.

[9] Gastaldi, L. and R. Nochetto, "Optimal L^∞-error estimates for nonconforming and mixed finite element methods of lowest order", *Numer. Math.*, 50, 587-611, 1987.

[10] Nakata, M., A. Weiser, and M. F. Wheeler, "Some superconvergence results for mixed finite element methods for elliptic problems on rectangular domains", in *The mathematics of finite elements and applications V*, J. R. Whiteman, ed., Academic Press, London, 1985.

[11] Nedelec, J. C., "Mixed finite elements in $\mathbb{R}^3$", *Numer. Math.*, 35, 315–341, 1980.

[12] Peaceman, D. W., *Fundamentals of numerical reservoir simulation*, Elsevier, Amsterdam, 1977.

[13] Raviart, P. A. and J. M. Thomas, "A mixed finite element method for 2nd order elliptic problems", in *Mathematical aspects of the finite element method, Lecture Notes in Math.*, Springer-Verlag, Berlin, 1977.

[14] Russell, T. F. and M. F. Wheeler, "Finite element and finite difference methods for continuous flows in porous media", in *The Mathematics of reservoir simulation*, R. E. Ewing, ed., SIAM, Philadelphia, 1983.

[15] Thomas, J. M., *Thèse de Doctorat d'état* à l'Université Pierre et Marie Curie, 1977.

[16] Weiser, A. and M. F. Wheeler, "On convergence of block-centered finite differences for elliptic problems", *SIAM J. Numer. Anal.*, 25, 351-375, 1988.

Chapter 16

Least-Squares Mixed Finite Element Methods for Steady Diffusion Problems

A.I. Pehlivanov[1] *and G.F. Carey*[1]

16.1 Introduction

Steady-state diffusion problems are of practical interest in a wide range of environmental transport problems, particularly those associated with flow and species transport in porous media. Standard Galerkin and mixed Galerkin methods are the primary finite element approaches for addressing these problems [12–14, 26]. However, there have also been studies and practical simulations involving other finite element schemes such as collocation [6] and hybrid methods [2, 27]. Recently, a class of least-squares mixed finite element methods have been developed and have become a topic of research interest (see [7, 8, 10, 11, 15, 17, 18, 20, 22–25, 29]).

In this least-squares class of methods we seek the best approximation to the solution of a mixed first-order system in a least-squares minimum residual sense. Since the flux components as well as the scalar potential appear as unknowns, the approach is capable of generating accurate nodal fluxes explicitly. Another interesting feature is that the least-squares system is symmetric positive and hence admits special sparse solution schemes such as Cholesky envelope solvers and preconditioned conjugate gradient

[1]TICAM/ASE–EM Dept., University of Texas at Austin, Austin, Texas 78712–1085

algorithms. Finally, this least-squares mixed formulation is not subject to the usual consistency (LBB) condition encountered in Galerkin mixed finite element formulations.

Nevertheless, there are still several open questions concerning the least-squares approach. For example, the nondimensionalization of the system equations and scaling are related issues that warrant study. The development of effective preconditioners for algebraic systems constructed using these methods has received little attention. For many porous flow or diffusion problems there are dramatic variations in material properties across the domain and this can lead to numerical conditioning problems.

This chapter is organized as follows: In Section 16.2 the governing mixed first-order system is specified. Then the least-squares mixed method is formulated in Section 16.3 and the relation with the classical mixed method is studied in Section 16.4. The finite element formulation is described in Section 16.5 and optimal L^2- and H^1-error estimates are derived. Finally, numerical experiments confirming the theoretical rates of convergence are presented in Section 16.6 together with some test problems with rapidly varying coefficients.

16.2 Governing Diffusion Equation

Consider the steady-state diffusion equation in conservative (self-adjoint) form

$$-\nabla \cdot (a(x)\,\nabla u) = f \quad \text{in } \Omega\,, \tag{16.1}$$

where Ω is a bounded domain in $\mathbb{R}^n$, $n = 2, 3$, with Lipschitz boundary Γ, $f(x)$ is a specified source term and $a(x)$ is a material property (diffusivity) coefficient satisfying

$$0 < \alpha_1 \leq a(x) \leq \alpha_2 \tag{16.2}$$

for α_1, α_2 constants. Specification of appropriate boundary conditions on Γ completes the classical statement of the diffusion problem. In the present work we will assume, for convenience, that homogeneous Dirichlet data $u = 0$ is specified on Γ.

In fluid flow problems in porous media u is the pressure of the fluid and the coefficient a is the ratio of the permeability of the medium and the viscosity of the fluid. Due to inhomogeneities in the porous medium, the coefficient a may vary rapidly and even be discontinuous. Appealing to Darcy's law, the flux or the Darcy velocity can be related to the pressure gradient by

$$\boldsymbol{\sigma} = -a(x)\,\nabla u \tag{16.3}$$

Using (16.3) in (16.1), the corresponding mixed first order system of differential equations is

$$\nabla \cdot \boldsymbol{\sigma} = f \quad \text{in} \quad \Omega \,, \tag{16.4}$$

$$\boldsymbol{\sigma} + a\,\nabla u = \mathbf{0} \quad \text{in} \quad \Omega \tag{16.5}$$

In fact, it is more typical that this transport system arises from conservation properties in this form and then (16.1) is obtained by substituting the constitutive relation (16.5) into (16.4).

In the present work we develop a least-squares finite element scheme for the mixed system (16.4)-(16.5) with homogeneous Dirichlet data

$$u = 0 \quad \text{on} \quad \Gamma \tag{16.6}$$

First, however, let us introduce some auxiliary conditions on the boundary and a curl property for the solution. These conditions have been shown to lead to a more accurate and robust method and follow naturally from the analysis.

Let $\boldsymbol{n} = (\nu_1, \ldots, \nu_n)$ be the unit outward normal vector to the boundary Γ. For $\boldsymbol{q} = (q_1, \ldots, q_n)$ introduce the exterior product

$$\Omega \subset \mathbb{R}^2 \quad : \quad \boldsymbol{n} \wedge \boldsymbol{q} = \nu_1 q_2 - \nu_2 q_1 \,,$$

$$\Omega \subset \mathbb{R}^3 \quad : \quad \boldsymbol{n} \wedge \boldsymbol{q} = (\nu_2 q_3 - \nu_3 q_2, \nu_3 q_1 - \nu_1 q_3, \nu_1 q_2 - \nu_2 q_1)$$

Then from the boundary condition $u = 0$ on Γ it follows that $\boldsymbol{n} \wedge \nabla u = 0$. Taking into account (16.5), this implies the property

$$\boldsymbol{n} \wedge a^{-1} \boldsymbol{\sigma} = 0 \quad \text{on} \quad \Gamma \,. \tag{16.7}$$

Later we shall use the operator $\nabla \times$ to denote, respectively,

$$\Omega \subset \mathbb{R}^2 \quad : \quad \nabla \times \boldsymbol{q} = \partial_1 q_2 - \partial_2 q_1 \,,$$

$$\Omega \subset \mathbb{R}^3 \quad : \quad \nabla \times \boldsymbol{q} = (\partial_2 q_3 - \partial_3 q_2, \partial_3 q_1 - \partial_1 q_3, \partial_1 q_2 - \partial_2 q_1) \,.$$

Also, when $\Omega \subset \mathbb{R}^2$ and $v \in H^1(\Omega)$ we denote $\nabla \times v = (-\partial_2 v, \partial_1 v)$. From (16.5) and the identity $\nabla \times \nabla v = \mathbf{0}$, we get

$$\nabla \times \mathbf{a}^{-1} \boldsymbol{\sigma} = 0 \tag{16.8}$$

16.3 Least-Squares Mixed Formulation

Now we are ready to define some associated function spaces which will be used in the least-squares formulation:

$$V = \left\{ v \in H^1(\Omega) : v = 0 \text{ on } \Gamma \right\}, \tag{16.9}$$

$$\widetilde{\boldsymbol{W}} = \left\{ \boldsymbol{q} \in L^2(\Omega)^n : \boldsymbol{\nabla} \cdot \boldsymbol{q} \in L^2(\Omega) \right\} \equiv H(\text{div}\,;\Omega), \tag{16.10}$$

$$\boldsymbol{W} = \left\{ \boldsymbol{q} \in \widetilde{\boldsymbol{W}} : \boldsymbol{\nabla} \times \mathbf{a}^{-1}\boldsymbol{q} \in L^2(\Omega)^{2n-3}\,, \boldsymbol{n} \wedge a^{-1}\boldsymbol{q} = 0 \text{ on } \Gamma \right\} \tag{16.11}$$

Let $(\cdot,\cdot)_{0,\Omega}$ be the standard inner product in $L^2(\Omega)$ or $L^2(\Omega)^n$. The least-squares minimization problem is: find $u \in V$, $\boldsymbol{\sigma} \in \boldsymbol{W}$ such that

$$J(u, \boldsymbol{\sigma}) = \inf_{v \in V, \boldsymbol{q} \in \boldsymbol{W}} J(v, \boldsymbol{q}) \tag{16.12}$$

where

$$\begin{aligned} J(v, \boldsymbol{q}) \;&=\; (\boldsymbol{\nabla} \times \mathbf{a}^{-1}\boldsymbol{q}, \boldsymbol{\nabla} \times \mathbf{a}^{-1}\boldsymbol{q})_{0,\Omega} \\ &\quad + (\boldsymbol{\nabla} \cdot \boldsymbol{q} - f, \boldsymbol{\nabla} \cdot \boldsymbol{q} - f)_{0,\Omega} \\ &\quad + (\boldsymbol{q} + a\,\boldsymbol{\nabla} v, a^{-1}(\boldsymbol{q} + a\,\boldsymbol{\nabla} v))_{0,\Omega} \end{aligned} \tag{16.13}$$

Note that a weight a^{-1} is applied to the square of $\boldsymbol{q} + a\,\boldsymbol{\nabla} v$.

Setting the first variation of J in (16.13) to zero, the corresponding weak statement can be written as: find $u \in V$, $\boldsymbol{\sigma} \in \boldsymbol{W}$ such that

$$a(u, \boldsymbol{\sigma}; v, \boldsymbol{q}) = (f, \boldsymbol{\nabla} \cdot \boldsymbol{q})_{0,\Omega} \quad \text{for all} \quad v \in V,\ \boldsymbol{q} \in \boldsymbol{W}\,, \tag{16.14}$$

where for notational convenience we have introduced the bilinear functional

$$a(u, \boldsymbol{\sigma}; v, \boldsymbol{q}) = \tilde{a}(u, \boldsymbol{\sigma}; v, \boldsymbol{q}) + (\boldsymbol{\nabla} \times \mathbf{a}^{-1}\boldsymbol{\sigma}, \boldsymbol{\nabla} \times \mathbf{a}^{-1}\boldsymbol{q})_{0,\Omega} \tag{16.15}$$

with

$$\tilde{a}(u, \boldsymbol{\sigma}; v, \boldsymbol{q}) = (\boldsymbol{\nabla} \cdot \boldsymbol{\sigma}, \boldsymbol{\nabla} \cdot \boldsymbol{q})_{0,\Omega} + (\boldsymbol{\sigma} + a\,\boldsymbol{\nabla} u, a^{-1}\boldsymbol{q} + \boldsymbol{\nabla} v)_{0,\Omega} \tag{16.16}$$

The basic coercivity estimate

$$\|v\|_{1,\Omega}^2 + \|\boldsymbol{q}\|_{0,\Omega}^2 + \|\boldsymbol{\nabla} \cdot \boldsymbol{q}\|_{0,\Omega}^2 \le C\tilde{a}(v, \boldsymbol{q}; v, \boldsymbol{q}) \tag{16.17}$$

holds for all $v \in V$, $\boldsymbol{q} \in \widetilde{\boldsymbol{W}}$ (see [25]). Adding the term $\|\boldsymbol{\nabla} \times \mathbf{a}^{-1}\boldsymbol{q}\|_{0,\Omega}^2$, we immediately obtain

$$\|v\|_{1,\Omega}^2 + \|\boldsymbol{q}\|_{0,\Omega}^2 + \|\boldsymbol{\nabla} \cdot \boldsymbol{q}\|_{0,\Omega}^2 + \|\boldsymbol{\nabla} \times \mathbf{a}^{-1}\boldsymbol{q}\|_{0,\Omega}^2 \le Ca(v, \boldsymbol{q}; v, \boldsymbol{q}) \tag{16.18}$$

for all $v \in V$, $\boldsymbol{q} \in \boldsymbol{W}$. Hence problem (16.14) has a unique solution $u \in V$, $\boldsymbol{\sigma} \in \boldsymbol{W}$. We would like to emphasize that the boundary condition $\boldsymbol{n} \wedge a^{-1}\boldsymbol{\sigma} = 0$ on Γ (see (16.7)) is not used in the proof of (16.17) and (16.18). However, if the boundary condition (16.7) is fulfilled and the domain Ω is convex (or the domain is a curvilinear polygon (polytope) with no concave angles), then the Friedrichs' inequality holds (see [28]):

$$\|\boldsymbol{q}\|_{1,\Omega}^2 \le C\left(\|\boldsymbol{q}\|_{0,\Omega}^2 + \|\boldsymbol{\nabla}\cdot\boldsymbol{q}\|_{0,\Omega}^2 + \|\boldsymbol{\nabla}\times \mathbf{a}^{-1}\boldsymbol{q}\|_{0,\Omega}^2\right) . \tag{16.19}$$

Hence, in this case we have a coercivity estimate in stronger norms

$$\|v\|_{1,\Omega}^2 + \|\boldsymbol{q}\|_{1,\Omega}^2 \le Ca(v,\boldsymbol{q};v,\boldsymbol{q}) . \tag{16.20}$$

Inequality (16.20) is crucial to the proof of optimal L^2-error estimates for the velocity approximation.

16.4 Relation to the Classical Mixed Method

First, recall the classical mixed method. We start from the system

$$\begin{aligned} \boldsymbol{\nabla}\cdot\boldsymbol{\sigma} &= f \quad \text{in } \Omega , &(16.21)\\ a^{-1}\boldsymbol{\sigma} + \boldsymbol{\nabla}u &= 0 \quad \text{in } \Omega &(16.22)\end{aligned}$$

Multiplying (16.21) by $-v \in L^2(\Omega)$, (16.22) by $\boldsymbol{q} \in H(\mathrm{div}\,;\Omega)$, and integrating by parts in the term $(\boldsymbol{\nabla}u, \boldsymbol{q})_{0,\Omega}$, we get the classical mixed formulation (see [26]): find $u \in L^2(\Omega)$, $\boldsymbol{\sigma} \in H(\mathrm{div}\,;\Omega)$ such that

$$(a^{-1}\boldsymbol{\sigma},\boldsymbol{q})_{0,\Omega} - (u,\boldsymbol{\nabla}\cdot\boldsymbol{q})_{0,\Omega} - (\boldsymbol{\nabla}\cdot\boldsymbol{\sigma},v)_{0,\Omega} = (-f,v)_{0,\Omega} \tag{16.23}$$

for all $v \in L^2(\Omega)$, $\boldsymbol{q} \in H(\mathrm{div}\,;\Omega)$.

Now, let us multiply (16.4) by $\boldsymbol{\nabla}\cdot\boldsymbol{q}$ (or, equivalently, apply the least-squares method to equation (16.4)). Then

$$(\boldsymbol{\nabla}\cdot\boldsymbol{\sigma},\boldsymbol{\nabla}\cdot\boldsymbol{q})_{0,\Omega} = (f,\boldsymbol{\nabla}\cdot\boldsymbol{q})_{0,\Omega} \quad \text{for all } \boldsymbol{q} \in H(\mathrm{div}\,;\Omega) \tag{16.24}$$

Adding (16.23) and (16.24), we obtain the weak formulation: find $u \in L^2(\Omega)$, $\boldsymbol{\sigma} \in H(\mathrm{div}\,;\Omega)$ such that

$$\begin{aligned}(a^{-1}\boldsymbol{\sigma},\boldsymbol{q})_{0,\Omega} - (u,\boldsymbol{\nabla}\cdot\boldsymbol{q})_{0,\Omega} - (\boldsymbol{\nabla}\cdot\boldsymbol{\sigma},v)_{0,\Omega} + (\boldsymbol{\nabla}\cdot\boldsymbol{\sigma},\boldsymbol{\nabla}\cdot\boldsymbol{q})_{0,\Omega}\\ = (-f,v)_{0,\Omega} + (f,\boldsymbol{\nabla}\cdot\boldsymbol{q})_{0,\Omega} \qquad (16.25)\end{aligned}$$

for all $v \in L^2(\Omega)$, $\boldsymbol{q} \in H(\mathrm{div}\,;\Omega)$. This approach was proposed in [4]. Below we comment on the finite element approximation for problems (16.23) and (16.25).

Suppose that $u \in V$, $v \in V$. Then (16.25) may be rewritten as

$$\begin{aligned}(a^{-1}\boldsymbol{\sigma},\boldsymbol{q})_{0,\Omega} + (\boldsymbol{\nabla} u,\boldsymbol{q})_{0,\Omega} + (\boldsymbol{\sigma},\boldsymbol{\nabla} v)_{0,\Omega} + (\boldsymbol{\nabla}\cdot\boldsymbol{\sigma},\boldsymbol{\nabla}\cdot\boldsymbol{q})_{0,\Omega} \\ = (-f,v)_{0,\Omega} + (f,\boldsymbol{\nabla}\cdot\boldsymbol{q})_{0,\Omega}\end{aligned} \tag{16.26}$$

for all $v \in V$, $\boldsymbol{q} \in \widetilde{\boldsymbol{W}}$ (see (16.9) and (16.10) for the definition of spaces).

On the other hand, the standard Galerkin formulation for problem (16.1) is

$$(a\,\boldsymbol{\nabla} u,\boldsymbol{\nabla} v)_{0,\Omega} = (f,v)_{0,\Omega} \quad \text{for all} \quad v \in V \tag{16.27}$$

Adding (16.26) and (16.27),

$$\tilde{a}(u,\boldsymbol{\sigma};v,\boldsymbol{q}) = (f,\boldsymbol{\nabla}\cdot\boldsymbol{q})_{0,\Omega} \quad \text{for all} \quad v \in V,\ \boldsymbol{q} \in \widetilde{\boldsymbol{W}} \tag{16.28}$$

Multiplying (16.8) by $\boldsymbol{\nabla} \times \mathbf{a}^{-1}\boldsymbol{q}$,

$$(\boldsymbol{\nabla} \times \mathbf{a}^{-1}\boldsymbol{\sigma},\boldsymbol{\nabla} \times \mathbf{a}^{-1}\boldsymbol{q})_{0,\Omega} = 0 \quad \text{for all} \quad \boldsymbol{q} \in \boldsymbol{W} \tag{16.29}$$

Then the weak formulation (16.14) follows from (16.28) and (16.29). Hence the least-squares mixed method for problem (16.1) may be viewed as a combination of the classical mixed method, a least-squares method for equations (16.4) and (16.8), and the standard Galerkin method.

Consider the classical mixed method (16.23). Let $V_h \subset L^2(\Omega)$, $\boldsymbol{W}_h \subset H(\mathrm{div}\,;\Omega)$ be finite element spaces. The approximation problem, corresponding to (16.23), is: find $u_h \in V_h$, $\boldsymbol{\sigma}_h \in \boldsymbol{W}_h$ such that

$$(a^{-1}\boldsymbol{\sigma}_h,\boldsymbol{q}_h)_{0,\Omega} - (u_h,\boldsymbol{\nabla}\cdot\boldsymbol{q}_h)_{0,\Omega} - (\boldsymbol{\nabla}\cdot\boldsymbol{\sigma}_h,v_h)_{0,\Omega} = (-f,v)_{0,\Omega} \tag{16.30}$$

for all $v_h \in V_h$, $\boldsymbol{q}_h \in \boldsymbol{W}_h$. In order to ensure stability and optimal error bounds, we have to satisfy the following consistency conditions concerning V_h and $\boldsymbol{W}_h$ (see [1,3,4,19]):

Condition C1: There exists a constant $\alpha > 0$ such that

$$(a^{-1}\boldsymbol{q}_h,\boldsymbol{q}_h)_{0,\Omega} \geq \alpha\left(\|\boldsymbol{q}_h\|_{0,\Omega}^2 + \|\boldsymbol{\nabla}\cdot\boldsymbol{q}_h\|_{0,\Omega}^2\right) \quad \text{for all} \quad \boldsymbol{q}_h \in X_h \tag{16.31}$$

where

$$X_h = \left\{\boldsymbol{q}_h \in \boldsymbol{W}_h \ :\ (\boldsymbol{\nabla}\cdot\boldsymbol{q}_h,v_h)_{0,\Omega} = 0 \ \ \forall v_h \in V_h\right\} \tag{16.32}$$

Condition C2: There exists a constant $\beta > 0$ such that

$$\sup_{\boldsymbol{q}_h \in \boldsymbol{W}_h} \frac{(\nabla \cdot \boldsymbol{q}_h, v_h)_{0,\Omega}}{\|\boldsymbol{q}_h\|_{H(\mathrm{div};\Omega)}} \geq \beta \|v_h\|_{0,\Omega} \quad \text{for all} \quad v_h \in V_h \tag{16.33}$$

The two conditions compete in the sense that Condition C2 requests a large enough (with respect to V_h) space $\boldsymbol{W}_h$ while Condition C1 requests a large enough (with respect to $\boldsymbol{W}_h$) space V_h. For example, the Raviart-Thomas spaces [26] satisfy both conditions at the same time. Further details are given in [4].

Now consider approximation of the augmented problem (16.25). Then Condition C1 (also known as "ellipticity on the kernel") becomes: There exists a constant $\alpha > 0$ such that

$$(a^{-1}\boldsymbol{q}_h, \boldsymbol{q}_h)_{0,\Omega} + \|\nabla \cdot \boldsymbol{q}_h\|_{0,\Omega}^2 \geq \alpha \left(\|\boldsymbol{q}_h\|_{0,\Omega}^2 + \|\nabla \cdot \boldsymbol{q}_h\|_{0,\Omega}^2 \right) \tag{16.34}$$

for all $\boldsymbol{q}_h \in X_h$. Obviously, Condition C1 is automatically satisfied in this case. In order to satisfy Condition C2 we have to choose $\boldsymbol{W}_h$ large enough and the form (16.25) now allows the use of continuous finite element spaces for the approximation of the velocity $\boldsymbol{\sigma}$ (see [4]).

In the next section we show that the least-squares mixed method does not impose any consistency conditions on the finite element spaces.

16.5 Least-Squares Mixed Method and Error Estimates

Let the domain Ω be partitioned into finite elements exactly. Denote the partition by T_h. Define the following finite element spaces:

$$V_h = \left\{ v_h \in C^0(\Omega) : v_h|_K \in P_k(K) \;\; \forall K \in T_h \,,\; v_h = 0 \text{ on } \Gamma \right\}, \tag{16.35}$$

$$\boldsymbol{W}_h = \left\{ \boldsymbol{q}_h \in C^0(\Omega)^n : \boldsymbol{q}_h|_K \in P_r(K)^n \quad \forall K \in T_h \,, \right.$$
$$\left. \boldsymbol{n} \wedge a^{-1}\boldsymbol{q}_h = 0 \;\text{ at the nodes on }\; \Gamma \right\} \tag{16.36}$$

Note that k and r are the polynomial degrees for the finite elements used in V_h and $\boldsymbol{W}_h$ respectively. The boundary condition $\boldsymbol{n} \wedge a^{-1}\boldsymbol{q}_h = 0$ is imposed only at the nodes on Γ because in the case of nonconstant coefficient $a(x)$ and/or when we use curved elements this boundary condition cannot be satisfied on the whole Γ. Hence, some (mild) nonconformity is introduced. Next, we comment on the boundary condition for $\boldsymbol{\sigma}_h$ at the corner nodes.

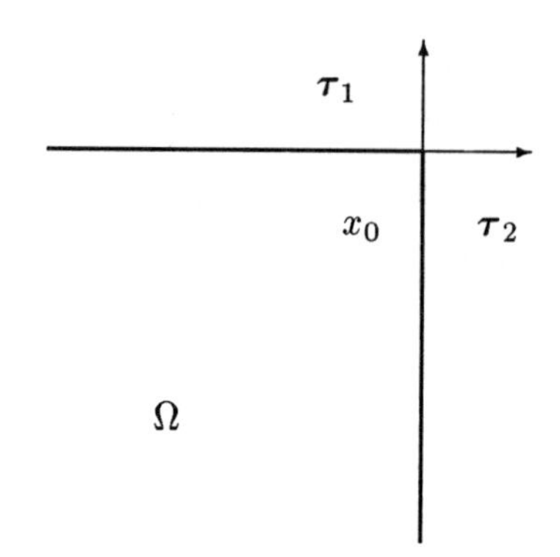

Figure 16.1: Boundary condition at the corner nodes

Suppose that the domain Ω is two-dimensional and we have the situation depicted in Figure 16.1, where x_0 is the corner node and $\boldsymbol{\tau}_1$, $\boldsymbol{\tau}_2$ are the corresponding tangential vectors. Then

$$\begin{aligned} \boldsymbol{n}_1 \wedge a^{-1}\boldsymbol{\sigma}_h(x_0) \equiv \boldsymbol{\tau}_1 \cdot a^{-1}\boldsymbol{\sigma}_h(x_0) &= -\partial_{\tau_1} u(x_0) , \\ \boldsymbol{n}_2 \wedge a^{-1}\boldsymbol{\sigma}_h(x_0) \equiv \boldsymbol{\tau}_2 \cdot a^{-1}\boldsymbol{\sigma}_h(x_0) &= -\partial_{\tau_2} u(x_0) . \end{aligned}$$

Here $\partial_{\tau_1} u(x_0)$ and $\partial_{\tau_2} u(x_0)$ are the tangential derivatives of u with respect to $\boldsymbol{\tau}_1$ and $\boldsymbol{\tau}_2$. These derivatives can be calculated from the boundary data (see also [5].) More details on the implementation may be found in [21].

The weak formulation is: find $u_h \in V_h$, $\boldsymbol{q}_h \in \boldsymbol{W}_h$ such that

$$a(u_h, \boldsymbol{\sigma}_h; v_h, \boldsymbol{q}_h) = (f, \boldsymbol{\nabla} \cdot \boldsymbol{q}_h)_{0,\Omega} \quad \text{for all} \quad v_h \in V_h,\ \boldsymbol{q}_h \in \boldsymbol{W}_h \tag{16.37}$$

To establish uniqueness, we first introduce the related space

$$\widetilde{\boldsymbol{W}}_h = \left\{ \boldsymbol{q}_h \in C^0(\Omega)^n : \boldsymbol{q}_h|_K \in P_r(K)^n \quad \forall K \in T_h \right\} \tag{16.38}$$

From (16.17) and the embeddings $V_h \subset V$, $\widetilde{\boldsymbol{W}}_h \subset \widetilde{\boldsymbol{W}}$, we immediately obtain

$$\|v_h\|_{1,\Omega}^2 + \|\boldsymbol{q}_h\|_{0,\Omega}^2 + \|\boldsymbol{\nabla} \cdot \boldsymbol{q}_h\|_{0,\Omega}^2 \le C\tilde{a}(v_h, \boldsymbol{q}_h; v_h, \boldsymbol{q}_h) \tag{16.39}$$

for all $v_h \in V_h$, $\boldsymbol{q}_h \in \widetilde{\boldsymbol{W}}_h$. Adding the term $\|\boldsymbol{\nabla} \times \mathbf{a}^{-1}\boldsymbol{q}_h\|_{0,\Omega}^2$,

$$\|v_h\|_{1,\Omega}^2 + \|\boldsymbol{q}_h\|_{0,\Omega}^2 + \|\boldsymbol{\nabla} \cdot \boldsymbol{q}_h\|_{0,\Omega}^2 + \|\boldsymbol{\nabla} \times \mathbf{a}^{-1}\boldsymbol{q}_h\|_{0,\Omega}^2 \le Ca(v_h, \boldsymbol{q}_h; v_h, \boldsymbol{q}_h) \tag{16.40}$$

for all $v_h \in V_h$, $\boldsymbol{q}_h \in \boldsymbol{W}_h$. Hence problem (16.37) has a unique solution $u_h \in V_h, \boldsymbol{q}_h \in \boldsymbol{W}_h$. Furthermore, in view of (16.39) and (16.40), we conclude that there is no consistency restriction on the degree of the polynomials in the respective finite element spaces. In other words, we do not have to select special elements to ensure stability. However, in order to get optimal error estimates certain relations between the element degree k for u_h (see (16.35)) and r for $\boldsymbol{\sigma}_h$ (see (16.36)) should be enforced. The cases $r = k$ and $r = k+1$ are now considered. The following error estimates are proved in [25] for the finite element approximation defined in (16.37):
(i) For $r = k$

$$\|u - u_h\|_{1,\Omega} + \|\boldsymbol{\sigma} - \boldsymbol{\sigma}_h\|_{H(\mathrm{div};\Omega)} \leq Ch^k \left(\|u\|_{k+1,\Omega} + \|\boldsymbol{\sigma}\|_{k+1,\Omega}\right) \tag{16.41}$$

If Ω is convex, then we have the stronger results

$$\|u - u_h\|_{1,\Omega} + \|\boldsymbol{\sigma} - \boldsymbol{\sigma}_h\|_{1,\Omega} \leq Ch^k \left(\|u\|_{k+1,\Omega} + \|\boldsymbol{\sigma}\|_{k+1,\Omega}\right), \tag{16.42}$$

$$\|u - u_h\|_{0,\Omega} + \|\boldsymbol{\sigma} - \boldsymbol{\sigma}_h\|_{0,\Omega} \leq Ch^{k+1} \left(\|u\|_{k+1,\Omega} + \|\boldsymbol{\sigma}\|_{k+1,\Omega}\right) \tag{16.43}$$

(ii) For $r = k + 1$, Ω convex

$$\|u - u_h\|_{0,\Omega} + \|\boldsymbol{\sigma} - \boldsymbol{\sigma}_h\|_{1,\Omega} \leq Ch^r \left(\|u\|_{r,\Omega} + \|\boldsymbol{\sigma}\|_{r+1,\Omega}\right) . \tag{16.44}$$

(iii) For $r = k + 1$, $k > 1$, $\Gamma \in C^{2,1}$ (see [16] for notations related to the boundary)

$$\|u - u_h\|_{-1,\Omega} + \|\boldsymbol{\sigma} - \boldsymbol{\sigma}_h\|_{0,\Omega} \leq Ch^{r+1} \left(\|u\|_{r,\Omega} + \|\boldsymbol{\sigma}\|_{r+1,\Omega}\right) \tag{16.45}$$

Note that the assumptions in (i), (ii), and (iii) regarding the domain Ω and the boundary Γ are used in the analysis but in most cases they are not sufficient to ensure the regularity of u and $\boldsymbol{\sigma}$ required by the norms in (16.41)-(16.45).

Now, let us go back to the least-squares functional (16.13) and the function space $\boldsymbol{W}$ defined in (16.11). There, the boundary condition (16.7) and the equation (16.8) are included, but the coercivity estimate (16.17) holds in the space $(V, \widetilde{\boldsymbol{W}})$ and does not depend on (16.7) and (16.8). Hence we may formulate a least-squares method without the curl-term and the auxiliary boundary condition for $\boldsymbol{\sigma}$. More specifically, the variational problem in this case is: find $u \in V$, $\boldsymbol{\sigma} \in \widetilde{\boldsymbol{W}}$ such that

$$\tilde{a}(u, \boldsymbol{\sigma}; v, \boldsymbol{q}) = (f, \nabla \cdot \boldsymbol{q})_{0,\Omega} \quad \text{for all } v \in V,\ \boldsymbol{q} \in \widetilde{\boldsymbol{W}} \tag{16.46}$$

The corresponding finite element method is: find $u_h \in V_h$, $\boldsymbol{\sigma}_h \in \widetilde{\boldsymbol{W}}_h$ such that

$$\tilde{a}(u_h, \boldsymbol{\sigma}_h; v_h, \boldsymbol{q}_h) = (f, \nabla \cdot \boldsymbol{q}_h)_{0,\Omega} \quad \text{for all} \quad v_h \in V_h, \ \boldsymbol{q}_h \in \widetilde{\boldsymbol{W}}_h \tag{16.47}$$

see (16.10) and (16.38). Note that the requirement $\boldsymbol{q}_h \in C^0(\Omega)^n$ in the definition of $\widetilde{\boldsymbol{W}}_h$ may be relaxed; in fact, we require that $\boldsymbol{q}_h \in H(\text{div}\,;\Omega)$. For example, the Raviart-Thomas elements may be used.

The following error estimates hold:

(i) For $r = k$

$$\|u - u_h\|_{1,\Omega} + \|\boldsymbol{\sigma} - \boldsymbol{\sigma}_h\|_{H(\text{div};\Omega)} \leq Ch^k \left(\|u\|_{k+1,\Omega} + \|\boldsymbol{\sigma}\|_{k+1,\Omega}\right) \tag{16.48}$$

(ii) For $r = k+1$

$$\|u - u_h\|_{0,\Omega} + \|\boldsymbol{\sigma} - \boldsymbol{\sigma}_h\|_{H(\text{div};\Omega)} \leq Ch^r \left(\|u\|_{r,\Omega} + \|\boldsymbol{\sigma}\|_{r+1,\Omega}\right) \tag{16.49}$$

Note that in the case $r = k+1$ we obtain the same error estimates as in the classical mixed method (see [14]).

16.6 Numerical Experiments

Let $\Omega \subset \mathbb{R}^2$ be the unit square. Consider problem (16.1) with non-homogeneous Dirichlet boundary conditions and exact solution $u = e^{2x_1^2 + x_2^2}$. We shall vary the coefficient $a(x)$. For the first set of experiments we take

$$a(x) = \frac{1}{1 + x_1^2} \tag{16.50}$$

Note that $1/2 \leq a(x) \leq 1$ for this test case.

For this problem our goal is to test the rates of convergence for the cases $r = k$ and $r = k+1$, $1 \leq k, r \leq 3$. First, we use the finite element formulation (16.37); i.e., the curl-term is used and the $\boldsymbol{\sigma}_h$-boundary condition is imposed. The results are presented in Table 16.1. The numerical tests of convergence confirm the theoretical estimates (16.41)–(16.45). Next, the results corresponding to neglecting the $\boldsymbol{\sigma}_h$-boundary condition are given in Table 16.2 and the case without the curl-term in Table 16.3. As we can see, there is a deterioration of the rates of convergence, especially for $\|\boldsymbol{\sigma} - \boldsymbol{\sigma}_h\|_{0,\Omega}$. Finally, the rates of convergence for the formulation (16.47) are given in Table 16.4. They confirm estimates (16.48) and (16.49).

Table 16.1: Experiments with curl-term and $\boldsymbol{\sigma}_h$-boundary condition

Norms	Rates of convergence				
	$k=1$ $r=1$	$k=1$ $r=2$	$k=2$ $r=2$	$k=2$ $r=3$	$k=3$ $r=3$
$\Vert u-u_h\Vert_{0,\Omega}$	2.01	1.99	2.99	2.99	4.01
$\vert u-u_h\vert_{1,\Omega}$	0.99	0.99	1.98	1.98	2.99
$\Vert\boldsymbol{\sigma}-\boldsymbol{\sigma}_h\Vert_{0,\Omega}$	1.96	2.36	3.03	4.04	4.04
$\vert\boldsymbol{\sigma}-\boldsymbol{\sigma}_h\vert_{1,\Omega}$	1.00	1.98	1.98	2.99	2.99
$\Vert\nabla\cdot(\boldsymbol{\sigma}-\boldsymbol{\sigma}_h)\Vert_{0,\Omega}$	1.02	2.00	2.00	3.00	3.01

Table 16.2: Experiments without $\boldsymbol{\sigma}_h$-boundary condition

Norms	Rates of convergence				
	$k=1$ $r=1$	$k=1$ $r=2$	$k=2$ $r=2$	$k=2$ $r=3$	$k=3$ $r=3$
$\Vert u-u_h\Vert_{0,\Omega}$	2.00	1.99	2.99	2.99	4.02
$\vert u-u_h\vert_{1,\Omega}$	0.99	0.99	1.98	1.98	2.99
$\Vert\boldsymbol{\sigma}-\boldsymbol{\sigma}_h\Vert_{0,\Omega}$	1.02	1.88	2.15	3.25	3.37
$\vert\boldsymbol{\sigma}-\boldsymbol{\sigma}_h\vert_{1,\Omega}$	0.65	1.40	1.58	2.58	2.74
$\Vert\nabla\cdot(\boldsymbol{\sigma}-\boldsymbol{\sigma}_h)\Vert_{0,\Omega}$	1.00	1.99	1.99	2.99	2.99

The least squares system has the form

$$\begin{pmatrix} M_{11} & M_{12} \\ M_{21} & M_{22} \end{pmatrix}\begin{pmatrix} \boldsymbol{\sigma}_h \\ u_h \end{pmatrix}=\begin{pmatrix} F \\ 0 \end{pmatrix} \tag{16.51}$$

We use the block diagonal matrix

$$\begin{pmatrix} M_{11} & 0 \\ 0 & M_{22} \end{pmatrix} \tag{16.52}$$

as a preconditioner for the conjugate gradient method. The preconditioned conjugate gradient scheme converges in 11 iterations for all cases in this set of experiments. This behavior is consistent with the coercivity estimates (16.17) and (16.18).

For the second set of experiments we set the coefficient $a(x)$ to be

$$a(x)=\frac{\epsilon}{\epsilon+x_1^2} \tag{16.53}$$

Table 16.3: Experiments without curl-term.

Norms	Rates of convergence				
	$k=1$ $r=1$	$k=1$ $r=2$	$k=2$ $r=2$	$k=2$ $r=3$	$k=3$ $r=3$
$\|u-u_h\|_{0,\Omega}$	2.00	1.99	2.99	2.99	4.01
$\|u-u_h\|_{1,\Omega}$	0.99	0.99	1.98	1.98	2.99
$\|\boldsymbol{\sigma}-\boldsymbol{\sigma}_h\|_{0,\Omega}$	1.47	1.78	2.10	2.73	3.14
$\|\boldsymbol{\sigma}-\boldsymbol{\sigma}_h\|_{1,\Omega}$	1.00	0.93	1.07	1.75	2.17
$\|\nabla\cdot(\boldsymbol{\sigma}-\boldsymbol{\sigma}_h)\|_{0,\Omega}$	0.98	1.99	1.99	2.98	2.98

Table 16.4: Experiments without curl-term or $\boldsymbol{\sigma}_h$ boundary conditions

Norms	Rates of convergence				
	$k=1$ $r=1$	$k=1$ $r=2$	$k=2$ $r=2$	$k=2$ $r=3$	$k=3$ $r=3$
$\|u-u_h\|_{0,\Omega}$	2.00	1.99	2.99	2.99	4.01
$\|u-u_h\|_{1,\Omega}$	0.99	0.99	1.98	1.98	2.99
$\|\boldsymbol{\sigma}-\boldsymbol{\sigma}_h\|_{0,\Omega}$	1.04	1.78	2.12	2.48	2.96
$\|\boldsymbol{\sigma}-\boldsymbol{\sigma}_h\|_{1,\Omega}$	0.53	0.78	1.06	1.46	2.06
$\|\nabla\cdot(\boldsymbol{\sigma}-\boldsymbol{\sigma}_h)\|_{0,\Omega}$	0.97	1.99	1.99	2.98	2.98

where the parameter $\epsilon \in (0,1]$ will be varied. For ϵ small, $0 < \epsilon << 1$,

$$\frac{\max a(x)}{\min a(x)} \approx \epsilon^{-1} \tag{16.54}$$

As a solution technique we use the block-ILU factorization method, recently proposed in [9]. This iterative method is well defined for positive definite block-tridiagonal systems and was extensively tested in [7]. For this set of experiments we use the least-squares formulation (16.37) with curl-term and $\boldsymbol{\sigma}_h$-boundary conditions. We modify the formulation in order to test the influence of a weight ω in the curl-term; i.e., we have

$$(\omega \nabla \times \mathbf{a}^{-1}\boldsymbol{\sigma}_h, \nabla \times \mathbf{a}^{-1}\mathbf{q}_h)_{0,\Omega} \tag{16.55}$$

as the curl-term. The domain Ω is partitioned uniformly into $2N^2$ linear triangular elements. The number of iterations for the case $\omega = 1$ and with different values for ϵ is presented in Table 16.5. As we see, the number of

Table 16.5: Experiments with weight $\omega = 1$

ϵ	Number of iterations		
	$N = 10$	$N = 15$	$N = 20$
1	7	9	12
0.2	12	19	24
0.1	16	28	37
0.05	20	41	58
0.01	27	76	122
0.005	29	84	141
0.001	30	90	149

Table 16.6: Experiments with weight $\omega = (a(x))^2$

ϵ	Number of iterations		
	$N = 10$	$N = 15$	$N = 20$
1	6	8	10
0.2	5	8	10
0.1	5	8	10
0.05	5	7	10
0.01	6	7	9
0.005	6	7	9
0.001	6	9	11

iterations increases when ϵ decreases. In order to overcome this effect, we set the weight $\omega = (a(x))^2$; i.e.,

$$\begin{aligned}
&(\omega\nabla \times \mathbf{a}^{-1}\boldsymbol{\sigma}_h, \nabla \times \mathbf{a}^{-1}\boldsymbol{q}_h)_{0,\Omega} \\
&\quad = (\nabla \times \boldsymbol{\sigma}_h - a^{-1}\nabla a \wedge \boldsymbol{\sigma}_h, \nabla \times \boldsymbol{q}_h - a^{-1}\nabla a \wedge \boldsymbol{q}_h)_{0,\Omega} \\
&\quad = (\nabla \times \boldsymbol{\sigma}_h, \nabla \times \boldsymbol{q}_h)_{0,\Omega} + \text{other terms} \qquad (16.56)
\end{aligned}$$

The results for this choice are listed in Table 16.6 and indicate that the number of iterations is bounded independent of ϵ. Apparently, this choice of the weight ω affects the constant in the Friedrichs' inequality (see (16.19)) which is crucial to the $H^1(\Omega)$-coercivity of the bilinear form.

Acknowledgements: This research has been supported in part by ARPA, TICAM, and the State of Texas.

References

[1] Babuska, I., "The finite element method with Lagrange multipliers", *Numer. Math.*, 20, 179–192, 1973.

[2] Babuska, I., J. T. Oden, and J. K. Lee, "Mixed-hybrid finite element approximations of second-order elliptic boundary value problems", *Comput. Methods Appl. Mech. Eng.*, 11, 175–206, 1977.

[3] Brezzi, F., "On the existence, uniqueness and approximation of saddle point problems arising from Lagrange multipliers", *RAIRO Sér. Anal. Numér.*, 8,R-2, 129–151, 1974.

[4] Brezzi, F., M. Fortin, and L. D. Marini, "Mixed finite element methods with continuous stresses", *Math. Models Meth. Appl. Sci.*, 3, 275–287, 1993.

[5] Carey, G. F., S. S. Chow, and M. R. Seager, "Approximate boundary-flux calculations", *Comput. Methods Appl. Mech. Engrg.*, 50, 107–120, 1985.

[6] Carey, G. F., D. Humphrey, and M. F. Wheeler, "Galerkin and collocation-Galerkin methods with superconvergence and optimal fluxes", *IJNME*, 17, 937–950, 1981.

[7] Carey, G. F., A. I. Pehlivanov, and P. S. Vassilevski, "Least-squares mixed finite element methods for non-selfadjoint elliptic problems: II. Performance of block-ILU factorization methods", to appear in *SIAM J. Sci. Comput.*

[8] Carey, G. F. and Y. Shen, "Convergence studies of least-squares finite elements for first order systems", *Comm. Appl. Numer. Methods*, 5, 427–434,1989.

[9] Chan, T. F. and P. S. Vassilevski, "A framework for block-ILU factorization using block size reduction", CAM Report 92–29(1992), Department of Mathematics, UCLA, to appear in *Math. Comp.*

[10] Chang, C. L., "A least-squares finite element method for the Helmholtz equation", *Comput. Methods Appl. Mech. Engrg.*, 83, 1–7, 1990.

[11] Chen, T.F., "On least-squares approximations to compressible flow problems", *Numer. Methods Partial Differential Equations*, 2, 207–228, 1986.

[12] Ciarlet, P. G., *The finite element method for elliptic problems,* North Holland, Amsterdam, New York, Oxford, 1978.

[13] Douglas, J., R. E. Ewing, and M. F. Wheeler, "The approximation of the pressure by a mixed method in the simulation of miscible displacement", *RAIRO Anal. Numer.*, 17, 17–33, 1983.

[14] Douglas, J. and J. E. Roberts, "Global estimates for mixed methods for second order elliptic equations", *Math. Comp.*, 44, 39–52, 1985.

[15] Fix, G. J., M. D. Gunzburger, and R. A. Nicolaides, "On mixed finite element methods for first order elliptic systems", *Numer. Math.*, 37, 29–48, 1981.

[16] Grisvard, P., *Elliptic problems in nonsmooth domains,* Pitman, Boston, London, Melbourne, 1985.

[17] Haslinger, J. and P. Neittaanmäki, "On different finite element methods for approximating the gradient of the solution to the Helmholtz equation", *Comput. Methods Appl. Mech. Engrg.*, 42, 131–148, 1984.

[18] Jespersen, D. C., "A least squares decomposition method for solving elliptic equations", *Math. Comp.*, 31, 873–880, 1984.

[19] Ladyzhenskaya, O. A., *The mathematical theory of viscous incompressible flows,* Gordon and Breach, London, 1969.

[20] Neittaanmäki, P. and J. Saranen, "On finite element approximation of the gradient for the solution of Poisson equation",*Numer. Math.*, 37, 333–337, 1981.

[21] Pehlivanov, A. I. and G. F. Carey, "Convergence studies of least-squares mixed finite element methods", in preparation.

[22] Pehlivanov, A. I. and G. F. Carey, "Error estimates for least-squares mixed finite elements", *RAIRO Anal. Numer.*, 28, 499–516, 1994.

[23] Pehlivanov, A. I., G. F. Carey, and R. D. Lazarov, " Least-squares mixed finite elements for second order elliptic problems", *SIAM J. Numer. Anal.*, 31, 1368–1377, 1994.

[24] Pehlivanov, A. I., G. F. Carey, R. D. Lazarov, and Y. Shen, "Convergence analysis of least-squares mixed finite elements", *Computing*, 51, 111–123, 1993.

[25] Pehlivanov, A. I., G. F. Carey, and P. S. Vassilevski, "Least-squares mixed finite element methods for non-selfadjoint elliptic problems: I. Error estimates", to appear in *Numer. Math.*

[26] Raviart, P. A. and J. M. Thomas, *A Mixed Finite Element Method for 2nd Order Elliptic Problems*, Lect. Notes in Math., Springer-Verlag, 606, 292–315, 1977.

[27] Raviart, P.A. and J. M. Thomas, "Primal hybrid finite element methods for second order elliptic problems", *Math. Comp.*, 31, 391–413, 1977.

[28] Saranen, J., "On an inequality of Friedrichs", *Math. Scand.*, 51, 310–322, 1982.

[29] Saranen, J., "Über die Approximation der Lösungen der Maxwellschen Randwertaufgabe mit der Methode der finiten Elemente", *Applicable Analysis*, 10, 15–30, 1980.

Chapter 17

Substructure Preconditioning for Porous Flow Problems

R.E. Ewing[1], *S. Maliassov*[1], *Y. Kuznetsov*[2] *and R.D. Lazarov*[1,3]

17.1 Introduction

In many engineering problems, such as petroleum reservoir simulation, groundwater contamination calculation, seismic exploration studies, etc., a very accurate velocity(flux) is needed especially in the presence of heterogeneities, orthotropy and large jumps in the material properties. More accurate approximation of the velocity can be achieved through the use of mixed finite element methods.

Let Ω be a convex polyhedral domain in $\mathbb{R}^3$, $f(x) \in L^2(\Omega)$ and $\boldsymbol{A}(x)$ be a sufficiently smooth three-by-three symmetric matrix-valued function on $\bar{\Omega}$ satisfying the uniform positive definiteness condition: there exists $\alpha > 0$ such that

$$\alpha^{-1}\boldsymbol{\xi}^T\boldsymbol{\xi} \le \sum_{i,j} \boldsymbol{\xi}^T \boldsymbol{A}(x)\boldsymbol{\xi} \le \alpha\boldsymbol{\xi}^T\boldsymbol{\xi}, \qquad \forall x \in \bar{\Omega}, \forall \boldsymbol{\xi} \in \mathbb{R}^3. \tag{17.1}$$

[1]Institute for Scientific Computation, Department of Mathematics, Texas A&M University, College Station, TX 77843.

[2]Institute of Numerical Mathematics, Russian Academy of Sciences, 14 Leninskiy Prospect, 117901 Moscow B-71, Russia.

[3]Institute of Mathematics, Bulgarian Academy of Sciences, 1113 Sofia, Bulgaria.

We consider the Dirichlet boundary value problem

$$\begin{aligned} \boldsymbol{q} + \boldsymbol{A}\nabla u &= \boldsymbol{0}, && \text{in } \Omega, \\ \nabla \cdot \boldsymbol{q} &= f, && \text{in } \Omega, \\ u &= 0, && \text{on } \partial\Omega, \end{aligned} \tag{17.2}$$

where $\partial\Omega$ is the boundary of Ω. In fluid flow in porous media $u(x)$ is referred to as pressure and $\boldsymbol{q}$ the Darcy's velocity vector. It is well known that (17.2) has a unique solution $u(x) \in H_0^1(\Omega) \cap H^2(\Omega)$, and the following elliptic regularity estimate holds true

$$\|u\|_{2,\Omega} \leq c\|f\|_{0,\Omega}, \tag{17.3}$$

where c is a constant dependent only on Ω and $\|\cdot\|_{0,\Omega}$ and $\|\cdot\|_{2,\Omega}$ are, respectively, the $L^2(\Omega)$ and $H^2(\Omega)$ Sobolev norms,

$$\|u\|_{0,\Omega} = \left(\int_\Omega u^2 dx\right)^{\frac{1}{2}}, \qquad \|u\|_{2,\Omega} = \left(\int_\Omega \sum_{|\alpha|\leq m} |\partial^\alpha u|^2 dx\right)^{\frac{1}{2}}. \tag{17.4}$$

The problem (17.2) can be discretized in various ways. Among the most popular methods are the finite volume method, the standard Galerkin finite element method and the mixed finite element method. Each of these methods has its advantages and disadvantages when applied to particular engineering problems. For example, for petroleum reservoir problems in geometrically simple domains and heterogeneous media the finite volume method has proven to be reliable, accurate and conserves mass locally. As shown in [21] the mixed finite element approach with special quadrature on rectangular elements is equivalent to a class of finite volume methods and gives superconvergent velocity calculations for smooth solutions. Based on this equivalence, an efficient multigrid solution procedure has been developed for structured grids [3]. However, in general the mixed finite element method leads to an algebraic saddle-point problem that is more difficult and more expensive to solve. Although some reliable preconditioning algorithms for these saddle point problems have been proposed and studied (see, e.g. [4,11,17,19]), their efficiency depends strongly on the geometry of the domain, on the coefficient matrix $\boldsymbol{A}(x)$ and on the type of finite elements.

An alternative approach can be taken by developing hybrid methods. This has been studied in the pioneering work of Arnold and Brezzi [2] where the continuity of the velocity vector normal to the boundary of each element is enforced by Lagrange multipliers. These Lagrange multipliers on the element boundaries correspond physically to the trace of the pressure $u(x)$.

We now explain briefly the main idea of this Lagrange formulation for the mixed finite element method. First, introduce the spaces

$$\boldsymbol{V} = H(div;\Omega) = \left\{\boldsymbol{q} \in \left(L^2(\Omega)\right)^3, \nabla\cdot\boldsymbol{q} \in L^2(\Omega)\right\}, \qquad W = L^2(\Omega); \tag{17.5}$$

Then the weak formulation of the system (17.2) is: find a pair $(\boldsymbol{q}, u) \in \boldsymbol{V}\times W$ such that

$$(\nabla\cdot\boldsymbol{q}, w) + (\boldsymbol{A}^{-1}\boldsymbol{q}, \boldsymbol{p}) - (u, \nabla\cdot\boldsymbol{p}) = (f, w), \quad \forall\, w \in W, \quad \boldsymbol{p} \in \boldsymbol{V} \tag{17.6}$$

Next, let $\bar{\boldsymbol{V}}_h \times W_h \subset \boldsymbol{V} \times W$ be a finite element space over the partition $\mathcal{T}_T$ of Ω into tetrahedra (or over the partition $\mathcal{T}_C$ into cubes) (see [6,16]). The requirement $\bar{\boldsymbol{V}}_h \subset \boldsymbol{V}$ implies that the normal component of the vector $\boldsymbol{q}$ is continuous across the interelement boundaries $\partial\mathcal{T}_T$. The construction in [2] is based on the idea of relaxing this continuity requirement and defining the space $\boldsymbol{V}_h = \left\{\boldsymbol{q} \in (L_h^2(\Omega))^3 : \boldsymbol{q}|_T \in \bar{\boldsymbol{V}}_h,\ T \in \mathcal{T}_T\right\}$. In order to apply the interelement continuity of the normal component of $\boldsymbol{q}$ we introduce the space of Lagrange multipliers

$$L_h = \left\{\lambda \in L^2(\partial\mathcal{T}_T) : \lambda|_{\partial T} \in \bar{\boldsymbol{V}}_h \cdot \nu \ \text{ for each } T \in \mathcal{T}_T\right\} \tag{17.7}$$

where ν is the normal vector to ∂T.

Now the approximation to (17.6) using Lagrange multipliers is formulated for the unknown triple $(\boldsymbol{q}_h, u_h, \lambda_h) \in \boldsymbol{V}_h \times W_h \times L_h$. We skip the details of the weak formulation on $\boldsymbol{V}_h \times W_h \times L_h$ referring to [1,2,5,8]. It suffices to remark that this then yields a linear system for the nodal unknowns in the usual way. In fact, if the vectors $\boldsymbol{Q}$, $\boldsymbol{U}$ and $\boldsymbol{\Lambda}$ correspond to the degrees of freedom in the representation of $\boldsymbol{q}_h$, u_h and λ_h with respect to the bases in $\boldsymbol{V}_h$, W_h and L_h, respectively, then the algebraic form of this approximation is [6]

$$\begin{pmatrix} \boldsymbol{M} & \boldsymbol{B} & \boldsymbol{C} \\ \boldsymbol{B}^T & \boldsymbol{0} & \boldsymbol{0} \\ \boldsymbol{C}^T & \boldsymbol{0} & \boldsymbol{0} \end{pmatrix} \begin{pmatrix} \boldsymbol{Q} \\ \boldsymbol{U} \\ \boldsymbol{\Lambda} \end{pmatrix} = \boldsymbol{F} \tag{17.8}$$

where $\boldsymbol{M}$ is a symmetric and positive definite matrix. Note that the matrices $\boldsymbol{M}$ and $\boldsymbol{B}$ are block diagonal since the unknown nodal values of $\boldsymbol{q}_h$ and u_h over a given finite element $\boldsymbol{T}$ are related to the nodal values on the adjacent element only through the matrix $\boldsymbol{C}$. Therefore, using element-by-element elimination (condensation) we can reduce this system to the form

$$\boldsymbol{S}\boldsymbol{\Lambda} = \boldsymbol{\Phi} \tag{17.9}$$

For the description of the structure and the particular form of the Schur complement $\boldsymbol{S}$ in the case of these particular finite element spaces we refer to [1,5,7,8].

The system (17.9) can also be obtained from application of the Galerkin method to (17.2) with nonconforming elements [2]. In particular, the lowest order Raviart-Thomas mixed element approximations are equivalent to the usual P_1-nonconforming finite element approximations augmented with P_3-bubbles. Such relationships have been studied recently for a large variety of mixed finite element spaces [1,5,7].

This equivalence between the hybrid mixed element and this nonconforming finite element method establishes a framework for preconditioning and/or solving the algebraic problem and for postprocessing the finite element solution. Schematically this framework includes the following three steps:

1. Forming the reduced algebraic problem for the Lagrange multiplier scheme, that is equivalent to the nonconforming problem;

2. Construction of efficient methods based on multigrid, multilevel or domain decomposition for solving or preconditioning the reduced problem;

3. Recovery of the solution $u(x)$ and the velocity $\boldsymbol{q}$ from the computed Lagrange multipliers.

Recent progress in each of the above steps (see, e.g. [10,18,20]) gives us an indication that the mixed finite element method can be used as an accurate and efficient tool for solving general elliptic problems of second order in domains with complicated geometry. The goal of this chapter is to construct, study and implement efficient preconditioners for the nonconforming finite element approximations of problem on arbitrary tetrahedral meshes.

17.2 Problem Formulation

We consider Ω to be a unit cube in $\mathbb{R}^3$ and $\boldsymbol{A}(x) = a(x)\,\boldsymbol{I}$ to be a scalar matrix. Let $\mathcal{T}_T$ be a regular partitioning [9] of Ω to tetrahedra T with a characteristic size $h = diam(T)$. Later in Section 17.3 we introduce a special partitioning of Ω in order to obtain better properties in the corresponding algebraic system (see Fig. 17.1).

We introduce the set Q_h of barycenters of all faces of the tetrahedral partition of Ω, and the set $\overset{\circ}{Q}_h$ of those barycenters that are strictly inside

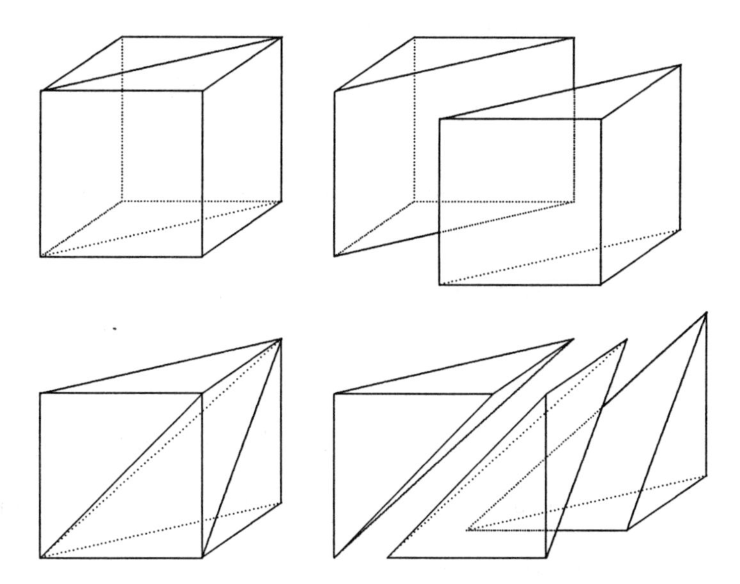

Figure 17.1: Partition of cube into prisms and tetrahedra.

Ω. The Crouzeix-Raviart nonconforming finite element space V_h consists of all piecewise-linear functions on $\mathcal{T}_T$ that vanish at barycenters of the boundary faces and are continuous at the barycenters of $\overset{\circ}{Q}_h$. Note that the space V_h is not a subspace of $H_0^1(\Omega)$.

Now we define the bilinear form on V_h

$$a_h(u,v) = \sum_{T \in \mathcal{T}_T} \int_T a(x) \nabla u \nabla v dx, \qquad \forall\, u, v \in V_h. \tag{17.10}$$

Thus the nonconforming discretization of problem (17.10) is: find $u_h \in V_h$ such that

$$a_h(u_h, v) = (f, v), \qquad \forall\, v \in V_h \tag{17.11}$$

where (f, v) denotes the L^2-inner product.

The natural degrees of freedom for the Crouzeix-Raviart nonconforming elements are the values at the barycenters of the faces of tetrahedral elements. Denote the vector of unknown values corresponding to a function $v_h \in V_h$ by $\mathbf{v}$ and assume that its dimension is N; i.e., $\boldsymbol{v} \in \mathbb{R}^N$. Note, that all unknowns on faces coincident with that part of the boundary with Dirichlet data are excluded.

Let $(\boldsymbol{u}, \boldsymbol{v})$ be a bilinear form defined on $\mathbb{R}^N$ by

$$(\boldsymbol{u}, \boldsymbol{v})_N = h^3 \sum_{x \in \overset{\circ}{Q}_h} u(x)v(x), \qquad u, v \in V_h. \tag{17.12}$$

It is clear that $(\cdot, \cdot)_N$ is equivalent to the L^2-inner product on V_h; i.e. there exists a constant $c > 0$ such that

$$c^{-1}\|v\|_0^2 \leq \|\boldsymbol{v}\| \leq c\|v\|_0^2, \qquad v \in V_h \tag{17.13}$$

where

$$\|\boldsymbol{v}\| = (\boldsymbol{v}, \boldsymbol{v})_N^{\frac{1}{2}}, \qquad \text{for } v \in V_h. \tag{17.14}$$

Then the discretization operator $\boldsymbol{A} : \mathbb{R}^N \to \mathbb{R}^N$ is defined by

$$(\boldsymbol{A}\boldsymbol{v}, \boldsymbol{w})_N = a_h(u, w), \qquad u, v \in V_h. \tag{17.15}$$

Similarly, we introduce the vector $\mathbf{F}$ as

$$(f, v) = (\boldsymbol{F}, \boldsymbol{v})_N, \qquad \forall\, v \in V_h. \tag{17.16}$$

Now, the problem (17.11) can be rewritten in matrix form as

$$\boldsymbol{A}\boldsymbol{u} = \boldsymbol{F} \tag{17.17}$$

where $\boldsymbol{A}$ is symmetric and positive definite.

17.3 Matrix Formulation and Its Properties

Our goal is to introduce an algebraic formulation of the approximate problem using a type of static condensation that eliminates certain local unknowns. In this way we can reduce substantially the size of the problem. For this approach we need a special partitioning of the domain into tetrahedra that have some regularity and preserve the simplicity of the algebraic problem.

First, we partition Ω into cubes with edges of size $h = 1/n$ and denote the cubes by $C^{(i,j,k)}$ where (x_{1i}, x_{2j}, x_{3k}) corresponds to the right rear upper corner of a cube. This partitioning is denoted by $\mathcal{T}_C$. Next, we divide each cube $C^{(i,j,k)}$ into two prisms $P_1 = P_1^{(i,j,k)}$ and $P_2 = P_2^{(i,j,k)}$ as shown in Figure 17.1 and denote this partitioning of Ω by $\mathcal{T}_P$. Finally, we divide each prism into three tetrahedra as shown in Figure 17.1 and denote this partitioning of Ω into tetrahedra by $\mathcal{T}_T$.

Let $P = P^{(i,j,k)} \in \mathcal{T}_P$ be a particular prism of the partition $\mathcal{T}_P$. Denote by V_h^P the subspace of restrictions of the functions in V_h onto P. These restrictions define vectors $\boldsymbol{u}_P$ that are restrictions of a vector $\boldsymbol{u} \in \mathbb{R}^N$. The dimension of V_h^P is denoted N^P. Obviously, for prisms with no faces on $\partial\Omega$ the dimension $N^P = 10$.

The local stiffness matrix $\boldsymbol{A}^P$ of prism $P \in \mathcal{T}_P$ is defined by

$$(\boldsymbol{A}^P \boldsymbol{u}_P, \boldsymbol{v}_P)_N = \sum_{T \subset P} \int_T a(x) \nabla u_h \cdot \nabla v_h \, dx \tag{17.18}$$

for any $P \in \mathcal{T}_P$. Then the global stiffness matrix is determined by assembling the local stiffness matrices in the usual manner. The following equality holds true for any $\boldsymbol{u}, \boldsymbol{v} \in \mathbb{R}^N$:

$$(\boldsymbol{A}\boldsymbol{u}, \boldsymbol{v})_N = \sum_{P \in \mathcal{T}_P} (\boldsymbol{A}^P \boldsymbol{u}_P, \boldsymbol{u}_P)_N. \tag{17.19}$$

Consider a prism P of an arbitrary cube that has no face on the boundary $\partial\Omega$. Enumerate the faces s_j, $j = 1, \ldots, 10$ of the tetrahedra in this prism as shown in Figure 17.2. Then the local stiffness matrix of this prism for the case $a(X) \equiv 1$ has the following form:

$$\boldsymbol{A}^P = \frac{3h}{2} \left[\begin{array}{cccccc|cccc} 1 & 0 & 0 & 0 & 0 & 0 & 0 & 0 & -1 & 0 \\ 0 & 1 & 0 & 0 & 0 & 0 & 0 & 0 & 0 & -1 \\ 0 & 0 & 1 & 0 & 0 & 0 & -1 & 0 & 0 & 0 \\ 0 & 0 & 0 & 1 & 0 & 0 & 0 & -1 & 0 & 0 \\ 0 & 0 & 0 & 0 & 1 & 0 & 0 & 0 & -1 & 0 \\ 0 & 0 & 0 & 0 & 0 & 1 & 0 & 0 & 0 & -1 \\ \hline 0 & 0 & -1 & 0 & 0 & 0 & 2 & 0 & -1 & 0 \\ 0 & 0 & 0 & -1 & 0 & 0 & 0 & 2 & 0 & -1 \\ -1 & 0 & 0 & 0 & -1 & 0 & -1 & 0 & 4 & -1 \\ 0 & -1 & 0 & 0 & 0 & -1 & 0 & -1 & -1 & 4 \end{array} \right]. \tag{17.20}$$

That is,

$$\boldsymbol{A}^P = \frac{3h}{2} \begin{bmatrix} \boldsymbol{A}_{11} & \boldsymbol{A}_{12} \\ \boldsymbol{A}_{21} & \boldsymbol{A}_{22} \end{bmatrix}. \tag{17.21}$$

Similarly, let us introduce the following matrix $\boldsymbol{B}^P$ defined on the same space V_h^P:

$$\boldsymbol{B}^P = \frac{3h}{2} \begin{bmatrix} \boldsymbol{A}_{11} & \boldsymbol{A}_{12} \\ \boldsymbol{A}_{21} & \boldsymbol{B}_{22} \end{bmatrix}, \qquad \boldsymbol{B}_{22} = \begin{bmatrix} 3 & -1 & -1 & 0 \\ -1 & 3 & 0 & -1 \\ -1 & 0 & 3 & 0 \\ 0 & -1 & 0 & 3 \end{bmatrix}. \tag{17.22}$$

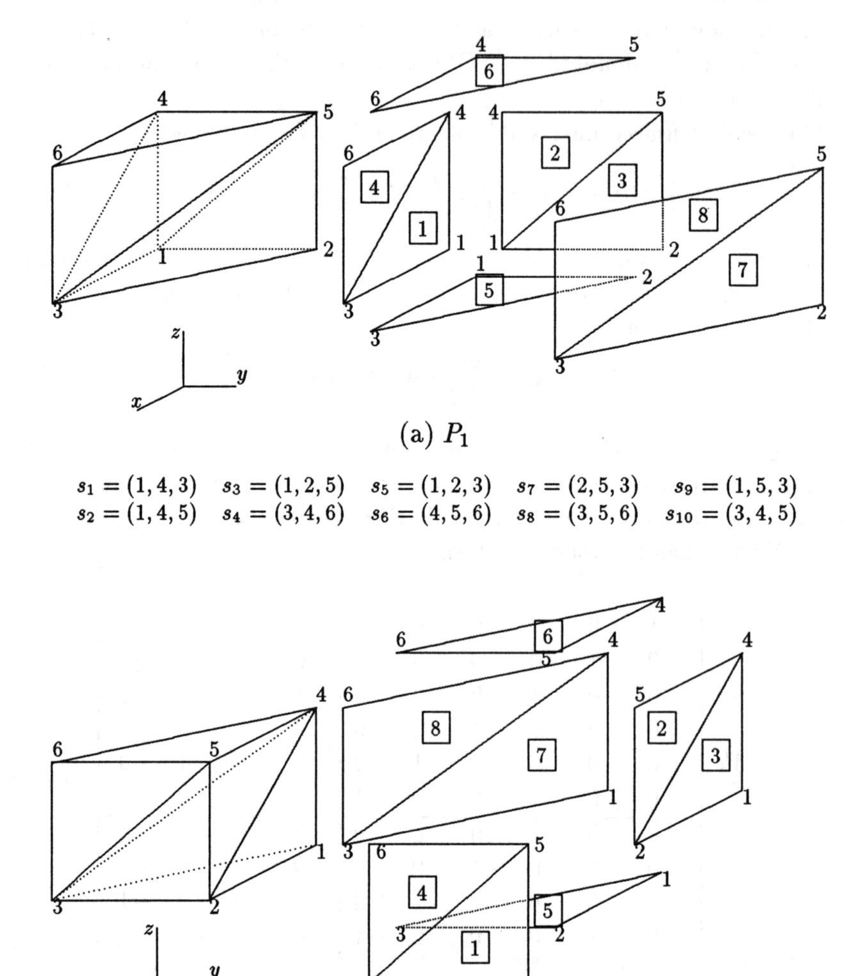

(a) P_1

$s_1 = (1, 4, 3)$ $s_3 = (1, 2, 5)$ $s_5 = (1, 2, 3)$ $s_7 = (2, 5, 3)$ $s_9 = (1, 5, 3)$
$s_2 = (1, 4, 5)$ $s_4 = (3, 4, 6)$ $s_6 = (4, 5, 6)$ $s_8 = (3, 5, 6)$ $s_{10} = (3, 4, 5)$

(b) P_2

$s_1 = (2, 3, 5)$ $s_3 = (1, 2, 4)$ $s_5 = (1, 2, 3)$ $s_7 = (1, 3, 4)$ $s_9 = (2, 3, 4)$
$s_2 = (2, 4, 5)$ $s_4 = (3, 5, 6)$ $s_6 = (4, 5, 6)$ $s_8 = (3, 4, 6)$ $s_{10} = (3, 4, 5)$

Figure 17.2: Local enumeration of faces in a prism.

It is easy to see that the following holds true:

PROPOSITION 1. $ker\boldsymbol{A}^P = ker\boldsymbol{B}^P$. □

Remark. If the prism $P \in \mathcal{T}_P$ has a face on $\partial\Omega$, the dimension of the matrix $\boldsymbol{A}^P$ will be less than 10 and the modification of $\boldsymbol{B}_{22}$ is obvious.

Finally, we define the $N \times N$ matrix $\boldsymbol{B}$ by

$$(\boldsymbol{B}\boldsymbol{u}, \boldsymbol{v})_N = \sum_{P \in \mathcal{T}_P} (\boldsymbol{B}^P \boldsymbol{u}_P, \boldsymbol{v}_P)_N \qquad \forall\, \boldsymbol{u}, \boldsymbol{v} \in \mathbb{R}^N. \tag{17.23}$$

Since matrix $\boldsymbol{B}$ will be used for preconditioning the original problem, (17.17) it is important to estimate the condition number of $\boldsymbol{B}^{-1}\boldsymbol{A}$. Using the fact that all element stiffness matrices are nonnegative and following [15], we easily get the estimates

$$\max_{u \in \mathbb{R}^N} \frac{(\boldsymbol{A}\boldsymbol{u}, \boldsymbol{u})}{(\boldsymbol{B}\boldsymbol{u}, \boldsymbol{u})} \leq \max_{\substack{P \in \mathcal{T}_P \\ (\boldsymbol{B}^P \boldsymbol{u}_P, \boldsymbol{u}_P) \neq 0}} \frac{(\boldsymbol{A}^P \boldsymbol{u}_P, \boldsymbol{u}_P)}{(\boldsymbol{B}^P \boldsymbol{u}_P, \boldsymbol{u}_P)}, \tag{17.24}$$

$$\min_{u \in \mathbb{R}^N} \frac{(\boldsymbol{A}\boldsymbol{u}, \boldsymbol{u})}{(\boldsymbol{B}\boldsymbol{u}, \boldsymbol{u})} \geq \min_{\substack{P \in \mathcal{T}_P \\ (\boldsymbol{B}^P \boldsymbol{u}_P, \boldsymbol{u}_P) \neq 0}} \frac{(\boldsymbol{A}^P \boldsymbol{u}_P, \boldsymbol{u}_P)}{(\boldsymbol{B}^P \boldsymbol{u}_P, \boldsymbol{u}_P)}. \tag{17.25}$$

In this way estimates of the maximal eigenvalue $\lambda_{\max}$ and the minimal eigenvalue $\lambda_{\min}$ of the eigenvalue problem

$$\boldsymbol{A}\boldsymbol{u} = \lambda \boldsymbol{B}\boldsymbol{u} \tag{17.26}$$

are obtained by solving the local problems

$$\boldsymbol{A}^P \boldsymbol{u}_P = \mu \boldsymbol{B}^P \boldsymbol{u}_P, \qquad (\boldsymbol{B}^P \boldsymbol{u}_P, \boldsymbol{u}_P) \neq 0, \qquad P \in \mathcal{T}_P. \tag{17.27}$$

Using superelement analysis it is easy to show that to get the minimal $\mu_{\min}$ and the maximal $\mu_{\max}$ eigenvalues of (17.27) one has to consider the worst case when the prism $P \in \mathcal{T}_P$ has no face on the boundary $\partial\Omega$, i.e., $P \cap \partial\Omega = \emptyset$. Then a direct calculation verifies the following result.

PROPOSITION 2. *The eigenvalues of problem (17.27) lie in the interval* $[2 - \sqrt{3},\ 2 + \sqrt{3}]$. □

The inequalities (17.24) and (17.25) then yield:

PROPOSITION 3. *The eigenvalues of problem (17.26) lie in the interval* $[2 - \sqrt{3}, 2 + \sqrt{3}]$ *and therefore*

$$\mathrm{cond}(\boldsymbol{B}^{-1}\boldsymbol{A}) \leq (2 + \sqrt{3})^2.$$

We emphasize that the condition number of the matrix $\boldsymbol{B}^{-1}\boldsymbol{A}$ is bounded by a constant independent of the step size of the mesh h.

Now we divide all unknowns in the system into two groups:

1. The first group consists of all unknowns corresponding to faces of the prisms in the partition $\mathcal{T}_P$, excluding, of course, the faces on $\partial\Omega$ (see Figure 17.2).

2. The second group consists of the unknowns corresponding to the faces of the tetrahedra that are internal to each prism (these are faces s_9 and s_{10} in Figure 17.2).

This splitting of the space $\mathbb{R}^N$ induces the following presentation of the vectors $\boldsymbol{v}^T = (\boldsymbol{v}_1^T, \boldsymbol{v}_2^T)$, where $\boldsymbol{v}_1 \in \mathbb{R}^{N_1}$ and $\boldsymbol{v}_2 \in \mathbb{R}^{N_2}$. Obviously, $N_1 = N - 4n^3$. Then matrix $\boldsymbol{B}$ can be presented in the block form

$$\boldsymbol{B} = \begin{bmatrix} \boldsymbol{B}_{11} & \boldsymbol{B}_{12} \\ \boldsymbol{B}_{21} & \boldsymbol{B}_{22} \end{bmatrix}, \qquad \dim \boldsymbol{B}_{11} = N_1. \tag{17.28}$$

Denote now by $\hat{\boldsymbol{B}}_{11} = \boldsymbol{B}_{11} - \boldsymbol{B}_{12}\boldsymbol{B}_{22}^{-1}\boldsymbol{B}_{21}$ the Schur complement of $\boldsymbol{B}$ obtained by elimination of the vector $\boldsymbol{v}_2$. Then $\boldsymbol{B}_{11} = \hat{\boldsymbol{B}}_{11} + \boldsymbol{B}_{12}\boldsymbol{B}_{22}^{-1}\boldsymbol{B}_{21}$, so the matrix $\boldsymbol{B}$ has the form

$$\boldsymbol{B} = \begin{bmatrix} \hat{\boldsymbol{B}}_{11} + \boldsymbol{B}_{12}\boldsymbol{B}_{22}^{-1}\boldsymbol{B}_{21} & \boldsymbol{B}_{12} \\ \boldsymbol{B}_{21} & \boldsymbol{B}_{22} \end{bmatrix}. \tag{17.29}$$

Note that for each prism $P \in \mathcal{T}_P$, the unknowns on the faces s_9 and s_{10} in (17.2) are connected through the equation $\boldsymbol{B}\boldsymbol{v} = \boldsymbol{F}$, only by the unknowns associated with this prism and therefore can be eliminated locally; i.e., the matrix $\boldsymbol{B}_{22}$ is block diagonal with 2×2 blocks and can be inverted locally (prism by prism). Thus matrix $\hat{\boldsymbol{B}}_{11}$ is easily computable.

17.4 Multilevel Substructuring Preconditioner

In this section we will propose two modifications of matrix $\boldsymbol{B}$ in (17.29) of the form

$$\tilde{\boldsymbol{B}} = \begin{bmatrix} \tilde{\boldsymbol{B}}_1 + \boldsymbol{B}_{12}\boldsymbol{B}_{22}^{-1}\boldsymbol{B}_{21} & \boldsymbol{B}_{12} \\ \boldsymbol{B}_{21} & \boldsymbol{B}_{22} \end{bmatrix} \tag{17.30}$$

and consider their computational properties.

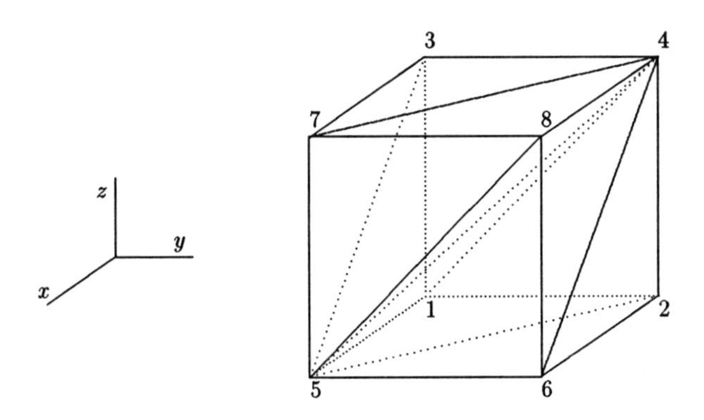

Figure 17.3: Cube $c^{(i,j,k)}$.

Group Partitioning of the Grid Points

Let us denote by $s_{r,l,m}^{(i,j,k)}$ the face of the cube $C^{(i,j,k)}$ with vertices r, l, m in Figure 17.3. For the sake of simplicity of exposition we introduce a special partitioning of the nodes of $\overset{\circ}{Q}_h$ into three groups:

1. First, we group the nodes on the faces

$$s_{2,4,5}^{(i,j,k)} \quad \text{and} \quad s_{4,5,7}^{(i,j,k)}, \qquad i,j,k = \overline{1,n}.$$

   We denote the unknowns at these nodes by $VI_\ell^{(i,j,k)}$, $\ell = 1,2$, $i,j,k = \overline{1,n}$.

2. Then, we take the nodes on the faces perpendicular to the x, y, and z axes:

$$\text{(i)} \quad s_{1,2,4}^{(i,j,k)}, \quad s_{1,3,4}^{(i,j,k)}, \quad i = \overline{2,n}, \quad j,k = \overline{1,n}.$$

   We denote the unknowns at these nodes by $Vx_\ell^{(i,j,k)}$, $\ell = 1,2$, $i = \overline{2,n}$, $j,k = \overline{1,n}$;

$$\text{(ii)} \quad s_{1,3,5}^{(i,j,k)}, \quad s_{5,3,7}^{(i,j,k)}, \quad j = \overline{2,n}, \quad i,k = \overline{1,n}. \tag{17.31}$$

   We denote the unknowns at these nodes by $Vy_\ell^{(i,j,k)}$, $\ell = 1,2$, $j = \overline{2,n}$, $i,k = \overline{1,n}$;

$$\text{(iii)} \quad s_{1,2,5}^{(i,j,k)}, \quad s_{2,5,6}^{(i,j,k)}, \quad i,j = \overline{1,n}, \quad k = \overline{2,n}.$$

We denote the unknowns at these nodes by $Vz_\ell^{(i,j,k)}$, $\ell = 1, 2$, $i, j = \overline{1, n}$, $k = \overline{2, n}$.

3. Finally, we take the remaining nodes on faces

$$s_{1,4,5}^{(i,j,k)}, \quad s_{3,4,5}^{(i,j,k)}, \quad s_{4,5,6}^{(i,j,k)}, \quad s_{4,5,8}^{(i,j,k)}, \quad i, j, k = \overline{1, n}.$$

We denote the unknowns at these nodes by $VA_\ell^{(i,j,k)}$, $\ell = \overline{1, n}$, $i, j, k = \overline{1, n}$.

Three Level Preconditioner: Variant I

Let us take an arbitrary cube $C^{(i,j,k)}$ that is partitioned into left and right prisms $P_p^{(i,j,k)}$, $p = 1, 2$ (see Figure 17.1). For notational convenience below, we omit indices '(i, j, k)' and 'p' in all cases where there is no ambiguity. In the local numeration (Figure 17.2) matrices $\boldsymbol{B}_1$ and $\boldsymbol{B}_2$ correspond to prisms having the form (17.20)-(17.22). We rewrite these matrices in the ordering (17.31) introduced above in this section:

$$\boldsymbol{B}_1 = \left(\frac{3h}{2}\right) \left[\begin{array}{rr|rrrrrr|rr}
3 & -1 & -1 & 0 & 0 & 0 & 0 & 0 & -1 & 0 \\
-1 & 3 & 0 & 0 & 0 & -1 & 0 & 0 & 0 & -1 \\
\hline
-1 & 0 & 1 & 0 & 0 & 0 & 0 & 0 & 0 & 0 \\
0 & 0 & 0 & 1 & 0 & 0 & 0 & 0 & 0 & -1 \\
0 & 0 & 0 & 0 & 1 & 0 & 0 & 0 & -1 & 0 \\
0 & -1 & 0 & 0 & 0 & 1 & 0 & 0 & 0 & 0 \\
0 & 0 & 0 & 0 & 0 & 0 & 1 & 0 & -1 & 0 \\
0 & 0 & 0 & 0 & 0 & 0 & 0 & 1 & 0 & -1 \\
\hline
-1 & 0 & 0 & 0 & -1 & 0 & -1 & 0 & 3 & 0 \\
0 & -1 & 0 & -1 & 0 & 0 & 0 & -1 & 0 & 3
\end{array}\right] \tag{17.32}$$

$$\boldsymbol{B}_2 = \left(\frac{3h}{2}\right) \left[\begin{array}{rr|rrrrrr|rr}
3 & -1 & 0 & 0 & -1 & 0 & 0 & 0 & -1 & 0 \\
-1 & 3 & 0 & -1 & 0 & 0 & 0 & 0 & 0 & -1 \\
\hline
0 & 0 & 1 & 0 & 0 & 0 & 0 & 0 & -1 & 0 \\
0 & -1 & 0 & 1 & 0 & 0 & 0 & 0 & 0 & 0 \\
-1 & 0 & 0 & 0 & 1 & 0 & 0 & 0 & 0 & 0 \\
0 & 0 & 0 & 0 & 0 & 1 & 0 & 0 & 0 & -1 \\
0 & 0 & 0 & 0 & 0 & 0 & 1 & 0 & -1 & 0 \\
0 & 0 & 0 & 0 & 0 & 0 & 0 & 1 & 0 & -1 \\
\hline
-1 & 0 & -1 & 0 & 0 & 0 & -1 & 0 & 3 & 0 \\
0 & -1 & 0 & 0 & 0 & -1 & 0 & -1 & 0 & 3
\end{array}\right].$$

The partitioning of nodes into groups (17.31) induces block matrices

$$\boldsymbol{B}_p = \begin{bmatrix} \boldsymbol{B}_{11,p} & \boldsymbol{B}_{12,p} \\ \boldsymbol{B}_{21,p} & \boldsymbol{B}_{22,p} \end{bmatrix}, \qquad p = 1,2, \tag{17.33}$$

where blocks $\boldsymbol{B}_{22,p}$ correspond to the unknowns of the third group and blocks $\boldsymbol{B}_{11,p}$ correspond to the unknowns of the first and second groups. We may eliminate the unknowns of the third group from each matrix $\boldsymbol{B}_p$, $p = 1,2$ locally on each prism. This yields the condensed form

$$\hat{\boldsymbol{B}}_{11,p} = \boldsymbol{B}_{11,p} - \boldsymbol{B}_{12,p}\boldsymbol{B}_{22,p}^{-1}\boldsymbol{B}_{21,p}, \qquad p = 1,2, \tag{17.34}$$

where $p = 1$ corresponds to a right prism and $p = 2$ to a left prism. Then

$$\hat{\boldsymbol{B}}_{11,1} = \left(\frac{3h}{2}\right) \left[\begin{array}{rr|rrrrrr} 8/3 & -1 & -1 & 0 & -1/3 & 0 & -1/3 & 0 \\ -1 & 8/3 & 0 & -1/3 & 0 & -1 & 0 & -1/3 \\ \hline -1 & 0 & 1 & 0 & 0 & 0 & 0 & 0 \\ 0 & -1/3 & 0 & 2/3 & 0 & 0 & 0 & -1/3 \\ -1/3 & 0 & 0 & 0 & 2/3 & 0 & -1/3 & 0 \\ 0 & -1 & 0 & 0 & 0 & 1 & 0 & 0 \\ -1/3 & 0 & 0 & 0 & -1/3 & 0 & 2/3 & 0 \\ 0 & -1/3 & 0 & -1/3 & 0 & 0 & 0 & 2/3 \end{array}\right], \tag{17.35}$$

$$\hat{\boldsymbol{B}}_{11,2} = \left(\frac{3h}{2}\right) \left[\begin{array}{rr|rrrrrr} 8/3 & -1 & -1/3 & 0 & -1 & 0 & -1/3 & 0 \\ -1 & 8/3 & 0 & -1 & 0 & -1/3 & 0 & -1/3 \\ \hline -1/3 & 0 & 2/3 & 0 & 0 & 0 & -1/3 & 0 \\ 0 & -1 & 0 & 1 & 0 & 0 & 0 & 0 \\ -1 & 0 & 0 & 0 & 1 & 0 & 0 & 0 \\ 0 & -1/3 & 0 & 0 & 0 & 2/3 & 0 & -1/3 \\ -1/3 & 0 & -1/3 & 0 & 0 & 0 & 2/3 & 0 \\ 0 & -1/3 & 0 & 0 & 0 & -1/3 & 0 & 2/3 \end{array}\right]. \tag{17.36}$$

Further, we define on each cube the matrices $\boldsymbol{B}_{1,p}$, $p = 1, 2$ as

$$\boldsymbol{B}_{1,1} = \left(\frac{3h}{2}\right) \left[\begin{array}{rr|rrrrrr} 8/3 & -1 & -1 & 0 & -1/3 & 0 & -1/3 & 0 \\ -1 & 8/3 & 0 & -1/3 & 0 & -1 & 0 & -1/3 \\ \hline -1 & 0 & 1 & 0 & 0 & 0 & 0 & 0 \\ 0 & -1/3 & 0 & 1/3 & 0 & 0 & 0 & 0 \\ -1/3 & 0 & 0 & 0 & 1/3 & 0 & 0 & 0 \\ 0 & -1 & 0 & 0 & 0 & 1 & 0 & 0 \\ -1/3 & 0 & 0 & 0 & 0 & 0 & 1/3 & 0 \\ 0 & -1/3 & 0 & 0 & 0 & 0 & 0 & 1/3 \end{array}\right], \tag{17.37}$$

$$\boldsymbol{B}_{1,2} = \left(\frac{3h}{2}\right) \left[\begin{array}{rr|rrrrrr} 8/3 & -1 & -1/3 & 0 & -1 & 0 & -1/3 & 0 \\ -1 & 8/3 & 0 & -1 & 0 & -1/3 & 0 & -1/3 \\ \hline -1/3 & 0 & 1/3 & 0 & 0 & 0 & 0 & 0 \\ 0 & -1 & 0 & 1 & 0 & 0 & 0 & 0 \\ -1 & 0 & 0 & 0 & 1 & 0 & 0 & 0 \\ 0 & -1/3 & 0 & 0 & 0 & 1/3 & 0 & 0 \\ -1/3 & 0 & 0 & 0 & 0 & 0 & 1/3 & 0 \\ 0 & -1/3 & 0 & 0 & 0 & 0 & 0 & 1/3 \end{array}\right]. \tag{17.38}$$

Both matrices $\hat{\boldsymbol{B}}_{11,p}$ and $\boldsymbol{B}_{1,p}$, $p = 1, 2$, are irreducible and have the same kernel; i.e., $\ker\hat{\boldsymbol{B}}_{11,p} = \ker\boldsymbol{B}_{1,p}$. It is easy to show that eigenvalues of the spectral problems (for both left and right prisms)

$$\hat{\boldsymbol{B}}_{11,p}\boldsymbol{u} = \mu\boldsymbol{B}_{1,p}\boldsymbol{u}, \quad \boldsymbol{u} \in \mathbb{R}^8, \qquad (\boldsymbol{B}_{1,p}\boldsymbol{u}, \boldsymbol{u}) \neq 0, \quad p = 1, 2, \tag{17.39}$$

belong to the interval $\mu \in [1, 3]$.

Now we define a new matrix on each prism

$$\tilde{\boldsymbol{B}}_p = \begin{bmatrix} \boldsymbol{B}_{1,p} + \boldsymbol{B}_{12,p}\boldsymbol{B}_{22,p}^{-1}\boldsymbol{B}_{21,p} & \boldsymbol{B}_{12,p} \\ \boldsymbol{B}_{21,p} & \boldsymbol{B}_{22,p} \end{bmatrix}, \qquad p = 1, 2. \tag{17.40}$$

In the case when cube C has nonempty intersection with $\partial\Omega$ all considerations are the same; matrices $\boldsymbol{B}_{1,p}$, $\boldsymbol{B}_{12,p}$, $\boldsymbol{B}_{21,p}$, $p = 1, 2$ will not have rows and columns corresponding to the nodes on the boundary.

Next, define the generalized eigenvalue problem for each prism P

$$\boldsymbol{B}_P\boldsymbol{u} = \xi\tilde{\boldsymbol{B}}_P\boldsymbol{u}, \qquad (\tilde{\boldsymbol{B}}_P\boldsymbol{u}, \boldsymbol{u}) \neq 0, \qquad P \in \mathcal{T}_P. \tag{17.41}$$

For inner prisms, that is for prisms which have no face on $\partial\Omega$, $\boldsymbol{u} \in \mathbb{R}^{10}$. Because the eigenvalues of the problems (17.39) belong to the interval

$[1,3]$, the same is true for the problems (17.41); thus we can formulate the following:

PROPOSITION 4. *Eigenvalues of the problems (17.41) belong to the interval* $[1,3]$. *Moreover, the eigenvalue problem on each prism*

$$A_P u = \nu \tilde{B}_P u, \qquad (\tilde{B}_P u, u) \neq 0, \qquad P \in \mathcal{T}_P, \tag{17.42}$$

has eigenvalues ν *that satisfy*

$$\nu \in [2-\sqrt{3}, 3(2+\sqrt{3})].$$

□

Now we define the symmetric positive-definite $N_1 \times N_1$ matrix $\tilde{B}_1$ by

$$(\tilde{B}_1 u_1, v_1) = \sum_{P \in \mathcal{T}_P} (\tilde{B}_P u_{1,P}, v_{1,P}), \tag{17.43}$$

where the vectors $v_1, u_1 \in \mathbb{R}^{N_1}$, and $u_{1,P}, v_{1,P}$ are the restrictions of the vectors u_1, v_1 to the prism P.

Along with matrix B in the form of (17.29), we introduce the matrix

$$\tilde{B} = \begin{bmatrix} \tilde{B}_1 + B_{12} B_{22}^{-1} B_{21} & B_{12} \\ B_{21} & B_{22} \end{bmatrix}. \tag{17.44}$$

Again, using superelement analysis and Proposition 4, it is easy to prove the following theorem.

THEOREM 1. *Matrix* $\tilde{B}$, *defined in (17.44), is spectrally equivalent to matrix* A*; i.e.,*

$$\alpha \tilde{B} \leq A \leq \beta \tilde{B},$$

with $\alpha = (2-\sqrt{3})$ *and* $\beta = 3(2+\sqrt{3})$. *Therefore,*

$$\operatorname{cond}(\tilde{B}^{-1} A) \leq 3(2+\sqrt{3})^2. \tag{17.45}$$

□

Instead of matrix B in the form of (17.29) we take the matrix $\tilde{B}$ in the form of (17.44) as the two-level preconditioner for matrix A. As we noted earlier, matrix B_{22} is block-diagonal and can be inverted locally on prisms.

Now consider the linear system

$$\tilde{B}_1 u = f \tag{17.46}$$

The matrix $\tilde{\boldsymbol{B}}_1$ can be written in the block form as

$$\tilde{\boldsymbol{B}}_1 = \begin{bmatrix} \boldsymbol{C}_{11} & \boldsymbol{C}_{12} \\ \boldsymbol{C}_{21} & \boldsymbol{C}_{22} \end{bmatrix}, \tag{17.47}$$

where the block $\boldsymbol{C}_{22}$ corresponds to the nodes from group (2), which are on the faces of tetrahedra perpendicular to the coordinate axis. It can be shown that matrix $\boldsymbol{C}_{22}$ is diagonal. In partitioning (17.31), we assume $\boldsymbol{u}$ and $\mathbf{f}$ ordered in the form

$$\boldsymbol{u} = \begin{bmatrix} \boldsymbol{u}_1 \\ \boldsymbol{u}_2 \end{bmatrix}, \qquad \boldsymbol{f} = \begin{bmatrix} \boldsymbol{f}_1 \\ \boldsymbol{f}_2 \end{bmatrix} \tag{17.48}$$

After elimination of the second group of unknowns,

$$\boldsymbol{u}_2 = \boldsymbol{C}_{22}^{-1}(\boldsymbol{f}_2 - \boldsymbol{C}_{21}\boldsymbol{u}_1)$$

we get the reduced system of linear equations

$$(\boldsymbol{C}_{11} - \boldsymbol{C}_{12}\boldsymbol{C}_{22}^{-1}\boldsymbol{C}_{21})\boldsymbol{u}_1 = \boldsymbol{f}_1 - \boldsymbol{C}_{12}\boldsymbol{C}_{22}^{-1}\boldsymbol{f}_2 = \tilde{\boldsymbol{f}}_1, \tag{17.49}$$

where the vector $\boldsymbol{u}_1$ and the block $\boldsymbol{C}_{11}$ correspond to the unknowns from the first group, which have only two unknowns per cube. The dimension of vectors $\boldsymbol{u}_1$ and $\boldsymbol{f}_1$ is equal to $2n^3$. Since we have introduced a two-level subdivision of matrix $\tilde{\boldsymbol{B}}_1$, the original matrix $\tilde{\boldsymbol{B}}$ can be considered as a three-level preconditioner.

Computational Scheme: Variant I

Let us write explicitly the equations from (17.46) in terms of the unknowns introduced in (17.31), i.e., in terms of

$$fI_\ell^{(i,j,k)}, \quad UI_\ell^{(i,j,k)}, \quad \ell = 1,2, \quad i,j,k = \overline{1,n};$$

$$fx_\ell^{(i,j,k)}, \quad Ux_\ell^{(i,j,k)}, \quad \ell = 1,2, \quad i = \overline{2,n}, \qquad j,k = \overline{1,n};$$

$$fy_\ell^{(i,j,k)}, \quad Uy_\ell^{(i,j,k)}, \quad \ell = 1,2, \quad j = \overline{2,n}, \qquad i,k = \overline{1,n};$$

$$fz_\ell^{(i,j,k)}, \quad Uz_\ell^{(i,j,k)}, \quad \ell = 1,2, \quad k = \overline{2,n}, \qquad i,j = \overline{1,n}.$$

To simplify the representation we consider the case $a(x) = 1$. Then

$$\begin{bmatrix} 16/3 & -2 \\ -2 & 16/3 \end{bmatrix} \boldsymbol{UI}^{(i,j,k)}$$

$$-(1-\delta_{in})\begin{bmatrix} 1/3 & 0 \\ 0 & 1 \end{bmatrix} \boldsymbol{Ux}^{(i,j,k)} - (1-\delta_{i1})\begin{bmatrix} 1 & 0 \\ 0 & 1/3 \end{bmatrix} \boldsymbol{Ux}^{(i-1,j,k)}$$

$$-(1-\delta_{jn})\begin{bmatrix} 1 & 0 \\ 0 & 1/3 \end{bmatrix} \boldsymbol{Uy}^{(i,j,k)} - (1-\delta_{j1})\begin{bmatrix} 1/3 & 0 \\ 0 & 1 \end{bmatrix} \boldsymbol{Uy}^{(i,j-1,k)}$$

$$-(1-\delta_{kn})\frac{1}{3}\boldsymbol{Uz}^{(i,j,k)} - (1-\delta_{k1})\frac{1}{3}\boldsymbol{Uz}^{(i,j,k-1)} = \left(\frac{2}{3h}\right) \boldsymbol{fI}^{(i,j,k)},$$

$$i,j,k = \overline{1,n} \qquad (17.50)$$

$$\frac{4}{3}\boldsymbol{Ux}^{(i,j,k)} - \begin{bmatrix} 1 & 0 \\ 0 & 1/3 \end{bmatrix} \boldsymbol{UI}^{(i-1,j,k)} -$$

$$\begin{bmatrix} 1/3 & 0 \\ 0 & 1 \end{bmatrix} \boldsymbol{UI}^{(i,j,k)} = \left(\frac{2}{3h}\right) \boldsymbol{fx}^{(i,j,k)}, \quad i = \overline{2,n},\ j,k = \overline{1,n},$$

$$\frac{4}{3}\boldsymbol{Uy}^{(i,j,k)} - \begin{bmatrix} 1/3 & 0 \\ 0 & 1 \end{bmatrix} \boldsymbol{UI}^{(i,j-1,k)} -$$

$$\begin{bmatrix} 1 & 0 \\ 0 & 1/3 \end{bmatrix} \boldsymbol{UI}^{(i,j,k)} = \left(\frac{2}{3h}\right) \boldsymbol{fy}^{(i,j,k)}, \quad j = \overline{2,n},\ i,k = \overline{1,n},$$

$$\frac{2}{3}\boldsymbol{Uz}^{(i,j,k)} - \frac{1}{3}\boldsymbol{UI}^{(i,j,k-1)} - \frac{1}{3}\boldsymbol{UI}^{(i,j,k)} = \left(\frac{2}{3h}\right) \boldsymbol{fz}^{(i,j,k)},$$

$$k = \overline{2,n},\ i,j = \overline{1,n} \qquad (17.51)$$

where the function $\delta_{ik} = 1$ if $i = k$ or 0 if $i \neq k$ is introduced to take into account the Dirichlet boundary conditions, and any vector $\boldsymbol{vr}^{(i,j,k)} = [vr_1^{(i,j,k)},\ vr_2^{(i,j,k)}]^T \in \mathbb{R}^2$.

After eliminating the unknowns $Ux_\ell^{(i,j,k)}, Uy_\ell^{(i,j,k)}, Uz_\ell^{(i,j,k)}$ from equations (17.50), we will have a block "seven-point" computational scheme with 2×2-blocks for the unknowns $UI_\ell^{(i,j,k)}$.

From (17.51) we have

$$\boldsymbol{Ux}^{(i,j,k)} = \frac{6}{12h}\boldsymbol{fx}^{(i,j,k)} + \frac{3}{4}\begin{bmatrix} 1 & 0 \\ 0 & 1/3 \end{bmatrix}\boldsymbol{UI}^{(i-1,j,k)} +$$
$$\frac{3}{4}\begin{bmatrix} 1/3 & 0 \\ 0 & 1 \end{bmatrix}\boldsymbol{UI}^{(i,j,k)}, \qquad i = \overline{2,n}, \quad j,k = \overline{1,n},$$

$$\boldsymbol{Uy}^{(i,j,k)} = \frac{6}{12h}\boldsymbol{fy}^{(i,j,k)} + \frac{3}{4}\begin{bmatrix} 1/3 & 0 \\ 0 & 1 \end{bmatrix}\boldsymbol{UI}^{(i,j-1,k)} +$$
$$\frac{3}{4}\begin{bmatrix} 1 & 0 \\ 0 & 1/3 \end{bmatrix}\boldsymbol{UI}^{(i,j,k)}, \qquad i,k = \overline{1,n}, \quad j = \overline{2,n},$$

$$\boldsymbol{Uz}^{(i,j,k)} = \frac{1}{h}\boldsymbol{fz}^{(i,j,k)} + \frac{1}{2}\boldsymbol{UI}^{(i,j,k-1)} + \frac{1}{2}\boldsymbol{UI}^{(i,j,k)},$$

$$i,j = \overline{1,n}, \quad k = \overline{2,n} \tag{17.52}$$

Substituting these expressions for $\boldsymbol{Ux}^{(i,j,k)}$, $\boldsymbol{Uy}^{(i,j,k)}$, and $\boldsymbol{Uz}^{(i,j,k)}$ into (17.50) we get

$$\begin{bmatrix} 16/3 & -2 \\ -2 & 16/3 \end{bmatrix}\boldsymbol{UI}^{(i,j,k)}$$

$$-(1-\delta_{i1})\left(\frac{1}{4}\boldsymbol{UI}^{(i-1,j,k)} + \begin{bmatrix} 3/4 & 0 \\ 0 & 1/12 \end{bmatrix}\boldsymbol{UI}^{(i,j,k)}\right)$$

$$-(1-\delta_{in})\left(\frac{1}{4}\boldsymbol{UI}^{(i+1,j,k)} + \begin{bmatrix} 1/12 & 0 \\ 0 & 3/4 \end{bmatrix}\boldsymbol{UI}^{(i,j,k)}\right)$$

$$-(1-\delta_{j1})\left(\frac{1}{4}\boldsymbol{UI}^{(i,j-1,k)} + \begin{bmatrix} 1/12 & 0 \\ 0 & 3/4 \end{bmatrix}\boldsymbol{UI}^{(i,j,k)}\right)$$

$$-(1-\delta_{jn})\left(\frac{1}{4}\boldsymbol{UI}^{(i,j+1,k)} + \begin{bmatrix} 3/4 & 0 \\ 0 & 1/12 \end{bmatrix}\boldsymbol{UI}^{(i,j,k)}\right)$$

$$-(1-\delta_{k1})\frac{1}{6}\left(\boldsymbol{UI}^{(i,j,k-1)}+\boldsymbol{UI}^{(i,j,k)}\right)-(1-\delta_{kn})\frac{1}{6}\left(\boldsymbol{UI}^{(i,j,k+1)}\right.$$

$$\left.+\boldsymbol{UI}^{(i,j,k)}\right) = \boldsymbol{F}^{(i,j,k)}, \qquad i,j,k=\overline{1,n}, \tag{17.53}$$

where

$$\boldsymbol{F}^{(i,j,k)} = \left(\frac{2}{3h}\right)\left\{\boldsymbol{fI}^{(i,j,k)}\right.$$
$$+(1-\delta_{i1})\frac{3}{4}\begin{bmatrix} 1/3 & 0 \\ 0 & 1 \end{bmatrix}\boldsymbol{fx}^{(i-1,j,k)} + (1-\delta_{in})\frac{3}{4}\begin{bmatrix} 1 & 0 \\ 0 & 1/3 \end{bmatrix}\boldsymbol{fx}^{(i+1,j,k)}$$
$$+(1-\delta_{j1})\frac{3}{4}\begin{bmatrix} 1 & 0 \\ 0 & 1/3 \end{bmatrix}\boldsymbol{fy}^{(i,j-1,k)} + (1-\delta_{jn})\frac{3}{4}\begin{bmatrix} 1/3 & 0 \\ 0 & 1 \end{bmatrix}\boldsymbol{fy}^{(i,j+1,k)}$$
$$\left.+(1-\delta_{k1})\frac{1}{2}\boldsymbol{fz}^{(i,j,k-1)} + (1-\delta_{kn})\frac{1}{2}\boldsymbol{fz}^{(i,j,k+1)}\right\}, \quad i,j,k=\overline{1,n} \tag{17.54}$$

Thus, to solve the linear system (17.46) with matrix $\tilde{\boldsymbol{B}}_1$, we first solve problem (17.53) with $(2n \times 2n)$-seven-block-diagonal matrix to obtain $\boldsymbol{u}_1$ and after that compute the vector $\boldsymbol{u}_2$ from (17.52).

Unfortunately, the matrix of linear system (17.53) has a rather complicated form which makes the solution rather difficult. Below we show that if on each prism instead of matrices (17.38) we introduce another matrix $\boldsymbol{B}_1^{(i,j,k)}$ then we will have a similar matrix (17.53). In this case we can use the method of separation of variables or another fast method to solve for $\boldsymbol{u}_1$.

Three-Level Preconditioner: Variant II

Instead of matrices (17.38), we introduce $\hat{\boldsymbol{B}}_1$ with

$$\hat{\boldsymbol{B}}_1 = \left(\frac{3h}{2}\right)\left[\begin{array}{rr|rrrrrr} 8/3 & -1 & -2/3 & 0 & -2/3 & 0 & -1/6 & -1/6 \\ -1 & 8/3 & 0 & -2/3 & 0 & -2/3 & -1/6 & -1/6 \\ \hline -2/3 & 0 & 2/3 & 0 & 0 & 0 & 0 & 0 \\ 0 & -2/3 & 0 & 2/3 & 0 & 0 & 0 & 0 \\ -2/3 & 0 & 0 & 0 & 2/3 & 0 & 0 & 0 \\ 0 & -2/3 & 0 & 0 & & 2/3 & 0 & 0 \\ -1/6 & -1/6 & 0 & 0 & 0 & 0 & 1/3 & 0 \\ -1/6 & -1/6 & 0 & 0 & 0 & 0 & 0 & 1/3 \end{array}\right] \tag{17.55}$$

This matrix is irreducible and has the same kernel as matrices $\boldsymbol{B}_{11,p}$, $p=1,2$.

PROPOSITION 5. *For any $P \in \mathcal{T}_P$ the eigenvalues of the spectral problems*

$$\boldsymbol{B}_{11,P}\boldsymbol{u} = \mu \hat{\boldsymbol{B}}_1 \boldsymbol{u}, \qquad (\hat{\boldsymbol{B}}_1, \boldsymbol{u}, \boldsymbol{u}) \neq 0 \tag{17.56}$$

belong to the interval

$$\mu \in \left[\frac{5}{11}(3-\sqrt{3}), \frac{3}{5}(3+\sqrt{3})\right]. \tag{17.57}$$

□

Again, we define the matrices $\hat{\boldsymbol{B}}_P$

$$\hat{\boldsymbol{B}}_P = \begin{bmatrix} \hat{\boldsymbol{B}}_1 + \boldsymbol{B}_{12,P}\boldsymbol{B}_{22,P}^{-1}\boldsymbol{B}_{21,P} & \boldsymbol{B}_{12,P} \\ \boldsymbol{B}_{21,P} & \boldsymbol{B}_{22,P} \end{bmatrix}, \qquad P \in \mathcal{T}_P \tag{17.58}$$

and consider the eigenvalue problems (17.41).

PROPOSITION 6. *The eigenvalues of problems (17.41) with the new block $\hat{\boldsymbol{B}}_1$ belong to the interval defined in (17.57). For the same reason, the eigenvalues of the spectral problems (17.42) for each prism satisfy*

$$\nu \in \left[\frac{5}{11}(3-\sqrt{3})(2-\sqrt{3}), \frac{3}{5}(3+\sqrt{3})(2+\sqrt{3})\right].$$

□

Now we define another symmetric positive-definite $N_1 \times N_1$-matrix $\hat{\boldsymbol{B}}_1$, using equality (17.43) with $\hat{\boldsymbol{B}}_P$ instead of $\tilde{\boldsymbol{B}}_P$. Then the new preconditioner $\hat{\boldsymbol{B}}$ is defined by

$$\hat{\boldsymbol{B}} = \begin{bmatrix} \hat{\boldsymbol{B}}_1 + \boldsymbol{B}_{12}\boldsymbol{B}_{22}^{-1}\boldsymbol{B}_{21} & \boldsymbol{B}_{12} \\ \boldsymbol{B}_{21} & \boldsymbol{B}_{22} \end{bmatrix}. \tag{17.59}$$

In the same way as before, using superelement analysis and Proposition 6, we have:

THEOREM 2. *Matrix $\hat{\boldsymbol{B}}$ defined in (17.59) with the new blocks $\hat{\boldsymbol{B}}_1$ in the form of (17.55) is spectrally equivalent to matrix $\boldsymbol{A}$; i.e.,*

$$\alpha \hat{\boldsymbol{B}} \leq \boldsymbol{A} \leq \beta \hat{\boldsymbol{B}} \tag{17.60}$$

with $\alpha = \frac{5}{11}(3-\sqrt{3})(2-\sqrt{3})$ and $\beta = \frac{3}{5}(3+\sqrt{3})(2+\sqrt{3})$. Therefore,

$$\operatorname{cond}(\hat{\boldsymbol{B}}^{-1}\boldsymbol{A}) \leq \nu, \tag{17.61}$$

where $\nu = 5(2+\sqrt{3})^2$. □

Let us now consider the linear system

$$\hat{\boldsymbol{B}}_1 \boldsymbol{u} = \boldsymbol{f}. \tag{17.62}$$

As in the treatment for matrix $\tilde{\boldsymbol{B}}_1$, matrix $\hat{\boldsymbol{B}}_1$ can be represented in block form as

$$\hat{\boldsymbol{B}}_1 = \begin{bmatrix} \boldsymbol{C}_{11} & \hat{C}_{12} \\ \hat{C}_{21} & \hat{C}_{22} \end{bmatrix}, \tag{17.63}$$

where block $\boldsymbol{C}_{11}$ coincides with the same block of matrix $\tilde{\boldsymbol{B}}_1$ (17.47), and matrix $\hat{C}_{22}$ is diagonal.

Computational Scheme: Variant II

Again, we write equations (17.62) explicitly in terms of the unknowns introduced in (17.31) for the case $a(x) = 1$:

$$\begin{bmatrix} 16/3 & -2 \\ -2 & 16/3 \end{bmatrix} \boldsymbol{UI}^{(i,j,k)} - (1-\delta_{i1})\frac{2}{3}\boldsymbol{Ux}^{(i-1,j,k)} - (1-\delta_{in})\frac{2}{3}\boldsymbol{Ux}^{(i,j,k)}$$

$$-(1-\delta_{j1})\frac{2}{3}\boldsymbol{Uy}^{(i,j-1,k)} - (1-\delta_{jn})\frac{2}{3}\boldsymbol{Uy}^{(i,j,k)}$$

$$-(1-\delta_{k1})\frac{1}{6}\begin{bmatrix} 1 & 1 \\ 1 & 1 \end{bmatrix} \boldsymbol{Uz}^{(i,j,k-1)} - (1-\delta_{kn})\frac{1}{6}\begin{bmatrix} 1 & 1 \\ 1 & 1 \end{bmatrix} \boldsymbol{Uz}^{(i,j,k)}$$

$$= \left(\frac{2}{3h}\right) \boldsymbol{fI}^{(i,j,k)}, \qquad i,j,k = \overline{1,n}, \tag{17.64}$$

$$\frac{4}{3}\boldsymbol{Ux}^{(i,j,k)} - \frac{2}{3}\boldsymbol{UI}^{(i-1,j,k)} - \frac{2}{3}\boldsymbol{UI}^{(i,j,k)} = \left(\frac{2}{3h}\right) \boldsymbol{fx}^{(i,j,k)},$$

$$i = \overline{2,n}, \quad j,k = \overline{1,n},$$

$$\frac{4}{3}\boldsymbol{Uy}^{(i,j,k)} - \frac{2}{3}\boldsymbol{UI}^{(i,j-1,k)} - \frac{2}{3}\boldsymbol{UI}^{(i,j,k)} = \left(\frac{2}{3h}\right) \boldsymbol{fy}^{(i,j,k)},$$

$$j = \overline{2,n}, \quad i,k = \overline{1,n},$$

$$\frac{2}{3}\boldsymbol{Uz}^{(i,j,k)} - \frac{1}{6}\begin{bmatrix} 1 & 1 \\ 1 & 1 \end{bmatrix}\boldsymbol{UI}^{(i,j,k-1)}$$
$$-\frac{1}{6}\begin{bmatrix} 1 & 1 \\ 1 & 1 \end{bmatrix}\boldsymbol{UI}^{(i,j,k)} = \frac{2}{3h}\boldsymbol{fz}^{(i,j,k)}, \quad k = \overline{2,n}, \quad i,j = \overline{1,n} \qquad (17.65)$$

Eliminating unknowns $Ux_\ell^{(i,j,k)}$, $Uy_\ell^{(i,j,k)}$, $Uz_\ell^{(i,j,k)}$, $\ell = 1,2$, from equations (17.64), we get a block "seven-point" scheme with 2×2-blocks for the unknowns $UI_\ell^{(i,j,k)}$, $\ell = 1,2$, $i,j,k = \overline{1,n}$. From (17.65) we have

$$\boldsymbol{Ux}^{(i,j,k)} = \frac{3}{4}\left(\frac{2}{3h}\right)\boldsymbol{fx}^{(i,j,k)} + \frac{1}{2}\boldsymbol{UI}^{(i-1,j,k)} + \frac{1}{2}\boldsymbol{UI}^{(i,j,k)},$$

$$i = \overline{2,n}, \quad j,k = \overline{1,n},$$

$$\boldsymbol{Uy}^{(i,j,k)} = \frac{3}{4}\left(\frac{2}{3h}\right)\boldsymbol{fy}^{(i,j,k)} + \frac{1}{2}\boldsymbol{UI}^{(i,j-1,k)} + \frac{1}{2}\boldsymbol{UI}^{(i,j,k)},$$

$$j = \overline{2,n}, \quad i,k = \overline{1,n},$$

$$\boldsymbol{Uz}^{(i,j,k)} = \frac{3}{2}\left(\frac{2}{3h}\right)\boldsymbol{fz}^{(i,j,k)} + \frac{1}{4}\begin{bmatrix} 1 & 1 \\ 1 & 1 \end{bmatrix}\boldsymbol{UI}^{(i,j,k-1)} +$$
$$+\frac{1}{4}\begin{bmatrix} 1 & 1 \\ 1 & 1 \end{bmatrix}\boldsymbol{UI}^{(i,j,k)}, \quad k = \overline{2,n}, \quad i,j = \overline{1,n}$$
$$(17.66)$$

Substituting (17.66) into (17.64), we obtain

$$\begin{bmatrix} 16/3 & -2 \\ -2 & 16/3 \end{bmatrix}\boldsymbol{UI}^{(i,j,k)} - (1-\delta_{i1})\frac{1}{3}\left(\boldsymbol{UI}^{(i-1,j,k)} + \boldsymbol{UI}^{(i,j,k)}\right)$$
$$-(1-\delta_{in})\frac{1}{3}\left(\boldsymbol{UI}^{(i+1,j,k)} + \boldsymbol{UI}^{(i,j,k)}\right)$$
$$-(1-\delta_{jn})\frac{1}{3}\left(\boldsymbol{UI}^{(i,j+1,k)} + \boldsymbol{UI}^{(i,j,k)}\right)$$
$$-(1-\delta_{j1})\frac{1}{3}\left(\boldsymbol{UI}^{(i,j-1,k)} + \boldsymbol{UI}^{(i,j,k)}\right)$$
$$-(1-\delta_{k1})\frac{1}{12}\begin{bmatrix} 1 & 1 \\ 1 & 1 \end{bmatrix}\left(\boldsymbol{UI}^{(i,j,k-1)} + \boldsymbol{UI}^{(i,j,k)}\right)$$

$$-(1-\delta_{kn})\frac{1}{12}\begin{bmatrix} 1 & 1 \\ 1 & 1 \end{bmatrix}\left(\boldsymbol{UI}^{(i,j,k+1)} + \boldsymbol{UI}^{(i,j,k)}\right) = \boldsymbol{F}^{(i,j,k)},$$

$$i,j,k=\overline{1,n} \qquad (17.67)$$

where

$$\boldsymbol{F}^{(i,j,k)} = \left(\frac{2}{3h}\right)\left\{\boldsymbol{fI}^{(i,j,k)} + (1-\delta_{i1})\frac{1}{2}\boldsymbol{fx}^{(i-1,j,k)} + (1-\delta_{in})\frac{1}{2}\boldsymbol{fx}^{(i,j,k)}\right.$$

$$+(1-\delta_{j1})\frac{1}{2}\boldsymbol{fy}^{(i,j-1,k)} + (1-\delta_{jn})\frac{1}{2}\boldsymbol{fy}^{(i,j,k)}$$

$$\left.+(1-\delta_{k1})\frac{1}{4}\begin{bmatrix} 1 & 1 \\ 1 & 1 \end{bmatrix}\boldsymbol{fz}^{(i,j,k-1)} + (1-\delta_{kn})\frac{1}{4}\begin{bmatrix} 1 & 1 \\ 1 & 1 \end{bmatrix}\boldsymbol{fz}^{(i,j,k)}\right\} \qquad (17.68)$$

To solve system (17.67) we introduce the rotation matrix

$$\boldsymbol{Q} = \frac{1}{\sqrt{2}}\begin{bmatrix} 1 & 1 \\ -1 & 1 \end{bmatrix} \qquad (17.69)$$

and new vectors $\boldsymbol{V}^{(i,j,k)} = (V_1^{(i,j,k)}\ V_2^{(i,j,k)})^T$, $i,j,k=\overline{1,n}$ such that

$$\boldsymbol{V}^{(i,j,k)} = \boldsymbol{QUI}^{(i,j,k)}, \qquad i,j,k=\overline{1,n} \qquad (17.70)$$

Then replacing $\boldsymbol{UI}^{(i,j,k)}$ in (17.67) by

$$\boldsymbol{UI}^{(i,j,k)} = \boldsymbol{Q}^T\boldsymbol{V}^{(i,j,k)}, \qquad i,j,k=\overline{1,n} \qquad (17.71)$$

and multiplying both sides of equation (17.67) by matrix $\boldsymbol{Q}$ we get the following problem for the unknowns $\boldsymbol{V}^{(i,j,k)}$:

$$\begin{bmatrix} 10/3 & 0 \\ 0 & 22/3 \end{bmatrix}\boldsymbol{V}^{(i,j,k)}$$

$$-(1-\delta_{i1})\frac{1}{3}\left(\boldsymbol{V}^{(i-1,j,k)} + \boldsymbol{V}^{(i,j,k)}\right) - (1-\delta_{in})\frac{1}{3}\left(\boldsymbol{V}^{(i+1,j,k)} + \boldsymbol{V}^{(i,j,k)}\right)$$

$$-(1-\delta_{j1})\frac{1}{3}\left(\boldsymbol{V}^{(i,j-1,k)} + \boldsymbol{V}^{(i,j,k)}\right) - (1-\delta_{jn})\frac{1}{3}\left(\boldsymbol{V}^{(i,j+1,k)} + \boldsymbol{V}^{(i,j,k)}\right)$$

$$-(1-\delta_{k1})\frac{1}{6}\begin{bmatrix} 1 & 0 \\ 0 & 0 \end{bmatrix}\left(\boldsymbol{V}^{(i,j,k-1)}+\boldsymbol{V}^{(i,j,k)}\right)$$

$$-(1-\delta_{kn})\frac{1}{6}\begin{bmatrix} 1 & 0 \\ 0 & 0 \end{bmatrix}\left(\boldsymbol{V}^{(i,j,k+1)}+\boldsymbol{V}^{(i,j,k)}\right)=Q\cdot\boldsymbol{F}^{(i,j,k)}=\tilde{\boldsymbol{F}}^{(i,j,k)},$$

$$i,j,k=\overline{1,n}. \tag{17.72}$$

It is easy to see that problem (17.72) is decomposed into the following two independent problems:

$$10V_1^{(i,j,k)}-(1-\delta_{i1})\left(V_1^{(i-1,j,k)}+V_1^{(i,j,k)}\right)-(1-\delta_{in})\left(V_1^{(i+1,j,k)}+V_1^{(i,j,k)}\right)$$

$$-(1-\delta_{j1})\left(V_1^{(i,j-1,k)}+V_1^{(i,j,k)}\right)-(1-\delta_{jn})\left(V_1^{(i,j+1,k)}+V_1^{(i,j,k)}\right)$$

$$-(1-\delta_{k1})\frac{1}{2}\left(V_1^{(i,j,k-1)}+V_1^{(i,j,k)}\right)-(1-\delta_{kn})\frac{1}{2}\left(V_1^{(i,j,k+1)}+V_1^{(i,j,k)}\right)$$

$$=3\tilde{F}_1^{(i,j,k)}, \qquad i,j,k=\overline{1,n} \tag{17.73}$$

and

$$-22V_2^{(i,j,k)}-(1-\delta_{i1})\left(V_2^{(i-1,j,k)}+V_2^{(i,j,k)}\right)-(1-\delta_{in})\left(V_2^{(i+1,j,k)}+V_2^{(i,j,k)}\right)$$

$$-(1-\delta_{j1})\left(V_2^{(i,j-1,k)}+V_2^{(i,j,k)}\right)-(1-\delta_{jn})\left(V_2^{(i,j+1,k)}+V_2^{(i,j,k)}\right)$$

$$=3\tilde{F}_2^{(i,j,k)}, \qquad i,j=\overline{1,n}, \quad \forall\, k=\overline{1,n} \tag{17.74}$$

That is, the linear system (17.67) of dimension $2n^3$ is reduced to one linear system of equations (17.73) of dimension n^3 and n linear systems of equations (17.74) of dimension n^2. For all these problems the method of separation of variables can be used. After we find the solution of these problems, vectors $\boldsymbol{UI}^{(i,j,k)}$ can be easily retrieved by using the relations (17.71).

The Method of Separation of Variables

Let us consider the method of separation of variables for problems (17.73) and (17.74). Problem (17.73) can be represented in the form

$$\boldsymbol{C}^{(3)}\boldsymbol{V}=\boldsymbol{f}, \qquad \boldsymbol{V},\boldsymbol{f}\in\mathbb{R}^{(n^3)} \tag{17.75}$$

with the matrix $\boldsymbol{C}^{(3)} = \frac{1}{2}\boldsymbol{C}_0 \otimes \boldsymbol{I}_0 \otimes \boldsymbol{I}_0 + \boldsymbol{I}_0 \otimes \boldsymbol{C}_0 \otimes \boldsymbol{I}_0 + \boldsymbol{I}_0 \otimes \boldsymbol{I}_0 \otimes \boldsymbol{C}_0$, where $\boldsymbol{I}_0$ is an $(n \times n)$-identity matrix and $(n \times n)$-matrix $\boldsymbol{C}_0$ has the form

$$\boldsymbol{C}_0 = \frac{1}{3}\begin{bmatrix} 3 & -1 & & & \\ -1 & 2 & -1 & & \\ & \ddots & \ddots & \ddots & \\ & & -1 & 2 & -1 \\ & & & -1 & 3 \end{bmatrix}. \tag{17.76}$$

If we represent matrix $\boldsymbol{C}_0$ in the form $\boldsymbol{C}_0 = \boldsymbol{Q}_0\boldsymbol{\Lambda}_0\boldsymbol{Q}_0^T$, where $\boldsymbol{\Lambda}_0$ is an $(n\times n)$-diagonal matrix and $\boldsymbol{Q}_0$ is an $(n \times n)$-orthogonal matrix ($\boldsymbol{Q}_0^{-1} = \boldsymbol{Q}_0^T$), then matrix $\boldsymbol{C}^{(3)}$ can be rewritten as

$$\boldsymbol{C}^{(3)} = \boldsymbol{Q}^{(3)}\boldsymbol{\Lambda}^{(3)}\boldsymbol{Q}^{(3)T}, \tag{17.77}$$

where

$$\boldsymbol{Q}^{(3)} = \boldsymbol{Q}_0 \otimes \boldsymbol{Q}_0 \otimes \boldsymbol{Q}_0,$$

$$\boldsymbol{\Lambda}^{(3)} = \frac{1}{2}\boldsymbol{\Lambda}_0 \otimes \boldsymbol{I}_0 \otimes \boldsymbol{I}_0 + \boldsymbol{I}_0 \otimes \boldsymbol{\Lambda}_0 \otimes \boldsymbol{I}_0 + \boldsymbol{I}_0 \otimes \boldsymbol{I}_0 \otimes \boldsymbol{\Lambda}_0.$$

Note that $\boldsymbol{Q}^{(3)}$ is an $(n^3 \times n^3)$-orthogonal matrix and $\boldsymbol{\Lambda}^{(3)}$ is an $(n^3 \times n^3)$-diagonal matrix.

Then the system (17.75) can be solved using the following steps

$$\tilde{\boldsymbol{f}} = \left[\boldsymbol{Q}^{(3)}\right]^T \boldsymbol{f}; \quad \boldsymbol{\Lambda}^{(3)}\boldsymbol{W} = \tilde{\boldsymbol{f}}; \quad \boldsymbol{V} = \boldsymbol{Q}^{(3)}\boldsymbol{W} \tag{17.78}$$

Similarly, problem (17.74) can be rewritten in the form

$$\boldsymbol{C}^{(2)}\boldsymbol{u} = \boldsymbol{b}, \qquad \boldsymbol{u}, \boldsymbol{b} \in \mathbb{R}^{(n^2)} \tag{17.79}$$

with the matrix

$$\boldsymbol{C}^{(2)} = \boldsymbol{K}_0 \otimes \boldsymbol{I}_0 + \boldsymbol{I}_0 \otimes \boldsymbol{K}_0,$$

where $(n \times n)$-matrix $\boldsymbol{K}_0$ has the form

$$\boldsymbol{K}_0 = \frac{1}{3}\begin{bmatrix} 10 & -1 & & & \\ -1 & 9 & -1 & & \\ & \ddots & \ddots & \ddots & \\ & & -1 & 9 & -1 \\ & & & -1 & 10 \end{bmatrix}. \tag{17.80}$$

Representing matrix $\boldsymbol{K}_0$ in the form

$$\boldsymbol{K}_0 = \boldsymbol{R}_0 \boldsymbol{D}_0 \boldsymbol{R}_0^T,$$

where $\boldsymbol{D}_0$ is an $(n \times n)$-diagonal matrix and $\boldsymbol{R}_0$ is an $(n \times n)$-orthogonal matrix, we can rewrite matrix $\boldsymbol{C}^{(2)}$ in the form

$$\boldsymbol{C}^{(2)} = \boldsymbol{Q}^{(2)} \boldsymbol{\Lambda}^{(2)} \boldsymbol{Q}^{(2)^T},$$

where $\boldsymbol{Q}^{(2)} = \boldsymbol{R}_0 \otimes \boldsymbol{R}_0$ and $\boldsymbol{\Lambda}^{(2)} = \boldsymbol{D}_0 \otimes \boldsymbol{I}_0 + \boldsymbol{I}_0 \otimes \boldsymbol{D}_0$. Then, the above algorithms are again applied:

$$\tilde{\boldsymbol{b}} = \left[\boldsymbol{Q}^{(2)}\right]^T \boldsymbol{b}, \quad \boldsymbol{\Lambda}^{(2)} \boldsymbol{W} = \tilde{\boldsymbol{b}}, \quad \boldsymbol{u} = \boldsymbol{Q}^{(2)} \boldsymbol{W}. \tag{17.81}$$

Preconditioned Conjugate Gradient Method

We solve system (17.17) by a preconditioned conjugate gradient method:

$$\begin{gathered} \boldsymbol{u}_0 = 0, \quad \boldsymbol{u}^{(k+1)} = \boldsymbol{u}^k - \frac{1}{P_k}\left[\tilde{\boldsymbol{B}}^{-1}\boldsymbol{\xi}^k - d_{k-1}(\boldsymbol{u}^k - \boldsymbol{u}^{k-1})\right], \\ \boldsymbol{\xi}^k = \boldsymbol{A}\boldsymbol{u}^k - \boldsymbol{f}, \; P_k = \frac{\|\boldsymbol{B}^{-1}\boldsymbol{\xi}^k\|_A}{\|\boldsymbol{\xi}^k\|^2_{\boldsymbol{B}^{-1}}} - d_{k-1}, \; d_k = P_k \frac{\|\boldsymbol{\xi}^{k+1}\|^2_{\boldsymbol{B}^{-1}}}{\|\boldsymbol{\xi}^k\|^2_{\boldsymbol{B}^{-1}}}, \\ k = 0, 1, \ldots, k_\varepsilon; \qquad k_{-1} = 0. \end{gathered} \tag{17.82}$$

It is well known that, for a given tolerance ε and convergence defined by

$$\|\boldsymbol{u}^{k_\varepsilon+1} - \boldsymbol{u}^*\|_A \leq \varepsilon \|\boldsymbol{u}^\circ - \boldsymbol{u}^*\|_A, \tag{17.83}$$

where $\boldsymbol{u}^* = \boldsymbol{A}^{-1}\boldsymbol{f}$ and $\boldsymbol{u}^\circ \in \mathbb{R}^N$ is any initial vector, the number of iterations K_ε satisfy

$$K_\varepsilon > \frac{\ell n \left(\frac{\varepsilon}{2}\right)}{\ell n \; q}, \tag{17.84}$$

where $q = (\sqrt{\nu} - 1)/(\sqrt{\nu} + 1)$ with the value of ν defined in (17.61).

THEOREM 3. *The number of operations for solving system (17.17) by method (17.82) with matrix $\hat{\boldsymbol{B}}$ defined in (17.59) with accuracy ε in the sense of (17.83) is given by the expression $cN^{4/3} \ln \left(\frac{2}{\varepsilon}\right)$, where the constant $c > 0$ does not depend on N.* □

For example, if $\varepsilon = 10^{-6}$ then $K_\varepsilon \leq 60$ iterations.

Table 17.1: Comparison performance for preconditioning

		CG WITHOUT PRECONDITIONING			CG WITH PRECONDITIONING		
n	N	n_{iter}	cond	time (sec)	n_{iter}	cond	time (sec)
4	672	40	66	0.18	22	9.84	0.22
8	5760	73	265	2.18	24	10.7	1.27
16	47616	130	1062	49.2	24	11.94	15.7
32	387072	200<	—	1248	25	12.2	163
40	758400				25	12.26	376
50	1485000				25	12.33	771

17.5 Results of the Numerical Experiments

The method of preconditioning, on the basis of multilevel substructuring as discussed above, was tested on the model problem

$$-\Delta u = f, \quad \text{in } \Omega = (0,1)^3 \subset \mathbb{R}^3, \qquad u|_{\partial\Omega} = 0 \tag{17.85}$$

with the nonconforming finite element method. The domain was divided into n^3 cubes (n in each direction) and each cube was partitioned into 6 tetrahedra. The total dimension of the original algebraic system was $N = 12n^3 - 6n^2$. The right hand side was generated randomly. The original algebraic problem has been solved by the conjugate gradient method in (17.82) with the preconditioner (17.59) to an accuracy $\varepsilon = 10^{-6}$. For comparison this problem has also been solved by the same method without preconditioning. The condition number of matrix $\boldsymbol{B}^{-1}\boldsymbol{A}$ was calculated from the relation between conjugate gradients and Lanczos algorithm [13]. The method was implemented in FORTRAN-77 in double precision. All experiments were carried out on a Sun Workstation. The results are summarized in Table 17.1.

References

[1] Arbogast, T. and Z. Chen, "On the implementation of mixed methods as nonconforming methods for second order elliptic problems", IMA Preprint #1172, to appear in *Math. Comp.*, 1995.

[2] Arnold, D.N. and F. Brezzi, "Mixed and nonconforming finite element methods: implementation, postprocessing and error estimates", *RAIRO, Model. Math. Anal. Numer.*, 19, 7–32, 1985.

[3] Bramble, J., R. Ewing, J. Pasciak, and J. Shen, "Analysis of the multigrid algorithms for cell-centered finite difference approximations", submitted to *Advances in Comput. Math*,1994.

[4] Bramble, J. and J.E. Pasciak, "A preconditioning technique for indefinite systems resulting from mixed approximations of elliptic problems", *Math. Comp.*, 50 , 1–17, 1988.

[5] Brenner, S.C., "A multigrid algorithm for the lowest-order Raviart-Thomas mixed triangular finite element method", *SIAM J. Numer. Anal.*, 29, 647–678, 1992.

[6] Brezzi, F. and M. Fortin, *Mixed and hybrid finite element methods*, Springer-Verlag, New York, Berlin, Heidelberg, 1991.

[7] Chen, Z. "Analysis of mixed methods using conforming and nonconforming finite element methods", *RAIRO, Math. Model. Numer. Anal.*, 27, 9–34, 1993. Raviart

[8] Chen, Z. "Multigrid algorithms for mixed methods for second order elliptic problems", *submitted to Math.Comp.*

[9] Ciarlet, P.G. *The finite element method for elliptic problems*, North-Holland, Amsterdam, New York, Oxford, 1978.

[10] Cowsar, L.C. "Domain decomposition methods for nonconforming finite elements spaces of Lagrange-type", *Preprint TR93-11*, March 1993.

[11] Ewing, R.E. and M.F. Wheeler, "Computational aspects of mixed finite element methods", in: *Numerical Methods for Scientific Computing*, R.S. Stepleman, ed., North-Holland, New York, 163–172, 1983.

[12] Falk, R. and J. Osborn, "Error estimates for mixed methods", *RAIRO, Model. Math. Anal. Numer.*, 14, 249–277, 1980

[13] Golub, G.M. and C.F. Van Loan, *Matrix computations*, Johns Hopkins University Press, 1989.

[14] Grisvard, P. *Elliptic problems in nonsmooth domains*, Pitman, Boston, Massachusetts, 1985.

[15] Kuznetsov, Y.A. "Multigrid domain decomposition methods", *Proceedings of 3rd international symposium on domain decomposition methods*, 1989 (T. Chan, R. Glowinski, J. Periaux, O. Widlund, Eds), SIAM, Philadelphia, 290–313,1990.

[16] Raviart, P.A. and J.M. Thomas, "A mixed finite element method for second order elliptic problems", in: *Mathematical aspects of the FEM*, Lecture Notes in Mathematics, 606, Springer-Verlag, Berlin and New York, 292–315, 1977.

[17] Rusten, T. and R. Winther, "A preconditioned iterative method for saddle-point problems", *SIAM J. Matrix Anal. Appl.*, 13, 887–904,1992.

[18] Sarkis, M. "Two-level Schwarz methods for nonconforming finite elements and discontinuous coefficients", *Preprint*, 1993.

[19] Vassilevski, P. and R. Lazarov, "Preconditioning saddle-point problems arising from mixed finite element discretization of elliptic equations", UCLA CAM Report 92-46, 1992.

[20] Vassilevski, P. and J. Wang, "Multilevel iterative methods for mixed finite element discretizations of elliptic problems", *Numer. Math.*, 63, 503–520,1992.

[21] Weiser, A. and M. Wheeler, "On convergence of block-centered finite-differences for elliptic problems", *SIAM J. Numer. Anal.*, 25, 351–375,1988.

Chapter 18

Error Estimates for Saturated Groundwater Flows

S.-S. Chow[1]

18.1 Introduction

In this chapter we briefly review some error estimates for the finite element solutions to some boundary value problems arising from steady state saturated groundwater flow applications.

In the first part we examine models for which Darcy's law provides an adequate description, and discuss the well known finite element error estimates for the standard variational formulation. In the second part we turn to situations where the groundwater flows do not satisfy Darcy's law and Forchheimer's law and Missbach's law are introduced. Some basic ideas, such as continuity and monotonicity properties of the operator associated with the boundary value problem, which lead to the establishment of various sub-optimal and optimal finite element error estimates are discussed. In the last part, results relating to the mixed method applied to non-Darcian flow problems will be discussed.

[1]Currently visiting Univ. of Texas of the Permian Basin, Odessa, TX.

18.2 Darcy's Law

Darcy's law [6] is most often used in the description of fluid motion in a statistically isotropic and homogeneous porous medium. For many practical applications, especially those in which inertial effects are negligible and the flow is in the prelaminar regime, the relationship between hydraulic gradient and the flow velocity is basically linear. Darcy's law, when coupled with mass conservation and appropiate boundary conditions, then provides an adequate mathematical model from which valuable information may be deduced.

More specifically, Darcy's law states that the flow velocity is proportional to the drop in hydraulic gradient, when the effect of gravity is neglected. Thus

$$V = -K\nabla h \tag{18.1}$$

where V is the flow velocity, h is the piezometric head and K is the permeability of the porous medium. The anisotropic form of Darcy's law has K as a tensor.

For steady state flows, we also have the mass conservation requirement

$$\nabla \cdot V = f(x) \tag{18.2}$$

where f is the source/sink term. Combining this with Darcy's law and assuming that head values are prescribed on the boundary $\partial\Omega$ of the region of interest $\Omega \subset I\!R^n, n \geq 1$, we obtain the familiar second order linear boundary value problem

$$\nabla \cdot (K\nabla h) = f(x) \text{ in } \Omega \tag{18.3}$$

$$h = h_0 \text{ on } \partial\Omega \tag{18.4}$$

Before stating the variational formulation for the above boundary value problem, we first introduce some notations. For a given domain $\mathcal{O}$, the Sobolev space $W^{k,p}(\mathcal{O})$, $k = \{0, \pm 1, \pm 2, \ldots\}$, $p \in [1, \infty]$, is equipped with the usual norm $\|\cdot\|_{k,p,\mathcal{O}}$ and seminorm $|\cdot|_{k,p,\mathcal{O}}$. We shall omit the index p when $p = 2$, the index $\mathcal{O}$ when $\mathcal{O} = \Omega$ and all the indices when $k = 0, p = 2, \mathcal{O} = \Omega$. We write $H^k(\mathcal{O})$ for $W^{k,2}(\mathcal{O})$ Since the distinction should be clear from the context, we use $(\cdot, \cdot)$ to denote both the $L^2(\mathcal{O})$ inner product and the $H^{-1}(\mathcal{O}) \times H_0^1(\mathcal{O})$ duality paring. We also use C and c (occasionally with subscripts) to denote arbitrary constants.

Variational Formulation

By multiplying the equation in the boundary value problem by a test function and applying integration by parts we obtain the variational problem: Find the head u with $u - h_0 \in H_0^1(\Omega)$ such that

$$a(u,v) = (f,v) \quad \text{for all } v \in H_0^1(\Omega) \tag{18.5}$$

where

$$a(u,v) = \int_\Omega K\nabla u \nabla v \, dx \tag{18.6}$$

It is well known [9] that the finite element approximation u_h to the above problem satisfies the error estimates

$$||u - u_h||_1 \leq Ch^k ||u||_{k+1} \tag{18.7}$$

$$||u - u_h|| \leq Ch^{k+1} ||u||_{k+1} \tag{18.8}$$

whenever u has sufficient regularity. Here k denotes the degree of the finite element.

These error estimates are obtained under two important properties of the bilinear form $a(\cdot,\cdot)$, namely continuity

$$|a(w,v)| \leq C_1 ||w||_1 ||v||_1 \tag{18.9}$$

and coercivity

$$a(v,v) \geq c_1 ||v||_1^2 \tag{18.10}$$

which when combined leads to the abstract error estimate

$$||u - u_h||_1 \leq C \inf ||u - v||_1, \quad for \ v \in H_0^1(\Omega). \tag{18.11}$$

Approximation theory for finite elements may then be applied to derive the error estimates listed above.

18.3 Non-Darcian Law

When Darcy's law is applied to the study of fluid flow in porous media, accurate descriptions of the flow conditions are obtained only in cases when the flow rate under consideration is in the prelaminar or creeping flow regime. For higher flow velocities, it has been recognized since the last century [11] that Darcy's law fails to remain valid. The inertial effects, neglected in Darcy's law, must now be taken into consideration. This inclusion of inertial effects is also important in porous media with large grain size. Thus in the study of flow through rockfill dams and flow in the area adjacent to a pumping well in a coarse-grained aquifer, it is necessary to employ a modified version of Darcy's law to describe the relation between the head loss and the velocity in order to obtain realistic solutions [24], [32], [20], [22].

For incompressible flows in which the inertial effects become significant, Forchheimer's law and Missbach's law are usually chosen to model the nonlinear effect: Let h denote the hydraulic head and q_s denote the velocity in the direction s, Forchheimer's law [15] may be stated as

$$-\frac{\partial h}{\partial s} = aq_s + bq_s^2 \tag{18.12}$$

while Missbach's law [26] is

$$-\frac{\partial h}{\partial s} = cq^{n+1} \tag{18.13}$$

where a, b, c, and n are positive parameters determined from experimental data. If we assume that the porous medium under consideration is statistically isotropic, we may modify these flow laws to a form that is invariant to coordinate transformation, with respectively,

$$-\nabla h = (a + b|\mathbf{q}|)\mathbf{q} \tag{18.14}$$

and

$$-\nabla h = C|\mathbf{q}|^n\mathbf{q} \tag{18.15}$$

where $\mathbf{q}$ is the superficial or average seepage velocity.

More generally, one may consider the nonlinear seepage law [27]

$$-\nabla h = (a + b|\mathbf{q}|^n)\mathbf{q} \tag{18.16}$$

where a, b, and n are positive constant parameters.

Noting that the function

$$s \mapsto \ell(s)s \equiv (a + b|s|^n)s$$

is a strictly increasing function on the set of all nonnegative numbers $\mathbb{R}^+$, its inverse $k(x,\cdot)\cdot$ may be defined as follows:

$$\text{for } \ a.e.\ x \ \text{ in } \ \Omega\ , \quad t = \ell(x,s)s \ \text{ iff } \ s = k(x,t)t\ . \tag{18.17}$$

Combining the general non-Darcy law with the mass balance equation (18.2) and assuming again that the head value is prescribed, we can make use of the above inverse relation to obtain a nonlinear boundary value problem of the form

$$-\nabla \cdot (k(|\nabla u|)\nabla u) = f(x) \text{ in } \Omega \quad (18.18)$$

$$u = u_0 \text{ on } \partial\Omega \quad (18.19)$$

When $a = 0$ or $n = 1$ in (18.16), the coefficient $k(\cdot)$ is known explicitly. However, in general the exact algebraic form of the principal coefficient is difficult, if not impossible, to determine. Nonetheless, its unique existence is assured due to the strictly increasing property of $\ell(\cdot)\cdot$, and, in the course of computing, one may numerically evaluate it from the known expression for ℓ. Boundary value problems with a nonlinear principal coefficient dependent on the gradient of the solution occur in many diverse applications apart from non-Darcian flows — some typical examples include gas flow in blast furnace, magnetostatic field in ferrous material, minimal surface equation and quasi-Newtonian flows.

Variational Formulation

The nonlinear boundary value problem may be put in the weak form in some appropiate Sobolev space V whose choice is dependent on the structure of the principal coefficient. Then (18.5) becomes

$$(Au, v) \equiv \int_\Omega k(|\nabla u|)\nabla u \cdot \nabla v = (f, v) \text{ for all } v \in V \quad (18.20)$$

As in the case of linear problems, error estimates for the finite element approximation may be obtained from an abstract error estimate and finite element approximation theory results. The abstract estimate is developed from the continuity property and the monotonicity property (which may be considered to be a generalization of the coercivity property) of the abstract operator A. These properties may be established by exploiting the algebraic structure of the principal coefficient. For example, let p be a fixed constant in (1,2] and related to n in (18.16) and suppose the nonlinear coefficient $k(\cdot)$ has the following properties:

1. $k(\cdot)$ is a positive continuous function on $\mathbb{R}^+$, with upper bound $\hat{k} > 0$.
2. $k(t)t$ is a strictly increasing function of t and vanishes at $t = 0$.
3. $k(\cdot)\cdot$ is Hölder continuous with exponent $p - 1$.

4. $k(t)t$ is continuously differentiable for all $t > 0$ and satisfies the inequality

$$(K_1 + K_2 t^{2-p})\frac{d}{dt}\Big(k(t)t\Big) \geq C_1 > 0 \quad \text{for all} \quad t > 0 \tag{18.21}$$

where K_1, K_2, and C_1 are positive constants.

Then the abstract operator is Hölder continuous with exponent $p-1$

$$||Aw - Av||^* \leq C||w - v||_{1,p}^{p-1} \text{ for all } v, w \in V, \tag{18.22}$$

where $||\cdot||^*$ denotes the dual norm of $||\cdot||_{1,p}$, and strongly monotone, i.e. for all $v, w \in V$,

$$(K_1 + K_2(||w||_{1,p} + ||v||_{1,p})^{2-p})(Aw - Av, w - v) \geq \alpha||w - v||_{1,p}^{p}. \tag{18.23}$$

These important properties of the abstract operator not only guarantee the unique solvability of the weak solution but also allow us to derive the abstract error estimates

$$||u - u_h||_{1,p} \leq C \inf\{||u - v_h||_{1,p}^{\frac{1}{3-p}}; v_h \in V_h\} \tag{18.24}$$

where V_h is an appropriately chosen conforming finite element subspace of V. Details of these results may be found in [7], [9].

It is obvious that the above results leads to suboptimal estimates even when the weak solution is regular.

Improved error estimates may be obtained by observing that the weak formulation and the finite element system can also be posed as functional minimization problems over V and V_h respectively, and that the derivative of the functional is closely related to the abstract operator described above [31]. More specifically, let

$$J(v) = \int_\Omega \int_0^{|\nabla v|} k(t)t\,dt\,dx - \int_\Omega f v\,dx \tag{18.25}$$

The weak solution $u \in V$ and the finite element approximation $u_h \in V_h$ are obtained by minimizing $J(\cdot)$ over V and V_h respectively. Consequently, with the aid of the mean value theorem it can be shown that for $p \in (1,2]$,

$$||u - u_h||_{1,p} \leq C \inf\{||u - v_h||_{1,p}^{\frac{p}{2}}; v_h \in V_h\} \tag{18.26}$$

This is obviously an improvement to the previous abstract error estimate. However, it is also clear that the resulting finite element error estimate is still sub-optimal.

For problems involving the Forchheimer law, optimal error estimates for the finite element approximations may be obtained for the case $a > 0$ provided that the weak solution has a stronger regularity property, viz. $u \in W^{1,\infty}(\Omega) \cap V$. Then for the problem discussed earlier, and for all $w_h \in V^h \cap W^{1,2}(\Omega)$,

$$||u - u_h||_{1,s} \leq C||\nabla(u - w_h)||_{L^2} \tag{18.27}$$

where $s = p$ or 2. Note that the norm appearing on the right hand side in the above abstract error estimate is always the same, regardless of the choice of s. The condition $a > 0$ allows us to conclude that the coefficient and hence the abstract operator are uniformly bounded above. The strengthened regularity requirement gives rise to the fact that the coefficient is bounded below. See [8] for details.

For seepage flow problems best described by Missbach's law, Barrett and Liu [3] showed how optimal finite element error estimates may be obtained. In short, they showed that the apparent degeneracy in (Au, w) is in fact removable given a rather strong regularity assumption on the weak solution u.

Error estimates for the finite element approximations in other norms for related problems with a gradient nonlinearity in the principal coefficient have also also been established for many specific cases, see e.g. [1], [10], [12], [14], [17], [18], [19], [21], [25], [28].

18.4 Mixed Formulation

In many applications such as contaminant transport modeling and petroleum reservoir simulation, the quantity of interest is the flow velocity rather than the head value, since it is the velocity variable that links the coupled equation system in the mathematical model. The velocity values computed by numerically differentiating the head values obtained from the finite element solution of (18.3) and (18.4) suffer a deterioration in accuracy and may thus lead to non-physical results when coupled with the transport equation. As an alternative, the mixed finite element method may be used to compute the head and velocity values simultaneously while providing a better accuracy for the velocity unknown. Furthermore, for problem (18.20) where the principal coefficient $k(\cdot)\cdot$ is not known explicitly, the mixed method also provides an excellent choice as the known function ℓ rather than k is being used in the formulation.

For flow problems where Darcy's law is applicable, error estimates for the finite element approximations are well known (see Fortin and Brezzi [16]).

For non-Darcian flows, some early sub-optimal results for the finite element error estimates were provided by Scheurer [29]. Improved results may be obtained using the techniques of Barrett and Liu ([4], [5]).

Variational Inequality

Variational inequality is a useful tool in dealing with a free surface formed in flows in dams. For Darcian flows, the "linear" variational inequality problem has been studied extensively by Baiocchi *et al* [2].

In [29], Scheurer considered the abstract problem closely related to the Missbach seepage law:

$$(Au, v-u) \geq (f, v-u) \text{ for all } v \in K,$$

$$\text{where } K = \{v \in V; Bv = g\}, B: \text{ linear}$$

Under standard boundedness and inf-sup conditions on the operator B he derived the error estimate

$$||u-u_h||_{1,p}^{p-1} \leq C \inf_{w_h} ||u-w_h|| + C \inf_{q_h} ||\lambda - q_h|| \qquad (18.28)$$

where λ is the Lagrange multiplier for the above problem. Consequently, for linear elements, the energy norm error estimate is

$$||u-u_h||_{1,p} = \mathcal{O}(h^{\frac{1}{p-1}}) \qquad (18.29)$$

Shen [30] considered a similar problem but with a boundary obstacle constraint and derived essentially the same error estimate for the linear finite element approximation.

More recently, Liu and Barrett [23] obtained improved error estimates based on some quasi-norm for the finite element solutions of the variational inequality problem.

18.5 Conclusion

Mathematical models for saturated non-Darcian groundwater flows often lead to a certain class of boundary value problems with a gradient nonlinearity in the principal coefficient. If the finite element method is employed, it is highly desirable to have error estimates for the computed result. In order to arrive at these estimates, it is necessary to exploit the algebraic structure of the nonlinear coefficient to establish monotonicity and continuity of the associated abstract operator. Under certain regularity assumptions on the weak solution, it may be shown that near-optimal or optimal results are obtained.

References

[1] Axelsson,O., "Error estimates for Galerkin method for quasilinear parabolic and elliptic differential equations", *Numer. Math.*, 28, 1–14, 1977.

[2] Baiocchi,C., F. Brezzi, and V. Comincioli, "Free boundary problems in fluid flow through porous media", in *Finite Elements in Fluids*, Volume 3, R. Gallagher *et al* (eds), 283–291, Wiley, Chichester, 1978.

[3] Barrett, J.W., and W.B. Liu, "Finite element approximation of the p-Laplacian", *Mathematics of Computation* , 61, 204, pg 523, 1993.

[4] Barrett, J. W., and W.B. Liu,"Finite element error analysis of a quasi-Newtonian flow obeying the Carreau or power law", *Numerische Mathematik*, 64, 4, pg 433, 1993.

[5] Barrett, J. W., and W.B. Liu, "Quasi-norm error bounds for the finite element approximation of a non-Newtonian flow", *Numerische Mathematik*, 68, 4, pg 437, 1994.

[6] Bear, J., *Dynamics of fluids in porous media* , American Elsevier Pub. Co., New York, 1972.

[7] Chow,S.-S., "Finite element error estimates for non-linear elliptic equations of monotone type", *Numerische Mathematik*, 54, 373–393, 1989.

[8] Chow, S.-S., "Finite element error estimates for a blast furnace gas flow problem", *SIAM J. Numerical Analysis*, 29, 3, 769–780, 1992.

[9] Ciarlet, P.G., *The finite element method for elliptic problems*, North-Holland, Amsterdam, 1978.

[10] Dobrowolski, M., and R. Rannacher, "Finite Element Methods for Non-linear Elliptic Systems of Second Order", *Math. Nachr.*, 94, 155–172, 1980.

[11] Dupuit, J., *Etudes Théoriques et Pratiques sur le Mouvement des Eaux*, Dunod, Paris, 1863.

[12] Feistauer, M., and A. Ženišek, "Finite element solution of nonlinear elliptic problems", *Numerische Mathematik*, 50, 451–475, 1987.

[13] Feistauer, M., and A. Ženišek, "Compactness method in the finite element theory of nonlinear elliptic problems", *Numerische Mathematik*, 52, 147–163, 1988.

[14] Fix, G.J., and B. Neta, "Finite element approximation of a nonlinear diffusion problem", *Comp. and Maths. with Applications*, 3, 287–298, 1977.

[15] Forchheimer, P., "Wasserbewegung durch Boden", *Z. Ver. dt. Ing.*, 45,1782–1788, 1901.

[16] Fortin, M., and F. Brezzi, *Mixed and hybrid finite element methods*, Springer-Verlag, New York, 1991.

[17] Frehse, J., "Eine gleichmäßige asymptotische Fehlerabschätzung zur Methode der finiten Elemente bei quasilinearen elliptischen Randwertproblemen", in *Theory of Nonlinear Operators: Constructive Aspects*, Tagungsband der Akademie der Wissenschaften, Berlin (DDR), 1976.

[18] Frehse, J., and R. Rannacher, "Asymptotic L^∞-error estimates for linear finite element approximations of quasilinear boundary value problems", *SIAM J. Numer. Anal.*, 15, 418–431, 1978.

[19] Frehse, J., and R. Rannacher, "Optimal uniform convergence for the finite element approximation of a quasi-linear elliptic boundary value problem", in *Formulation and Computational Algorithms in Finite Element Analysis*, Chapter 27, 793–812, MIT, Cambridge, MA, 1976.

[20] Hannoura, A.A., and F.B.J. Barends, "Non-Darcy flow; A state of the art, in flow and transport in porous media", *Proceedings of Euromech 143*, A. Verruijt *et al* (eds), Rotterdam, 37–51, 1981.

[21] Johnson, C. and V. Thomée, "Error estimates for a finite element approximation of a minimal surface", *Math. Comp.*, 29, 343–349, 1975.

[22] Joseph,D.D., D.A. Nield and G. Papanicolaou, "Nonlinear equation governing flow in a saturated porous medium",*Water Resources Research*, 18, 4, 1049–1052, 1982 and 19, 2, 591, 1983.

[23] Liu, W. B., and J.W. Barrett, "Quasi-norm error bounds for the finite element approximation of some degenerate quasilinear elliptic equations and variational inequalities", *RAIRO. Mathematical Modelling and Numerical Analysis*, 28, 6, pg 725, 1994.

[24] McCorquodale, J.A., "Variational approach to non-Darcy flow", *Jnl. of Hydraulics Div.*, ASCE, 96, HY11, 2265–2278, 1970.

[25] Mittelmann, H.D., "On optimal pointwise estimates for a finite element solution of nonlinear boundary value problems", *SIAM J. Numer. Analysis*, 14,773–778, 1977.

[26] Missbach, A., *Listy Cukrova*, 55, pg. 293, 1937.

[27] Norrie, D.H., and G. de Vries, "A survey of the finite element applications in fluid mechanics", in *Finite Elements in Fluid*, Vol. 3, Gallagher et al. (eds.), Wiley, London, 363–395, 1978.

[28] Rannacher, R. "Some asymptotic error estimates for finite element approximation of minimal surfaces", *RAIRO*, 11, 181–196, 1977.

[29] Scheurer, B., "Existence et approximation de points selles pour certains problèmes non linéaires", *RAIRO Analyse numérique*, 11, 7, 369–400, 1977.

[30] Shen, S.-M., "Finite element approximations for a variational inequality with a nonlinear monotone operator", in *Numerical Methods for partial differential equations*, Lecture Notes in Mathematics, No. 1297, Y.L. Zhu and B.Y. Guo (eds), Springer-Verlag, Berlin, 1987.

[31] Tyukhtin, V.B., "The rate of convergence of approximation methods of solution of one-sided variational problems, I.", *Vestnik Leningrad Univ. Mat. Mekh. Astronom.*, No. 13, 111–113, July 1982 (in Russian).

[32] Volker,R.E., "Nonlinear flow in porous media by finite elements", *J. Hydraulics Div.*, ASCE, 95, HY6, 2093–2114, 1969.

Chapter 19

Waste Encapsulation by *In Situ* Vitrification

R.T. McLay[1], G.F. Carey[1] and R.J. MacKinnon[2]

19.1 *In Situ* Vitrification

In the past, it was common practice for high technology industries to dispose of hazardous waste from manufacturing processes in landfills and there are numerous such sites throughout the U.S. Since these buried wastes can migrate from the landfills and eventually pollute surrounding soil and ground water, there is a need for *in situ* treatment. One such technology, *in situ* vitrification (ISV), is being studied as a possible treatment of many classes of buried hazardous wastes (see [3,9]).

The basic idea in the ISV process is to use electrical resistance heating to melt soil and buried waste *in situ*, creating a glass-like material that resolidifies and permanently immobilizes the waste materials. This process can be briefly described as follows (see Figure 19.1): A square array of four electrodes is positioned with the ends just below the ground surface. Electrically conductive paths are then established between electrodes by placing a mixture of glass frit and graphite in shallow channels at the soil surface. A current is then passed between diagonally opposed electrodes, and the

[1]The University of Texas at Austin
[2]Brookhaven National Laboratories

resulting joule heat melts the soil and wastes. Non-volatile materials are encapsulated, while organic materials thermally decompose. Gaseous products are either dissolved in the melt or migrate upwards through and around the molten zone. A confinement hood, placed over the vitrified zone, collects these off-gases as they evolve from the melt. The gaseous effluents combust with the air present inside the hood and are subsequently drawn into a treatment system where contaminants are removed. This process continues until the desired melt depth is reached, at which time the electrical power is removed and the molten zone is allowed to solidify. Upon solidification, the transformed soil and waste mixture becomes a stable glass-like material. The final dimensions of the melt zone can range up to nine meters wide and nine meters deep.

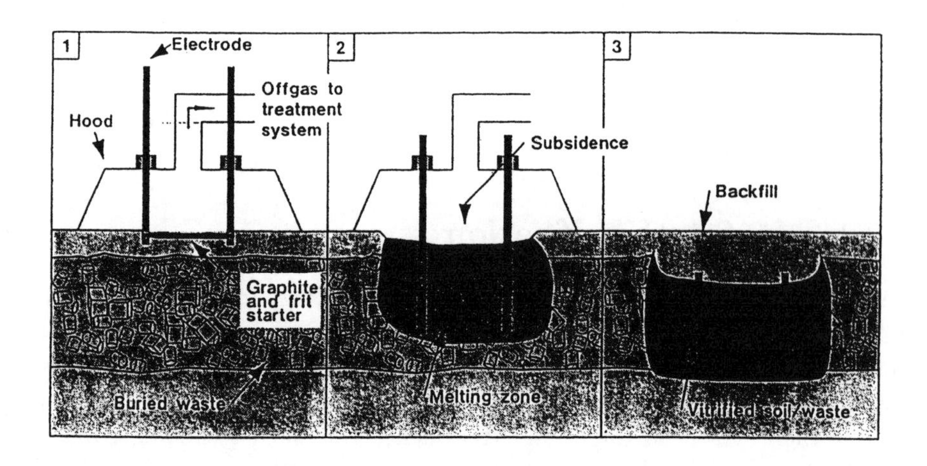

Figure 19.1: Characterization of ISV process

The physical processes controlling electrical heating and heat transfer are very closely coupled: First, the electrical conductivity depends strongly on the temperature; second, the temperature distribution is strongly governed by the electrical heating source. Furthermore, the extent of the melt zone and the buoyancy forces in the melt depend on the melt temperature. Convective heat transfer in the melt is determined by the fluid velocity field and, in particular, the structure of recirculation zones. Hence, a viable model should be capable of solving the coupled electric field, temperature, and viscous flow equations as well as determining the moving melt boundary location. The present work summarizes a finite element formulation and describes some related numerical experiments. The numerical work is restricted to a representative axisymmetric test problem that includes the important physical processes and yet remains computationally tractable. Potential applications of the model include sensitivity analyses to help assess the effect of uncertainty in material properties and to motivate further laboratory experiments.

19.2 Formulation

The problem domain is partitioned into two regions: the liquid melt Ω_m, and the surrounding solid soil Ω_s separated by Γ_m (Figure 19.2). Electrodes are identified by Γ_e, for e ranging from one to the number of electrodes. The free surface boundaries for the melt and the solid are denoted by Γ_{ms} and Γ_{ss}. Finally the far-field boundary is denoted Γ_f.

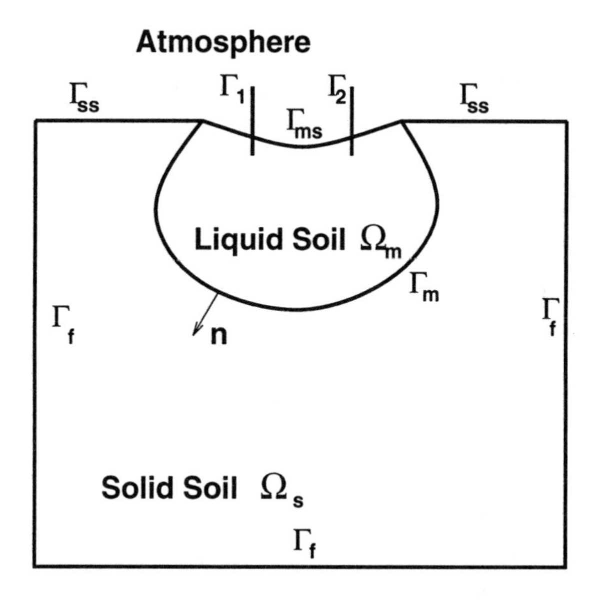

Figure 19.2: Idealized schematic of the ISV process

The transfer of heat can be characterized by

$$\rho c_p \frac{\partial T}{\partial t} + \rho c_p \boldsymbol{u} \cdot \nabla T - \nabla \cdot (k(T) \nabla T) = q(T, \phi) \text{ in } \Omega_m \cup \Omega_s \qquad (19.1)$$

where c_p is the specific heat, ρ is the density, k is the thermal conductivity, T is the temperature, $\boldsymbol{u}$ is the velocity of the viscous flow, q is the electrical heat source, and ϕ is the electric potential. Heat energy contributions due to chemical reactions and viscous dissipation are assumed negligible.

The electrical heat source, q, in (19.1) depends on electric potential ϕ and temperature T and may be modeled as

$$q(\phi, T) = \sigma_E(T)\nabla\phi \cdot \nabla\phi \quad \text{in} \quad \Omega_m \cup \Omega_s \tag{19.2}$$

where the dependence of electrical conductivity σ_E on temperature T is determined for the material by experiment. For example, the dependence in Figure 19.3 is assumed for the subsequent studies and is based on the experiments reported in [3].

The phase (melt) boundary moves with normal velocity V satisfying the Stefan condition

$$[[k\nabla T \cdot \boldsymbol{n}]] = LV \quad \text{on} \quad \Gamma_m \tag{19.3}$$

together with

$$T = T_{\text{melt}} \quad \text{on} \quad \Gamma_m \tag{19.4}$$

where L is the latent heat coefficient for phase change of the material, T_{melt} is the melting temperature, the notation $[[\,\cdot\,]]$ denotes the jump, and $\boldsymbol{n}$ is the unit normal to the phase boundary surface. The far-field boundary and ground surface conditions are

$$\begin{aligned} k\nabla T \cdot \boldsymbol{n} &= 0 \quad \text{on} \quad \Gamma_f \\ k\nabla T \cdot \boldsymbol{n} &= h_T(T - T_0) \quad \text{on} \quad \Gamma_{ss} \end{aligned} \tag{19.5}$$

and

$$k\nabla T \cdot \boldsymbol{n} = \epsilon\sigma(T^4 - T_0^4) \quad \text{on} \quad \Gamma_{ms} \tag{19.6}$$

where h_T is the heat-transfer coefficient from soil to air, T_0 is the external temperature, ϵ is the emissivity, and σ is the Stefan-Boltzmann constant.

The parabolic PDE (19.1) is to be integrated numerically to determine the temperature field over a period of several days in the ISV simulation. During the ISV process, the melt boundary moves slowly (2 to 10 cm/hr) and the relaxation time for the electric field is short compared to that for heat-transfer. This implies that the electric field rapidly adjusts to a steady state within each timestep of the heat-transfer solution. Consequently, it suffices that we solve the elliptic potential problem

$$\nabla \cdot (\sigma_E(T)\nabla\phi) = 0 \quad \text{in} \quad \Omega_m \cup \Omega_s \tag{19.7}$$

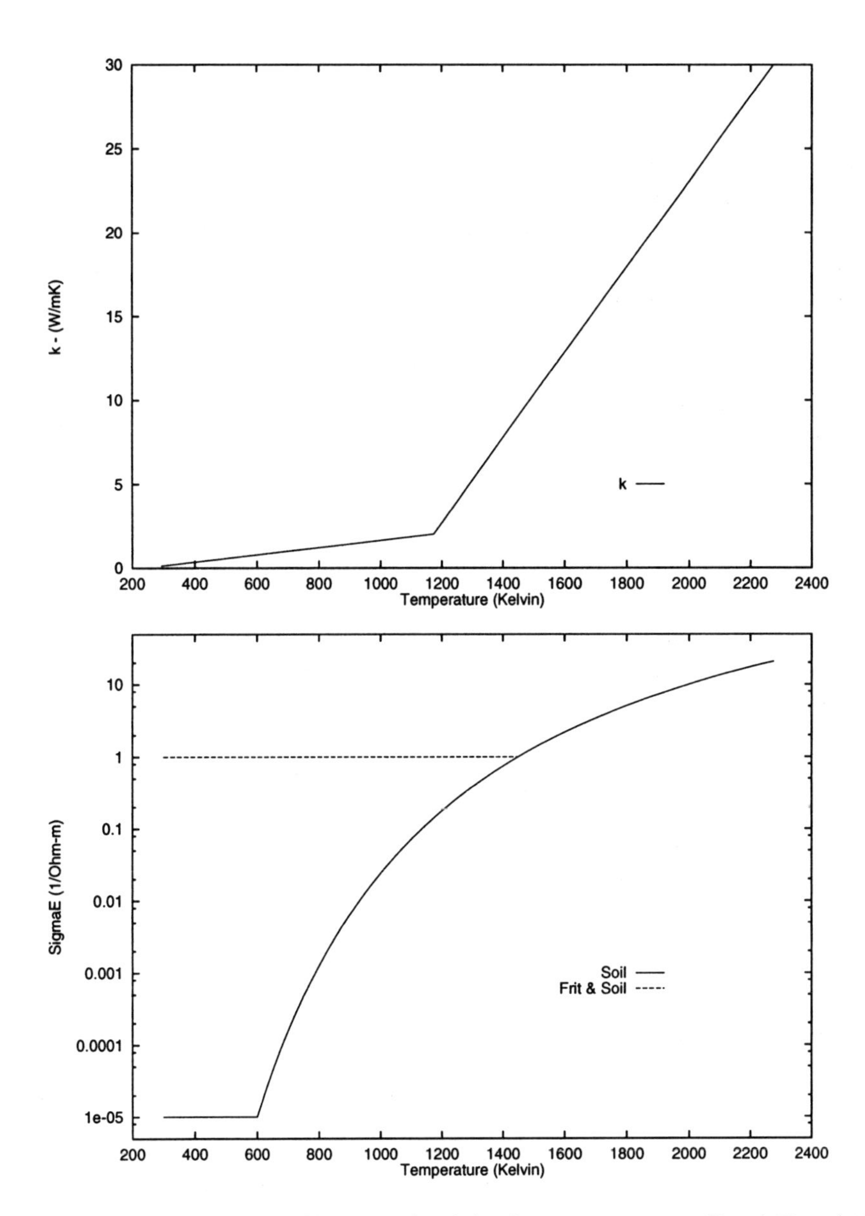

Figure 19.3: Dependence of heat conductivity k on temperature T and Electrical conductivity σ_E on temperature T (from [3])

The corresponding boundary conditions for this problem are

$$\phi = 0 \text{ on } \Gamma_f \tag{19.8}$$

$$\nabla\phi\cdot\boldsymbol{n} = 0 \text{ on } \Gamma_{ms}\cup\Gamma_{ss} \tag{19.9}$$

$$\phi = \tilde{\phi}^e \text{ on } \Gamma_e \tag{19.10}$$

where $\tilde{\phi}^e$ is the potential drop between electrode e and the far-field boundary.

The foregoing heat-transfer problem is coupled to the flow in the melt zone. The melt mixture may be non-Newtonian, but molten glass and sand can be modeled reasonably well by using the Navier-Stokes equations for viscous flow of an incompressible fluid having a temperature-dependent viscosity. Such models have proven effective for finite-element modeling of heat transfer and viscous flow in other applications (e.g., see [1,4,7,8,11].)

Introducing the well-known Boussinesq approximation [2], the viscous flow equations are

$$\frac{\partial\boldsymbol{u}}{\partial t}+\boldsymbol{u}\cdot\nabla\boldsymbol{u}-\frac{1}{\rho}\nabla\cdot\boldsymbol{\tau} = \beta(T-T^*)\boldsymbol{g}+\boldsymbol{g} \text{ in } \Omega_m \tag{19.11}$$

$$\nabla\cdot\boldsymbol{u} = 0 \text{ in } \Omega_m \tag{19.12}$$

where $\boldsymbol{u}$ is the velocity, p is the pressure, T^* is a reference temperature (at which buoyancy forces are zero), β is the coefficient of thermal expansion of the fluid, $\boldsymbol{g}=-g\boldsymbol{i}_z$ is the vertical force due to gravity, and $\boldsymbol{\tau}$ is the stress tensor given by

$$\tau_{ik} = -p\delta_{ik}+\rho\nu\left(\frac{\partial u_i}{\partial x_k}+\frac{\partial u_k}{\partial x_i}\right) \tag{19.13}$$

where ν is the kinematic viscosity.

"No slip" boundary conditions are assumed at the melt boundary and electrode surface; hence

$$\boldsymbol{u}=\boldsymbol{0} \text{ on } \Gamma_m \text{ and } \Gamma_j \tag{19.14}$$

Atmospheric pressure p_{atm} acts on the melt surface. That is,

$$\boldsymbol{\tau}\cdot\boldsymbol{n}=p_{\text{atm}} \text{ on } \Gamma_{ms} \tag{19.15}$$

The above equations, together with initial data constitute the mathematical statement of the viscous flow problem which is coupled to the previous heat transfer and electric potential systems.

19.3 Finite-Element Approximation

A variational form for the coupled problem can be constructed by introducing weighted-residual projections for the governing partial differential equations and applying the divergence theorem in the standard manner [6]. Then, for the coupled problem on a fixed domain, we have the weak statement: for $t > 0$ the solution $(\boldsymbol{u}, p, T, \phi)$ satisfies

$$\int_{\Omega_m} \left(\frac{\partial \boldsymbol{u}}{\partial t} \cdot \boldsymbol{v} + \boldsymbol{u} \cdot \nabla \boldsymbol{u} \cdot \boldsymbol{v} + \nu \nabla \boldsymbol{u} : \nabla \boldsymbol{v} - \frac{p}{\rho} \nabla \cdot \boldsymbol{v} \right) dx$$

$$+ \int_{\Gamma_{ms}} \frac{1}{\rho} \boldsymbol{\tau} \cdot \boldsymbol{n} \cdot \boldsymbol{v} \; ds = \int_{\Omega} \beta (T - T_0) \boldsymbol{g} \cdot \boldsymbol{v} \; dx + \int_{\Omega} \boldsymbol{g} \cdot \boldsymbol{v} dx \tag{19.16}$$

$$\int_{\Omega} b \nabla \cdot \boldsymbol{u} \; dx = 0 \tag{19.17}$$

$$\int_{\Omega_m \cup \Omega_s} \left(\rho c_p \frac{\partial T}{\partial t} w + \rho c_p \boldsymbol{u} \cdot \nabla T w + k \nabla T \cdot \nabla w \right) dx$$

$$- \int_{\partial \Omega_m \cup \partial \Omega_s} k \nabla T \cdot \boldsymbol{n} w \; ds = \int_{\Omega} q w \; dx \tag{19.18}$$

$$\int_{\Omega_m \cup \Omega_s} \sigma_E \nabla \phi \cdot \nabla \omega dx = 0 \tag{19.19}$$

for all admissible test functions $\boldsymbol{v}, b, w$, and ω. In (19.16), $\nabla \boldsymbol{u} : \nabla \boldsymbol{v}$ is the dyadic product $\sum_{i,j} \frac{\partial u_i}{\partial x_j} \frac{\partial v_i}{\partial x_j}$ and w, ω, $\boldsymbol{v}$ or $\boldsymbol{v} \cdot \boldsymbol{n}$ are zero on those parts of the boundary $\partial \Omega$ where T, ϕ, $\boldsymbol{u}$ or $\boldsymbol{u} \cdot \boldsymbol{n}$, respectively, are specified as essential data. Flux-type conditions can be incorporated directly as natural boundary conditions in (19.16)-(19.19). For example, if a radiative flux is given on Γ_{ms}, the corresponding contribution to the surface integral in (19.18) is

$$\int_{\Gamma_{ms}} k \nabla T \cdot \boldsymbol{n} w ds = \int_{\Gamma_{ms}} \epsilon \sigma (T^4 - T_0^4) w ds \tag{19.20}$$

where w is an arbitrary admissible test function on Γ_{ms}.

Introducing semidiscrete finite-element expansions for $\boldsymbol{u}$, p, T, and ϕ, we obtain a nonlinear system

$$\boldsymbol{M}_1 \frac{d\boldsymbol{U}}{dt} + \boldsymbol{g}_1(\boldsymbol{U}) + \boldsymbol{A}_1(\boldsymbol{T})\boldsymbol{U} + \boldsymbol{B}\boldsymbol{p} = \boldsymbol{f}_1(\boldsymbol{T}) \tag{19.21}$$

$$\boldsymbol{B}^T \boldsymbol{U} = \boldsymbol{0} \tag{19.22}$$

$$\boldsymbol{M}_2 \frac{d\boldsymbol{T}}{dt} + \boldsymbol{g}_2(\boldsymbol{U}, \boldsymbol{T}) + \boldsymbol{A}_2(\boldsymbol{T}) = \boldsymbol{f}_2(\boldsymbol{\phi}, \boldsymbol{T}) \tag{19.23}$$

$$\boldsymbol{A}_3(\boldsymbol{T}) \boldsymbol{\phi} = \boldsymbol{C}(\boldsymbol{T}) \tag{19.24}$$

where $\boldsymbol{U}$ is the vector of nodal velocity components, $\boldsymbol{T}$ is the vector of nodal temperatures, $\boldsymbol{p}$ is the nodal pressure vector and $\boldsymbol{\phi}$ is the vector of nodal potential values. Differencing (19.21) to (19.24) implicitly with respect to time through the time step Δt_n yields a coupled nonlinear system of algebraic equations to be solved for the nodal solution vector $(\boldsymbol{U}, \boldsymbol{p}, \boldsymbol{T}, \boldsymbol{\phi})$.

In the present numerical algorithm, the energy, electric potential, and momentum equations are iteratively decoupled by successive approximation of $\boldsymbol{T}$, $\boldsymbol{\phi}$, and $\boldsymbol{U}$, respectively, with initial iterates given by the solution at the previous step. Hence, this algorithm requires repeated solution of the (smaller) time-discretized nonlinear Navier-Stokes problem in (19.21) and (19.22), the linearized energy equation in (19.23) and the electric potential problem of (19.24). Successive approximation and Newton-Raphson iteration have been implemented for the non-linear systems with frontal elimination of the resulting sparse algebraic linear systems. The number of iterations can be varied according to the size of timestep. In the following numerical studies, we use a backward Euler implicit scheme to integrate (19.21) and (19.23) with timestep Δt_n.

The melt isotherm $1428K$ identifies the location of the phase boundary, and we set $\beta = 0$ in the solid subregion where $T < 1428K$. The latent-heat contribution at the phase-change boundary is accounted for by evaluating a temperature-dependent heat capacity as in [10]. This approach yields an artificial "mushy" zone between melt and solid. The solution proceeds by time stepping from the initial state. A detailed algorithm is given in [5].

19.4 Numerical Studies

In these studies we consider an axisymmetric problem in an annular domain with a centrally located electrode. This problem captures several important features of the coupled physical processes for an exploratory study and permits efficient computation. The problem domain has an inner (electrode) radius of $0.15m$ and outer radius of either $2.0m$ or $3.0m$ depending on the test case. There is an electrically conducting "frit" layer of depth $0.15m$ at the surface, and the domain is modeled to a depth of $1m$. The boundary conditions are: $k\nabla T \cdot \boldsymbol{n} = h_T(T - T_0)$ on Γ_{ms}; $k\nabla T \cdot \boldsymbol{n} = \epsilon\sigma(T^4 - T_0^4)$ on Γ_{ss}; $\nabla\phi \cdot \boldsymbol{n} = 0$, $\boldsymbol{u} \cdot \boldsymbol{n} = 0$ on $\Gamma_{ms} \cup \Gamma_{ss}$; $k\nabla T \cdot \boldsymbol{n} = \nabla\phi \cdot \boldsymbol{n} = 0$ and $\boldsymbol{u} = \boldsymbol{0}$ on Γ_f; $\boldsymbol{u} = \boldsymbol{0}$, $\phi = \tilde{\phi}^1(t)$, $k\nabla T \cdot \boldsymbol{n} = 0$ on Γ_1. The material properties are taken to be $\rho = 2170kg/m^3$ $c_p = 1046J/kgK$, with $k(T)$ and $\sigma_E(T)$ given in Figure 19.3. The frit layer and soil below are taken to have identical properties except that in the frit layer $\sigma_E = 1.0(1/\Omega m)$ for all temperatures less than $1428K$ (i.e., melting). At the fluid surface $z = 0$, the radiative heat transfer condition applies with reference exterior temperature T_0 taken as $300K$ and $h_T = \epsilon\sigma = 5.1 \times 10^{-8}$. (We have taken $\epsilon = 0.09$ as the emissivity in the present calculations. Experimental results for ϵ are needed for the field model.) The heat flux at the inner and outer boundaries and the base is assumed to be zero (insulated). The base and top are assumed also to be electrically insulated so that $\nabla\phi \cdot \boldsymbol{n} = 0$ at $z = 0$ and $z = z_b$. At the inner boundary the potential is taken to be $\phi = 500$ volts, with $\phi = 0$ at the outer boundary for the desired potential drop. The initial temperature is $T = 300K$.

The problem is solved on a finite-element discretization of biquadratic elements for velocity, temperature and potential fields and bi-linear pressures. The mesh is graded into the surface regions to a resolution necessary to model the flow behavior and temperature gradients. Given the uncertainty in material parameters, it is debatable whether finer grids will provide any further practical information. A constant timestep of $\Delta t = 200s$ was adequate for the test cases considered and facilitated comparisons.

The first case considered corresponds to $\beta = 10^{-4}$. In this problem, the behavior at early time is essentially conductive but as time increases a convective cell develops. The presence of a second counter-rotating cell is observed by $t = 72,000s$, and strengthens as t increases to $1.12 \times 10^5 s$ (Figure 19.4). The effect of this convective pair of cells on the heat transfer is evident in the non-convex form of the J contour ($T = 1450K$). The effect of convective mixing and radiation at the surface is to retard the depth of heating and melting. As time continues, the cells grow and the strength of the second convective roll diminishes. By $t = 1.62 \times 10^5 s$ there is basically one large clockwise cell still evident.

The effect of convection is to promote mixing, and this enhances cooling at the top surface. We emphasize that $\epsilon = 0.09$ is used in the radiation boundary condition, and cooling at the top surface will be more pronounced if the calculations are repeated with higher values of ϵ. The size of the volumetric expansion coefficient β should be measured accurately in future laboratory experiments, since this strongly influences the flow and the thermal penetration. Accurate measurement of surface boundary conditions is also needed. The conduction-only solution obviously fails to capture the effect of thermal mixing in the ISV process.

The next problem deals with the effect of buried conductors on the coupled heat transfer, electric field, and viscous flow problem for ISV processes. The analysis follows the formulation described in the previous sections for ISV modeling in an axisymmetric domain. The domain is $2m$ wide and $1m$

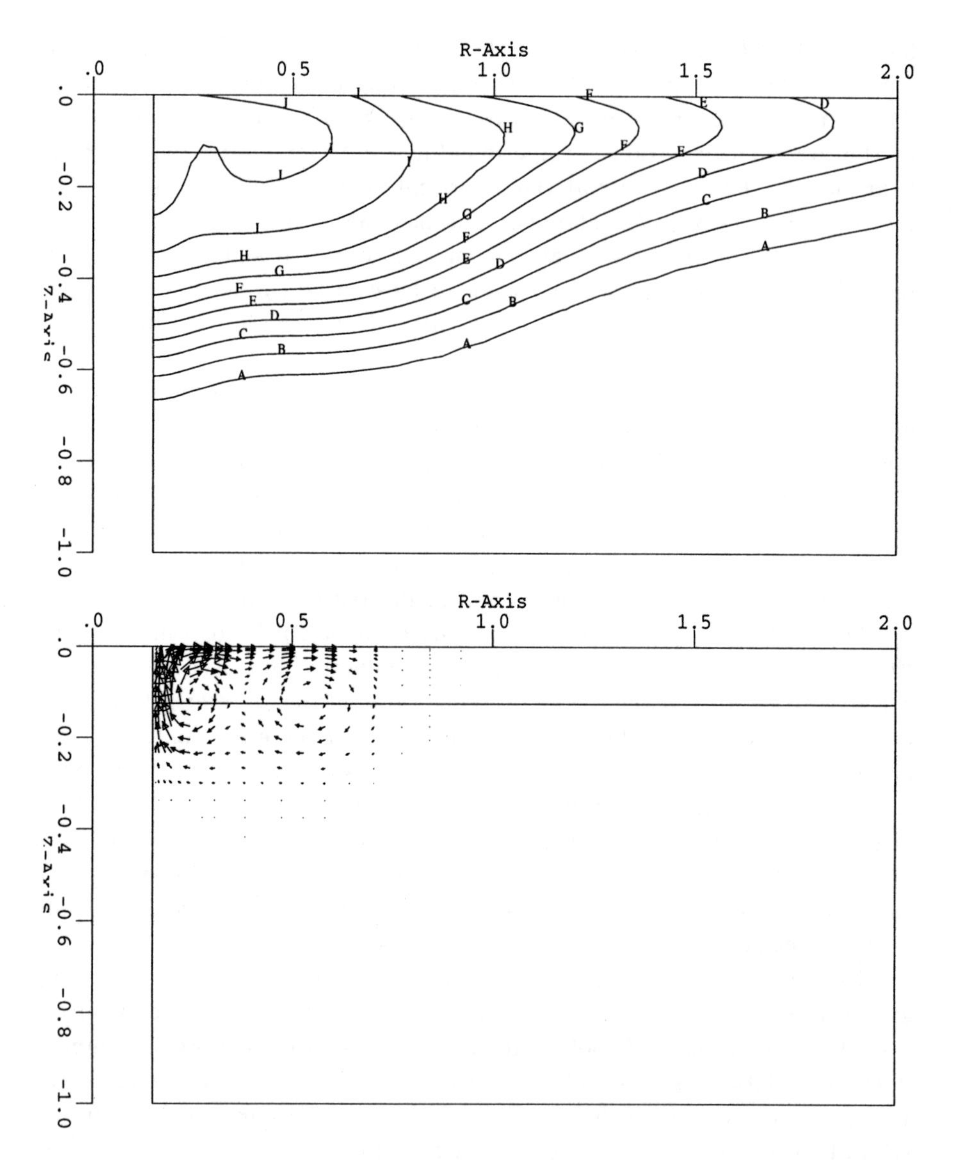

Figure 19.4: Isotherms and velocity vectors at $t = 112,000s$, $V_{\max} = 1.23mm/s$ ($\beta = 10^{-4}$), $T_A = 416K$, $T_J = 1462K$ for domain $.15m < r < 2m$, $-1m < z < 0.0m$.

deep with an inner electrode at $r = 0.15m$ as before. The applied voltage and other boundary conditions are the same as in the previous study. Material properties also correspond to those in the previous calculations. In particular, we have set $\beta = 0$ for $T < 1428K$, $\beta = 10^{-4}$ for $T > 1428K$, and $\epsilon = 0.09$ in the top surface radiative boundary condition with T_0 again equal to $300K$. For the simulations, the "buried block" is taken to have the following properties that are assumed constant: $\rho = 7,200kg/m^3$, $k = 15.5W/m^2K$, $c_p = 753J/kgK$, and $\sigma_E = 100/\Omega m$. No-slip boundary conditions are assumed at the surface of an interior block.

These properties differ significantly from those of the surrounding soil and will permit us to obtain a general understanding of the effect of local variations in properties on the overall heat transfer and melt growth in the ISV process. The values above for ρ and c_p are similar to those for metal (e.g., steel). Since the buried object may contain other materials, we have taken the electrical conductivity σ_E to be much higher than that of the surrounding soil but still considerably less than that of solid metal ($O(10^6)/\Omega m$). The problem can be solved with σ_E locally corresponding to the pure metal case but the essential physics remain unchanged. (The main complication is that the E-field problem becomes numerically less well-conditioned as σ_E becomes locally very large.)

The main objective is to ascertain the effect of the buried inclusion on the heat transfer and viscous flow processes. The previous calculations in the absence of an inclusion provide a baseline for comparison. Consider the block located near the electrode at $0.34 < r < 1.0$, $-0.25 < z < -0.45$ (Figure 19.5). We first compare the early time behavior (not shown) and confirm that the effect of the block is not apparent. This is consistent with what would be expected, since the block is embedded in essentially non-conducting material at this stage. However, by about $t = 22,000s$ the effect of the deformation of the temperature contours (with respect to the baseline study involving no inclusion) is apparent near the top left corner of the block. Since the block is a good thermal conductor, it heats readily and relatively uniformly. On the other hand, the adjacent soil has poor thermal conductivity and there will be a steep gradient near the block. We see this effect in the concentration of contours in the soil near the block corner. The corner singularity may also have some effect on the local thermal behavior, but this does not alter the qualitative result. (Since the block also has high electrical conductivity the E-field will behave similarly). The melt zone and recirculation cell are remote from the inclusion. The growth of the thermal and viscous flow fields at $t = 52,000s$ is shown in Figure 19.5. The effect of the inclusion is to inhibit the growth of the flow region relative to the baseline study. The peak temperature and velocity are higher than in the baseline case. (Convection is smaller with the smaller melt region and therefore there is less heat transfer to the surface in the present case).

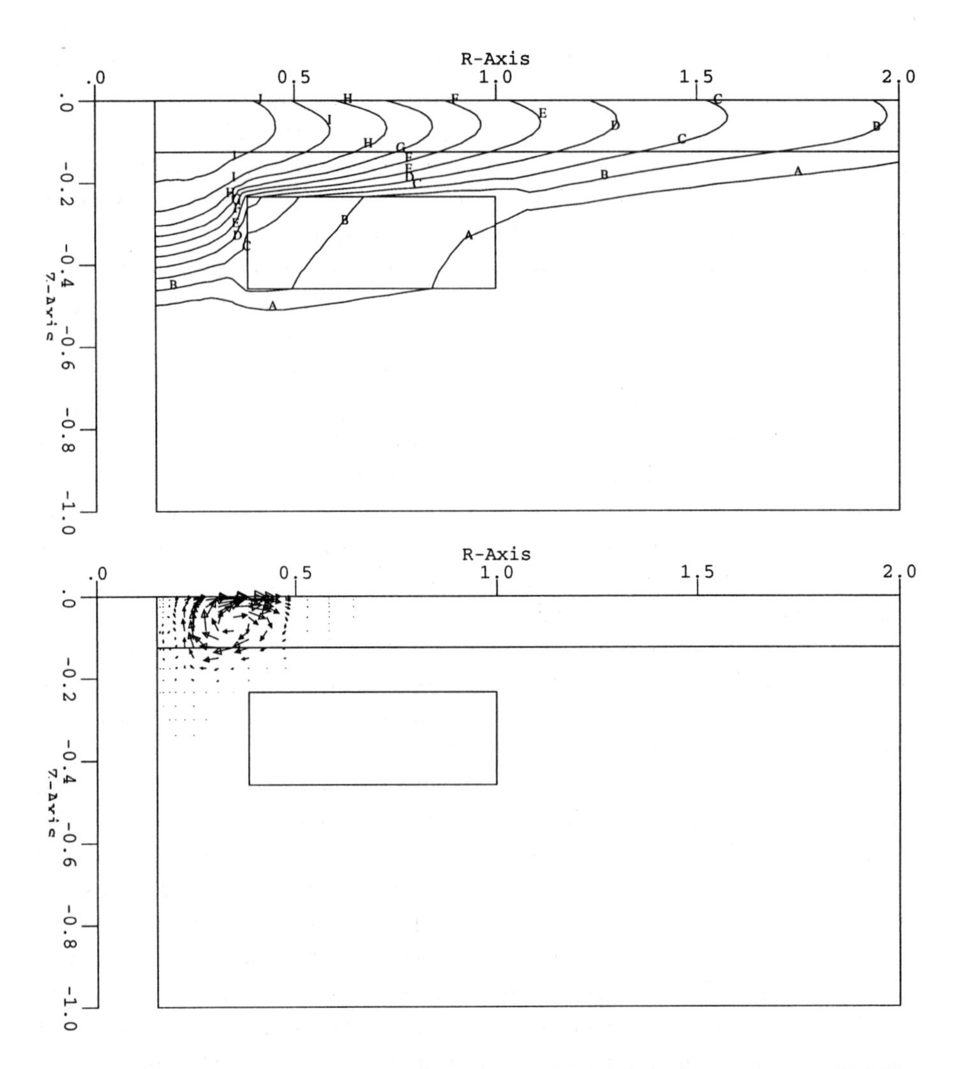

Figure 19.5: Isotherms and velocity vectors for buried object at $t = 52,000s$, $V_{\max} = 0.42mm/s$, $T_A = 408K$, $T_J = 1407K$ for domain $.15m < r < 2m$, $-1m < z < 0.0m$.

Locally higher temperatures imply, in turn, that the buoyancy force is larger and hence velocity is increased. Similar behavior is evident at later times, as seen in the results for $t = 2.55 \times 10^5 s$ in Figure 19.6. Note the wide uniform zone in Figure 19.6a containing the block. The no-slip condition on the block surface inhibits the growth of the melt cell.

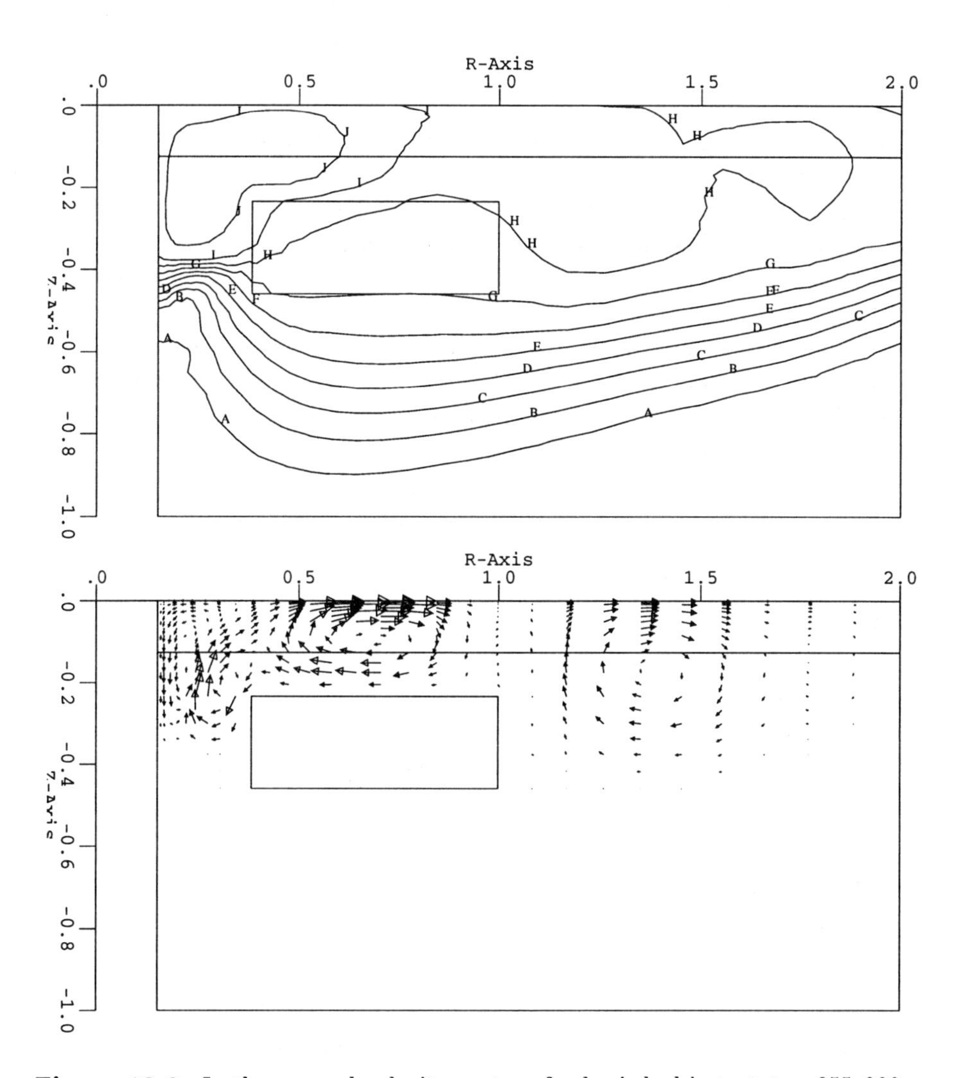

Figure 19.6: Isotherms and velocity vectors for buried object at $t = 255,000s$, $V_{\max} = 19.8mm/s$, $T_A = 452K$, $T_J = 1830K$ for domain $.15m < r < 2m$, $-1m < z < 0.0m$.

19.5 Conclusion

Practical ISV applications involve mixtures of materials such as buried waste products and soils of nonuniform composition. Modeling and simulation rely on knowledge of material properties and certain constitutive assumptions that are less reliable than in similar manufacturing processes. Nevertheless, under reasonable assumptions a model can be developed that will permit a better understanding of the process.

A finite-element model for the analysis of thermal-electric and flow interactions in the ISV process has been presented. Associated numerical studies designed to examine electrical heating, melt convection, buried conductors, and variable voltage effects have also been provided. These results can be summarized as follows: (a) convective mixing reduces peak melt temperatures and enhances cooling at the top surface; (b) both the volumetric expansion coefficient and the surface boundary conditions strongly influence melt flow and shape; (c) a conductive, buried inclusion tends to inhibit convection and melt growth and produce steep temperature gradients near the inclusion; and (d) a time-varying voltage may impact melt shape.

Acknowledgments: This research has been supported in part by the Department of Energy.

References

[1] Argyris, J., J. S. Doltsinis, P. M. Pimenta and H. Wüstenberg, "Finite element solution of viscous flow problems", in R. H. Gallagher *et al.* (eds.), *Finite Elements and Fluids*, 89–114, Wiley, 1986.

[2] Boussinesq, J., *Théorie Analytique de la Chaleur: mise en harmonie avec la thermodynamique et avec la théorie mécanique de la lumière*, Gauthier-Villars, 1903.

[3] Buelt, J. L., C. Timmerman, K. Oma, V. Fitzpatrick and J. Carter, "*In situ* vitrification of transuranic wastes: An updated systems evaluation and application assessment", Tech. Rep. PNL-4800, Suppl. 1 UC-70, Pacific Northwest Laboratory, 1987.

[4] Carey, G. F. and R. T. McLay, "Viscous flow and heat transfer with moving phase boundary", Paper 85-WA/HT-48 ASME Winter Annual Meeting, Miami, FL, December 1985.

[5] Carey, G. F., R. T. McLay and R. J. MacKinnon, "Finite-element modelling of *In Situ* vitrification", *In Situ*, 172, 201–226, 1993.

[6] Carey, G. F. and J. T. Oden, *Finite Elements: Fluid Mechanics*, 6, Prentice Hall, 1986.

[7] Gartling, D. K., "Finite element analysis of volumetrically heated fluids in an axisymmetric enclosure", in R. H. Gallagher *et al.* (eds.), *Finite Elements in Fluids*, 4, 233–250, Wiley, 1982.

[8] Lewis, R. W., K. Morgan and P. M. Roberts, "Determination of thermal stresses in solidification problems", in Pittman *et al.* (eds.), *Numerical Analysis of Forming*, 405–431, Wiley, 1984.

[9] MacKinnon, R. J., P. Murray, R. Johnson, D. Hagrman, C. Slater and E. Marwil, "*In situ* vitrification model development and implementation plan", Tech. Rep. EGG-WM-9036, Idaho National Engineering Laboratory, 1990.

[10] Morgan, K., R. W. Lewis and O. C. Zienkiewicz, "An improved algorithm for heat conduction problems with phase change", *Int. J. Num. Meth. Eng.*, 124, 1191–1195, 1978.

[11] Wang, H. P. and R. T. McLay, "Automatic remeshing scheme for modeling hot forming processes", *J. Fluid Eng.*, 108, 465–469, 1986.

Chapter 20
Numerical Simulation of Multiphase Flow in Groundwater Hydrology

Z. Chen[1]

20.1 Introduction

Soil venting is a new technology designed to remediate volatile waste from an unsaturated zone. The unsaturated zone is an important area where surface processes and the groundwater are tied together and much of our municipal water supply is drawn. For contaminants to reach and pollute the groundwater they often move through the unsaturated zone. Soil venting is a technology that attempts to remove contaminants from the soil before they can seriously pollute the groundwater supply.

A soil venting process works by pumping air through part of an unsaturated zone contaminated by a volatile contaminant, leading the contaminant to volatilize so that it can be removed by the gaseous phase flow. Early studies of this contaminant cleanup technology show that it can be efficient and economical [15]. However, although it is a practical technology, few theories are available to assess its performance. Also, soil venting is chemically and physically complicated. It involves flow and transport in the unsaturated zone. Hence numerical simulation of these processes is a crucial step in understanding and designing soil venting applications.

[1] Department of Mathematics, Texas A&M University

In this chapter we consider the modeling of flow and transport for an unsaturated zone, which is considered as a multiphase system. The multiphase system consists of the solid, air, and water phases. We first develop a nonlinear differential system for describing this system based on the established physics of porous medium flow and transport. The two-phase flow equations are written in a fractional flow formulation, i.e., in terms of a saturation and a global pressure. The fractional flow approach is motivated by petroleum reservoir simulation [5]. The main reason for this approach is that efficient and effective numerical methods can be devised to take advantage of many physical properties inherent in the flow equations.

Next, we consider numerical methods for the nonlinear differential system that describes fluid flow and contaminant transport in the unsaturated zone. The saturation equation is approximated by a standard finite element method, while the pressure equation is treated by a mixed finite element method. It is well known that for these applications the advective transport dominates the diffusive effects in incompressible flow. In the air-water system studied here, the transport again dominates the entire process. Hence it is important to obtain good approximate velocities. This motivates the use of the mixed method [13]. An explicit scheme is used to integrate the transport equation. The scheme utilizes a discontinuous finite element method for the advective part of the equation together with a mixed finite element approach for the diffusive part. This approach allows accurate approximation of steep solution gradients, is (trivially) conservative and the resulting algorithm is parallelizable.

20.2 Flow and Transport Equations

The unsaturated zone is considered as a multiphase system and equations are written for the air and water phases. A macroscopic description of this system exploits fluid and solid properties defined over the porous medium continuum $\Omega \subset \mathbb{R}^d$, $d \leq 3$. The usual equations describing two-phase flow in Ω are given by the mass balance equation and Darcy's law for each of the fluid phases [1]

$$\frac{\partial(\phi \rho_\alpha s_\alpha)}{\partial t} + \nabla \cdot (\rho_\alpha u_\alpha) = f_\alpha, \qquad x \in \Omega,\ t > 0, \tag{20.1}$$

$$u_\alpha = -\frac{k k_{r\alpha}}{\mu_\alpha}(\nabla p_\alpha - \rho_\alpha g), \qquad x \in \Omega,\ t > 0 \tag{20.2}$$

where $\alpha = w$ denotes the water phase, $\alpha = a$ indicates the air phase, ϕ and k are the porosity and absolute permeability of the porous system, ρ_α,

s_α, p_α, u_α, and μ_α are the density, saturation, pressure, volumetric velocity, and viscosity of the α-phase, f_α is the source/sink term, $k_{r\alpha}$ is the relative permeability of the α-phase, and g is the gravitational, downward-pointing, constant vector. These equations are augmented by the volume balance law:

$$s_a + s_w = 1, \tag{20.3}$$

and the definition of the capillary pressure

$$p_c(s_w) = p_a - p_w \tag{20.4}$$

The contaminant can be transported in either air or water phase so a mass balance equation may be included for each phase. However, we here consider the case where water is the wetting fluid so that air does not have any contact with the solid phase and the contaminant instantaneously establishes an equilibrium distribution between the water and air phases. Then the concentration in each phase is proportional to the concentration in the other phase by Henry's law. With these two assumptions, the contaminant transport equation for the volumetric concentration of the water phase c_w is

$$\frac{\partial(\phi_w c_w)}{\partial t} + \nabla \cdot (U_w c_w - \mathcal{D}_w \nabla c_w) + \Lambda_w c_w = F_w, \tag{20.5}$$

where

$$\phi_w = \phi(s_w + H s_a) + H_s, \qquad U_w = u_w + H u_a, \tag{20.6}$$

$$\mathcal{D}_w = \phi(s_w D_w + s_a H D_a), \quad \Lambda_w = \phi(\chi_w s_w + H \chi_a s_a), \tag{20.7}$$

H and H_s are the equilibrium phase partitioning constants between an air/water system and a soil/water system, respectively, $D_w(u_w)$ is the hydrodynamic diffusion/dispersion tensor of the water phase, $D_a(u_a)$ is defined similarly for the air phase, χ_α $(\alpha = w, a)$ denotes the reaction rate for phase α, and F_w indicates the source/sink term.

The most commonly encountered boundary conditions are of first-type and second-type $(\alpha = w, a)$:

$$p_\alpha = p_{\alpha D}(x,t), \text{ on } \Gamma_1^p, \; u_\alpha \cdot \nu = d_\alpha(x,t), \text{ on } \Gamma_2^p \tag{20.8}$$

$$c_w = c_{wD}(x,t), \text{ on } \Gamma_1^c, \; \partial c_w/\partial\nu = c_{wN}(x,t), \text{on } \Gamma_2^c \tag{20.9}$$

for $t > 0$, where $p_{\alpha D}$, d_α, c_{wD}, and c_{wN} are given functions, $\partial\Omega = \Gamma_1^\pi \cup \Gamma_2^\pi$ with Γ_1^π and Γ_2^π disjoint $(\pi = p, c)$, and ν is the outer unit normal to $\partial\Omega$. Other types of boundary conditions can be used [11].

20.3 A Fractional Flow Formulation

The fractional flow approach is often used by petroleum engineers [5], since efficient numerical methods can be devised to take advantage of many physical properties inherent in the flow equations. It is convenient to introduce the phase mobilities

$$\lambda_\alpha = k_{r\alpha}/\mu_\alpha, \quad \alpha = a, w, \tag{20.10}$$

and the total mobility

$$\lambda = \lambda_a + \lambda_w. \tag{20.11}$$

Then we define the global pressure [5] with $s = s_w$ as

$$\begin{aligned} p =& \frac{1}{2}(p_a + p_w) + \frac{1}{2}\int_{s_c}^{s} \frac{\lambda_a - \lambda_w}{\lambda} \frac{dp_c}{d\xi} d\xi \\ =& p_w + \int_0^{p_c(s)} \left(\frac{\lambda_a}{\lambda}\right) \left(p_c^{-1}(\xi)\right) d\xi, \end{aligned} \tag{20.12}$$

where $p_c(s_c) = 0$. As usual, we may assume that ρ_α depends on p [5]. Then the total velocity is given by

$$u = -k\lambda \left(\nabla p - G(s,p)\right), \tag{20.13}$$

where

$$G(s,p) = \frac{\lambda_a \rho_a + \lambda_w \rho_w}{\lambda} g. \tag{20.14}$$

Now it can be easily seen that

$$u_w = q_w u + k\lambda_a q_w \nabla p_c - k\lambda_a q_w \tilde{\rho}, \tag{20.15}$$

$$u_a = q_a u - k\lambda_w q_a \nabla p_c + k\lambda_w q_a \tilde{\rho}, \tag{20.16}$$

where $q_\alpha = \lambda_\alpha/\lambda$, $\alpha = a, w$, and $\tilde{\rho} = (\rho_a - \rho_w)g$. Consequently,

$$u = u_a + u_w. \tag{20.17}$$

Following [10], (20.2) can be manipulated using (20.12)–(20.17) to yield the pressure equation

$$\nabla \cdot u = -\frac{\partial \phi}{\partial t} - \sum_{\alpha=w}^{a} \frac{1}{\rho_\alpha} \left(\phi s_\alpha \frac{\partial \rho_\alpha}{\partial t} + u_\alpha \cdot \nabla \rho_\alpha - f_\alpha\right), \tag{20.18}$$

and the saturation equation

$$\phi\frac{\partial s_w}{\partial t}+\nabla\cdot\left(q_w u + k\lambda_a q_w(\nabla p_c - \tilde{\rho})\right)$$
$$= -s_w\frac{\partial \phi}{\partial t} - \frac{1}{\rho_w}\left(\phi s_w \frac{\partial \rho_w}{\partial t} + u_w\cdot\nabla\rho_w - f_w\right). \qquad (20.19)$$

The water phase is usually assumed to be incompressible. Then (20.18) and (20.19) can be simplified. Also, the term $u_a \cdot \nabla\rho_a$ is effectively quadratic in velocity, which is small in almost all of the domain [5], [20], and can be neglected. After we introduce the notation

$$c(s,p) = \frac{\phi s}{\rho_a}\frac{d\rho_a}{dp}, \qquad D(s) = -k\lambda_a q_w \frac{dp_c}{ds}, \qquad (20.20)$$

$$f(p) = \frac{f_a}{\rho_a} + \frac{f_w}{\rho_w}, \qquad a(s) = k\lambda, \qquad (20.21)$$

$$\tilde{f}_w = \frac{f_w}{\rho_w}, \qquad b(s,p) = -k\lambda_a q_w \tilde{\rho}, \qquad (20.22)$$

(20.18) and (20.19) can be now written as

$$c(s,p)\frac{\partial p}{\partial t} + \nabla\cdot u = f(p) - \frac{\partial \phi}{\partial t}, \qquad (20.23)$$

$$u = -a(s)\left(\nabla p - G(s,p)\right), \qquad (20.24)$$

$$\phi\frac{\partial s}{\partial t} - \nabla\cdot\left(D(s)\nabla s - q_w u - b(s,p)\right) = \tilde{f}_w - s\frac{\partial \phi}{\partial t}. \qquad (20.25)$$

The boundary conditions for the pressure-saturation equations become

$$p = p_D(x,t), \text{ on } \Gamma_1^p, \; u\cdot\nu = \tilde{d}(x,t), \text{on } \Gamma_2^p \qquad (20.26)$$

$$s = s_D(x,t), \text{ on } \Gamma_1^p, \; (D(s)\nabla s - q_w u - b(s,p))\cdot\nu = -d_w(x,t), \text{ on } \Gamma_2^p, \qquad (20.27)$$

for $t > 0$, where s_D and p_D are the transforms of p_{wD} and p_{aD} by (20.4) and (20.12), and $\tilde{d} = d_a + d_w$.

We also specify the initial conditions

$$p(x,0) = p^0(x), \qquad x \in \Omega, \tag{20.28}$$
$$s(x,0) = s^0(x), \qquad x \in \Omega. \tag{20.29}$$

Finally, the initial concentration is given by

$$c_w(x,0) = c_w^0(x), \qquad x \in \Omega. \tag{20.30}$$

To summarize, the resulting model consists of (20.5), (20.9), (20.23)–(20.30).

20.4 Finite Element Approximation

We first present the finite element approximation procedure for the flow part. The saturation equation is approximated by a standard Galerkin finite element method, while the pressure equation is treated by a mixed finite element method. First introduce the spaces

$$H(\text{div},\Omega) = \{v \in (L^2(\Omega))^d : \nabla \cdot v \in L^2(\Omega),\ d = 2,3\}, \tag{20.31}$$
$$V_\eta = \{v \in H(\text{div},\Omega) : v \cdot \nu = \eta \text{ on } \Gamma_2^p\}, \tag{20.32}$$
$$M = \{w \in H^1(\Omega) : w = 0 \text{ on } \Gamma_1^p\}. \tag{20.33}$$

For $0 < h_p < 1$ and $0 < h < 1$, let $\mathcal{J}_{h_p}$ and $\mathcal{J}_h$ denote two quasi-regular partitions of Ω into elements, say, simplexes, rectangular parallelepipeds, and/or prisms. In both partitions, we also require that adjacent elements completely share their common edge or face and assume that each exterior edge or face has imposed on it either Dirichlet or Neumann conditions, but not both. Let $M_h \subset W^{1,\infty}(\Omega) \cap M$ be a standard C^0-finite element space associated with $\mathcal{J}_h$ and let $V_{\eta,h} \times W_h = V_{\eta,h_p} \times W_{h_p} \subset V_\eta \times L^2(\Omega)$ be some standard mixed finite element space for second order elliptic problems defined over $\mathcal{J}_{h_p}$ (see, e.g., [2, 3, 4, 11, 19, 21].) Finally, let $\{t^n\}_{n=0}^{n_T}$ be a quasi-partition of $J = (0,T)$, $(T > 0)$ with $t^0 = 0$ and $t^{n_T} = T$, and set $\Delta t^n = t^{n+1} - t^n$, $\Delta t = \max\{\Delta t^n, 0 \le n \le n_T - 1\}$, and

$$\psi^n = \psi(t^n), \quad \partial\psi^n = (\psi^n - \psi^{n-1})/\Delta t^n. \tag{20.34}$$

We are now in a position to introduce our finite element procedure.

The fully-discrete finite element method is given as follows. The approximation procedure for the pressure is defined by the mixed method for a pair of maps $\{u_h^n, p_h^n\} \in V_{\tilde{d}^n,h} \times W_h$, $n = 1, 2, \cdots, n_T$ such that

$$\int_\Omega \alpha(s_h^{n-1}) u_h^n \cdot v \, d\Omega - \int_\Omega \nabla \cdot v p_h^n \, d\Omega \tag{20.35}$$
$$= \int_\Omega G(s_h^{n-1}, p_h^{n-1}) \cdot v \, d\Omega - \int_{\Gamma_1^p} p_D^n v \cdot \nu \, ds \quad \forall v \in V_{0,h},$$

$$\int_\Omega c(s_h^{n-1} p_h^{n-1}) \partial p_h^n \psi \, d\Omega + \int_\Omega \nabla \cdot u_h^n \psi \, d\Omega$$
$$= \int_\Omega \left(f(p_h^{n-1}) - \frac{\partial \phi^{n-1}}{\partial t} \right) \psi \, d\Omega, \quad \forall \psi \in W_h \tag{20.36}$$

and the finite element approximation for the saturation $s_h^n \in M_h + s_D^n$, $n = 1, 2, \cdots, n_T$ satisfies

$$\int_\Omega \phi \partial s_h^n w \, d\Omega + \int_\Omega (D(s_h^{n-1}) \nabla s_h^n - q_w(s_h^{n-1}) u_h^n - b(s_h^{n-1}, p_h^n)) \nabla w \, d\Omega$$
$$= \int_\Omega (\tilde{f}_w^n - s_h^n \frac{\partial \phi^n}{\partial t}) w \, d\Omega - \int_{\Gamma_2^p} d_w^n w \, ds, \qquad \forall w \in M_h. \tag{20.37}$$

The initial conditions are

$$p_h(\cdot, 0) = p_h^0, \quad s_h(\cdot, 0) = s_h^0 \tag{20.38}$$

where p_h^0 and s_h^0 are some approximations in W_h and M_h of p^0 and s^0, respectively.

For $n = 1, 2, \cdots, n_T$, equations (20.35) and (20.37) are integrated as follows: First, using s_h^{n-1}, p_h^{n-1}, and (20.35), evaluate $\{u_h^n, p_h^n\}$. Since it is linear, (20.35),(20.36) has a unique solution for each n [6], [16]. Next, using s_h^{n-1}, $\{u_h^n, p_h^n\}$, and (20.37), calculate s_h^n. Again, (20.37) has a unique solution for each n [23]. While the backward Euler scheme is presently used in (20.36) and (20.37), the Crank-Nicolson scheme and more accurate time stepping procedures can be used[14].

We now consider a finite element approximation scheme for the transport part. The scheme uses a combination of a discontinuous finite element method for the advective part and a mixed finite element method for the diffusive part. This scheme is trivially conservative and fully parallelizable. To see how to combine these two methods for the transport equation, let

Ω be the unit rectangle $[0,1]^2$. Let $\{x_{i+1/2}\}_{i=0}^{n_x} \times \{y_{j+1/2}\}_{j=0}^{n_y} \times \{z_{k+1/2}\}_{k=0}^{n_z}$ be a partition of Ω with $x_{1/2} = y_{1/2} = z_{1/2} = 0$ and $x_{n_x+1/2} = y_{n_y+1/2} = z_{n_z+1/2} = 1$. Then, set $I_i^x = (x_{i-1/2}, x_{i+1/2})$, $I_j^y = (y_{j-1/2}, y_{j+1/2})$, $I_k^z = (z_{k-1/2}, z_{k+1/2})$, $\Delta x_i = x_{i+1/2} - x_{i-1/2}$, $\Delta y_j = y_{j+1/2} - y_{j-1/2}$, and $\Delta z_k = z_{k+1/2} - z_{k-1/2}$. Associated with this partition, we introduce the spaces

$$\begin{aligned} Q_{\eta,h} = \{v \in H(\text{div};\Omega):\ & v|_{I_i^x \times I_j^y \times I_k^z} = (a_{i,j,k}^1 + a_{i,j,k}^2 x, a_{i,j,k}^3 + a_{i,j,k}^4 y, \\ & a_{i,j,k}^5 + a_{i,j,k}^6 z),\ a_{i,j,k}^l \in \mathbb{R},\ i = 1,\cdots,n_x, \\ & j = 1,\cdots,n_y,\ k = 1,\cdots,n_z,\ v\cdot\nu|_{\partial\Gamma_2^c} = \eta\}, \end{aligned} \tag{20.39}$$

$$\begin{aligned} R_h = \{w \in L^\infty(\Omega):\ & w|_{I_i^x \times I_j^y \times I_k^z} \in P^0(I_i^x \times I_j^y \times I_k^z), \quad i = 1,\cdots,n_x, \\ & j = 1,\cdots,n_y,\ k = 1,\cdots,n_z\}. \end{aligned} \tag{20.40}$$

If $v \in Q_{\eta,h}$, $v_{i+1/2,j,k}$ denotes $v(x_{i+1/2}, y_j, z_k)$; $v_{i,j+1/2,k}$ and $v_{i,j,k+1/2}$ can be similarly defined. If $w \in R_h$, then $w_{i,j,k}$ represents the constant value $w(x,y,z)$, $(x,y,z) \in I_i^x \times I_j^y \times I_k^z$. For notational and expositional convenience, let $\Delta x_0 = \Delta x_1$, $\Delta x_{n_x+1} = \Delta x_{n_x}$, $\Delta x_{i+1/2} = (\Delta x_i + \Delta x_{i+1})/2$, $\Delta y_0 = \Delta y_1$, $\Delta y_{n_y+1} = \Delta y_{n_y}$, $\Delta y_{j+1/2} = (\Delta y_j + \Delta y_{j+1})/2$, $\Delta z_0 = \Delta z_1$, $\Delta z_{n_z+1} = \Delta z_{n_z}$, and $\Delta z_{k+1/2} = (\Delta z_k + \Delta z_{k+1})/2$. Finally, define the notation $v^+ = \max\{v, 0\}$ and $v^- = \min\{v, 0\}$.

The subscript h is omitted below when no ambiguity occurs. Then the approximate solution $(c_w)_h^n \in R_h$ is required to satisfy, for $n = 0,\cdots,n_T - 1$, $i = 1,\cdots,n_x$, $j = 1,\cdots,n_y$, and $k = 1,\cdots,n_z$:

$$\begin{aligned} & \frac{(\phi_w c_w)_{i,j,k}^{n+1} - (\phi_w c_w)_{i,j,k}^n}{\Delta t^n} + \frac{\varphi_{1,i+1/2,j,k}^n - \varphi_{1,i-1/2,j,k}^n}{\Delta x_i} \\ & + \frac{\varphi_{2,i,j+1/2,k}^n - \varphi_{2,i,j-1/2,k}^n}{\Delta y_j} + \frac{\varphi_{3,i,j,k+1/2}^n - \varphi_{3,i,j,k-1/2}^n}{\Delta z_k} \\ & - \frac{1}{\Delta x_i}\left(\sigma_{1,i+1/2,j,k}^n - \sigma_{1,i-1/2,j,k}^n\right) \\ & - \frac{1}{\Delta y_j}\left(\sigma_{2,i,j+1/2,k}^n - \sigma_{2,i,j-1/2,k}^n\right) \\ & - \frac{1}{\Delta z_k}\left(\sigma_{3,i,j,k+1/2}^n - \sigma_{3,i,j,k-1/2}^n\right) + (\Lambda_w)_{i,j}^n (c_w)_{i,j}^n = (F_w)_{i,j}^n, \end{aligned} \tag{20.41}$$

where

$$\varphi^n_{1,i-1/2,j,k} = (c_w)^n_{i-1,j,k}(U_w)^{n+}_{1,i-1/2,j,k} + (c_w)^n_{i,j,k}(U_w)^{n-}_{1,i-1/2,j,k}, \tag{20.42}$$

$$\varphi^n_{2,i,j-1/2,k} = (c_w)^n_{i,j-1,k}(U_w)^{n+}_{2,i,j-1/2,k} + (c_w)^n_{i,j,k}(U_w)^{n-}_{2,i,j-1/2,k}, \tag{20.43}$$

$$\varphi^n_{3,i,j,k-1/2} = (c_w)^n_{i,j,k-1}(U_w)^{n+}_{3,i,j,k-1/2} + (c_w)^n_{i,j,k}(U_w)^{n-}_{3,i,j,k-1/2}, \tag{20.44}$$

and the function

$$\sigma^n_h = (\sigma^n_1, \sigma^n_2, \sigma^n_3) \in Q_{c^n_{wN},h} \tag{20.45}$$

is the solution of

$$\int_\Omega \sigma_h(t^n) v\, d\Omega = -\int_\Omega (c_w)_h(t^n)\nabla \cdot v\, d\Omega + \int_{\Gamma^c_1} c_{wD}(t^n) v \cdot \nu\, ds, \quad \forall v \in Q_{0,h}. \tag{20.46}$$

After the mass matrix has been mass-lumped [22], this implies

$$\sigma^n_{1,i-1/2,j,k} = (\mathcal{D}_w)^n_{i-1/2,j,k}((c_w)^n_{i,j,k} - (c_w)^n_{i-1,j,k})/\Delta x_{i-1/2}, \tag{20.47}$$

$$\sigma^n_{2,i,j-1/2,k} = (\mathcal{D}_w)^n_{i,j-1/2,k}((c_w)^n_{i,j,k} - (c_w)^n_{i,j-1,k})/\Delta y_{j-1/2}, \tag{20.48}$$

$$\sigma^n_{3,i,j,k-1/2} = (\mathcal{D}_w)^n_{i,j,k-1/2}((c_w)^n_{i,j,k} - (c_w)^n_{i,j,k-1})/\Delta z_{k-1/2}. \tag{20.49}$$

The initial condition is

$$(c_w)_h(\cdot, 0) = (c^0_w)_h, \quad x \in \Omega, \tag{20.50}$$

where $(c^0_w)_h$ is some approximation in R_h of c^0_w. Finally, the Neumann boundary condition (20.9) is discretized by the usual reflection principle, and on Γ^c_1 $(c_w)_h$ is defined by c_{wD}.

Note that the lowest-order Raviart-Thomas mixed method [21] over rectangles has been used in (20.41). Since the elements in $Q_{\eta,h}$ have continuous normal components on interelement edges, the numerical fluxes $\varphi^n_{1,i-1/2,j,k}$, $(\mathcal{D}_w)^n_{i-1/2,j,k}$, etc. are well defined. Furthermore, if appropriate approximations of the coefficient $(U_w)_h$ are introduced and the mass-lumping technique is used as above, the conservative scheme (20.41) can be deduced from the discontinuous finite element method [11], [12] or from the finite volume method [18], [17] combined with the mixed finite element method [22].

A convergence analysis has been given in [7], [9], and [8] for the finite element procedure (20.35), (20.37), and (20.41). It has been shown that error estimates of optimal order in the L^2-norm and almost optimal order in the L^∞-norm can be obtained in the nondegenerate case where the capillary diffusion coefficient $D(s)$ and the hydrodynamic diffusion/dispersion tensor $\mathcal{D}_w(s,u)$ are assumed to be uniformly positive. In the degenerate case where these coefficients can be zero, the choice of norm depends on the severity of their degeneracy, with almost optimal order convergence for non-severe degeneracy.

20.5 Numerical Results

Consider the problem

$$\frac{\partial}{\partial t}(\Phi u) + \nabla \cdot (Vu - D\nabla u) = -Ku, \qquad \text{on } [0,1], t > 0 \tag{20.51}$$

with $\Phi = 1$, $V = -0.5$, and $K = 0$. Boundary and initial conditions are given by

$$u(t,0) = 0, \quad u(t,1) = 1, \quad t > 0, \tag{20.52}$$

$$u(x,0) = 0, \quad x \in [0,1] \tag{20.53}$$

The exact and approximate 'nonviscous' solution (i.e., in the case of $D = 0$) and the 'viscous' solution with $D = 10^{-3}$ at $T = 0.5$ are displayed in Figure 20.1. Notice that the largest error in the approximation of u occurs around the location of the discontinuity $x = 0.75$. As expected, this implies that the presence of a discontinuity has an effect on the convergence. We also see that the 'nonviscous' solution provides a good approximation to the 'viscous' solution.

In the second example we consider problem (20.51) on $\Omega = (0,1)^2$. The coefficients are $\Phi = 1$, $V = (\cos(\frac{8}{\pi}), \sin(\frac{8}{\pi}))$, $k = 0$, $D = 10^{-3}$. and the boundary and initial data are, respectively,

$$u_D = \begin{cases} 1, & x = 0, \frac{1}{2} < y < 1, \\ 0, & \text{elsewhere}, \end{cases}$$
$$u(x,y,0) = \begin{cases} 1, & 0 \le x \le 1, \frac{1}{2} < y \le 1, \\ 0, & \text{elsewhere}. \end{cases} \tag{20.54}$$

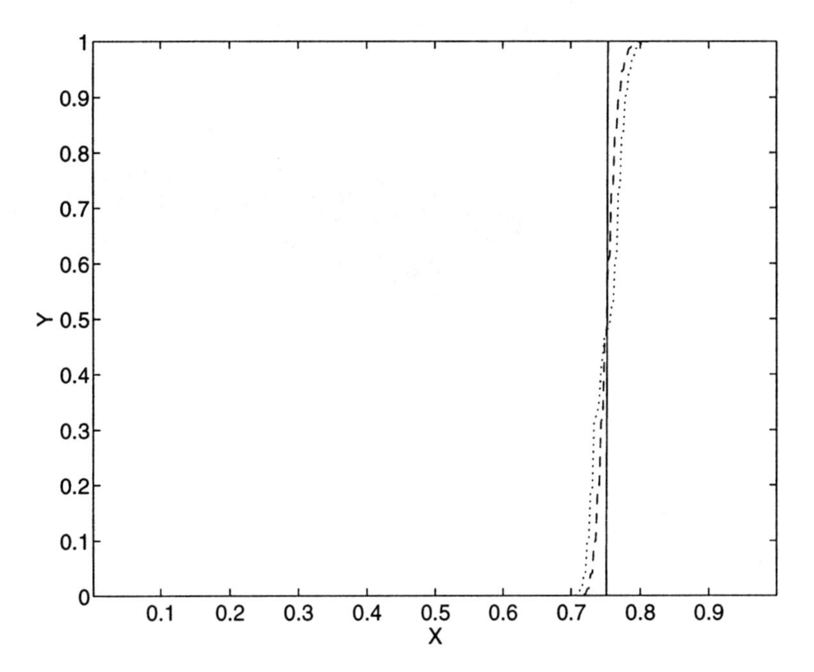

Figure 20.1: The '—', '- - -', and '. . .' represent u with $D = 0$ and the approximate solution u_h with $D = .0001$ and $D = 0$.

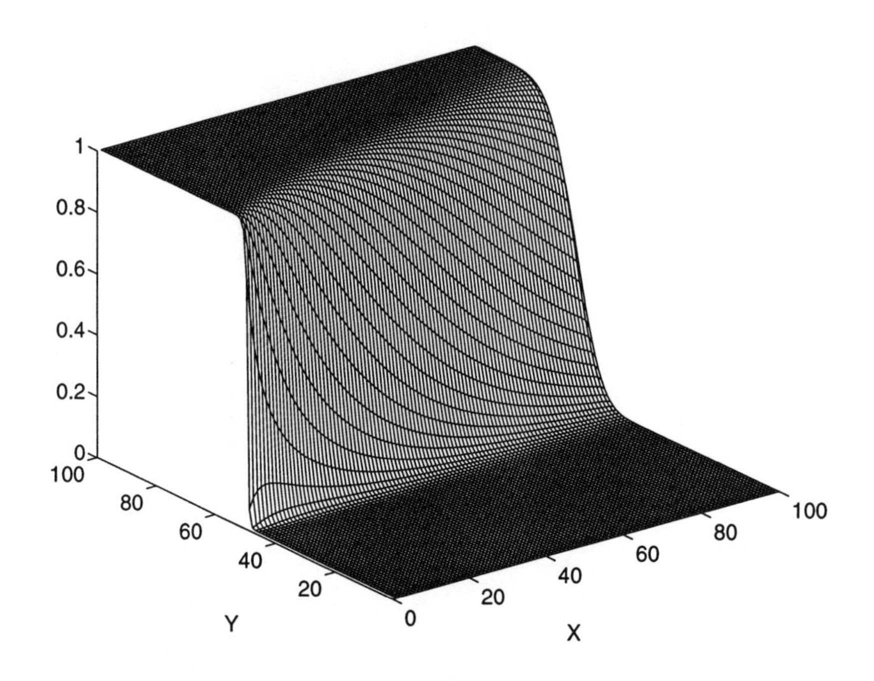

Figure 20.2: The approximate solution u_h on $(0,1)^2$.

The approximate solution of this problem obtained using the previous method with $\Delta x = \Delta y = 10^{-2}$ at time $T = 2$ is shown in Figure 20.2. The graph clearly shows that the method can capture the shock around the location $y = 1/2$.

20.6 Conclusions

We are presently working on a parallel groundwater flow and transport code on the Intel-Paragon. Numerical experiments are being carried out.

References

[1] Bear, J., *Dynamics of Fluids in Porous Media*, Dover, New York, 1972.

[2] Brezzi, F., J. J. Douglas, and L. Marini, "Two families of mixed finite elements for second order elliptic problems", *Numer. Math.*, 47, 217–235, 1985.

[3] Brezzi, F., J. J. Douglas, M. Fortin and L. Marini, "Efficient rectangular mixed finite elements in two and three space variables", *RAIRO Modèl. Math. Anal. Numér*, 21, 581–604, 1987.

[4] Brezzi, F., J. D. Jr, R. Durán and M. Fortin, "Mixed finite elements for second order elliptic problems in three variables", *Numer. Math.*, 51, 237–250, 1987.

[5] Chavent, G. and J. Jaffré, *Mathematical Models and Finite Elements for Reservoir Simulation*, North-Holland, Amsterdam, 1978.

[6] Chen, Z., "Analysis of mixed methods using conforming and nonconforming finite element methods", *RAIRO Modèl. Math. Anal. Numér.*, 27, 9–34, 1993.

[7] Chen, Z., M. Espedal and R. Ewing, "Continuous-time finite element analysis of multiphase flow in groundwater hydrology", *Applications of Mathematics*, 1995, to appear.

[8] Chen, Z. and R. Ewing, "Fully-discrete finite element analysis of multiphase flow in groundwater hydrology", in preparation.

[9] Chen, Z. and R. Ewing, "Stability and convergence of a finite element method for reactive transport in ground water", *ISC-94-07-Math Preprint*, 1994, (submitted).

[10] Chen, Z., R. Ewing and M. Espedal, "Multiphase flow simulation with various boundary conditions", in e. a. A. Peters (ed.), *Numerical Methods in Water Resources, Vol. 2*, 925–932, Kluwer Academic Publishers, Netherlands, 1994.

[11] Chen, Z. and J. J. Douglas, "Approximation of coefficients in hybrid and mixed methods for nonlinear parabolic problems", *Mat. Aplic. Comp.*, 10, 137–160, 1991.

[12] Cockburn, B. and C. Shu, "TVB Runge-Kutta local projection discontinuous Galerkin finite element method for scalar conservation laws II: general framework", *Math. Comp.*, 52, 411–435, 1989.

[13] Douglas, J., R. Ewing and M. Wheeler, "The approximation of the pressure by a mixed method in the simulation of miscible displacement", *RAIRO Anal. Numér.*, 17, 17–33, 1983.

[14] Ewing, R. and T. Russell, "Efficient time-stepping methods for miscible displacement problems in porous media", *SIAM J. Numer Anal.*, 19, 1–66, 1982.

[15] Fine, P. and B. Yaron, "Outdoor experiments on enhanced volatalization by venting of kerosene compound from soil", *J. of Contaminant Hydrology*, 12, 355–374, 1993.

[16] Johnson, C. and V. Thomée, "Error estimates for some mixed finite element methods for parabolic type problems, *RAIRO Anal. Numér.*", 15, 41–78, 1981.

[17] Lazarov, R. D., I. D. Mishev and P. S. Vassilevski, "Finite volume methods with local refinement for convection-diffusion problems", Tech. Rep. 92-50, Department of Mathematics, UCLA, 1992.

[18] McCormick, S., *Multilevel Adaptive Methods for Partial Differential Equations*, SIAM, Philadelphia, 1989.

[19] Nedelec, J., "Mixed finite elements in e^3", *Numer. Math.*, 35, 315–341, 1980.

[20] Peaceman, D., *Fundamentals of Numerical Reservoir Simulation*, Elsevier, New York, 1977.

[21] Raviart, P. and J. Thomas, *A mixed finite element method for second order elliptic problems*, Lecture Notes in Math. **606**, Springer, Berlin, 1977.

[22] Russell, T. and M. Wheeler, "Finite element and finite difference methods for continuous flows in porous media", in *The Mathematics of Reservoir Simulation*, 35–106, SIAM, Philadelphia, 1983.

[23] Thomée, V., *Galerkin Finite Element Methods for Parabolic Problems*, Lecture Notes in Math. **1054**, Springer-Verlag, Berlin, New York, 1984.

Index